NEUROMATIC

CLASS 200 | NEW STUDIES IN RELIGION

EDITED BY Kathryn Lofton AND John Lardas Modern

Kindred Spirits: Friendship and Resistance at the Edges of Modern Catholicism
by Brenna Moore

Ripples of the Universe: Spirituality in Sedona, Arizona
by Susannah Crockford

Making a Mantra: Tantric Ritual and Renunciation in the Jain Path to Liberation
by Ellen Gough

The Privilege of Being Banal: Art, Secularism, and Catholicism in Paris
by Elayne Oliphant

Experiments with Power: Obeah and the Remaking of Religion in Trinidad
by J. Brent Crosson

The Lives of Objects: Material Culture, Experience, and the Real in the History of Early Christianity
by Maia Kotrosits

Make Yourselves Gods: Mormons and the Unfinished Business of American Secularism
by Peter Coviello

Hunted: Predation and Pentecostalism in Guatemala
by Kevin Lewis O'Neill

NEUROMATIC

or, A Particular History of Religion and the Brain

JOHN LARDAS MODERN

The University of Chicago Press
Chicago and London

The University of Chicago Press, Chicago 60637
The University of Chicago Press, Ltd., London

Published 2021
Printed in the United States of America

30 29 28 27 26 25 24 23 22 21   1 2 3 4 5

ISBN-13: 978-0-226-79718-2 (cloth)
ISBN-13: 978-0-226-79962-9 (paper)
ISBN-13: 978-0-226-79959-9 (e-book)
DOI: https://doi.org/10.7208/chicago/9780226799599.001.0001

Library of Congress Cataloging-in-Publication Data

Names: Modern, John Lardas, 1971– author.
Title: Neuromatic; or, a particular history of religion and the brain / John Lardas Modern.
Other titles: Particular history of religion and the brain | Class 200: new studies in religion.
Description: Chicago ; London : University of Chicago Press, 2021. | Series: Class 200: new studies in religion | Includes bibliographical references and index.
Identifiers: LCCN 2020056567 | ISBN 9780226797182 (cloth) | ISBN 9780226799629 (paperback) | ISBN 9780226799599 (ebook)
Subjects: LCSH: Brain—Religious aspects. | Neurosciences—Religious aspects. | Cognitive neuroscience. | Neurosciences—History. | Religion and science.
Classification: LCC BL65.B73 M63 2021 | DDC 612.8/2—dc23
LC record available at https://lccn.loc.gov/2020056567

⊗ This paper meets the requirements of ANSI/NISO Z39.48-1992 (Permanence of Paper).

Somewhere a machine began to hum and I distrusted the man and woman above me.

RALPH ELLISON, *Invisible Man* (1952)

In memory of Dorothy Dale Flay (1913–1991)

CONTENTS

FIGURES

NEUROMATIC

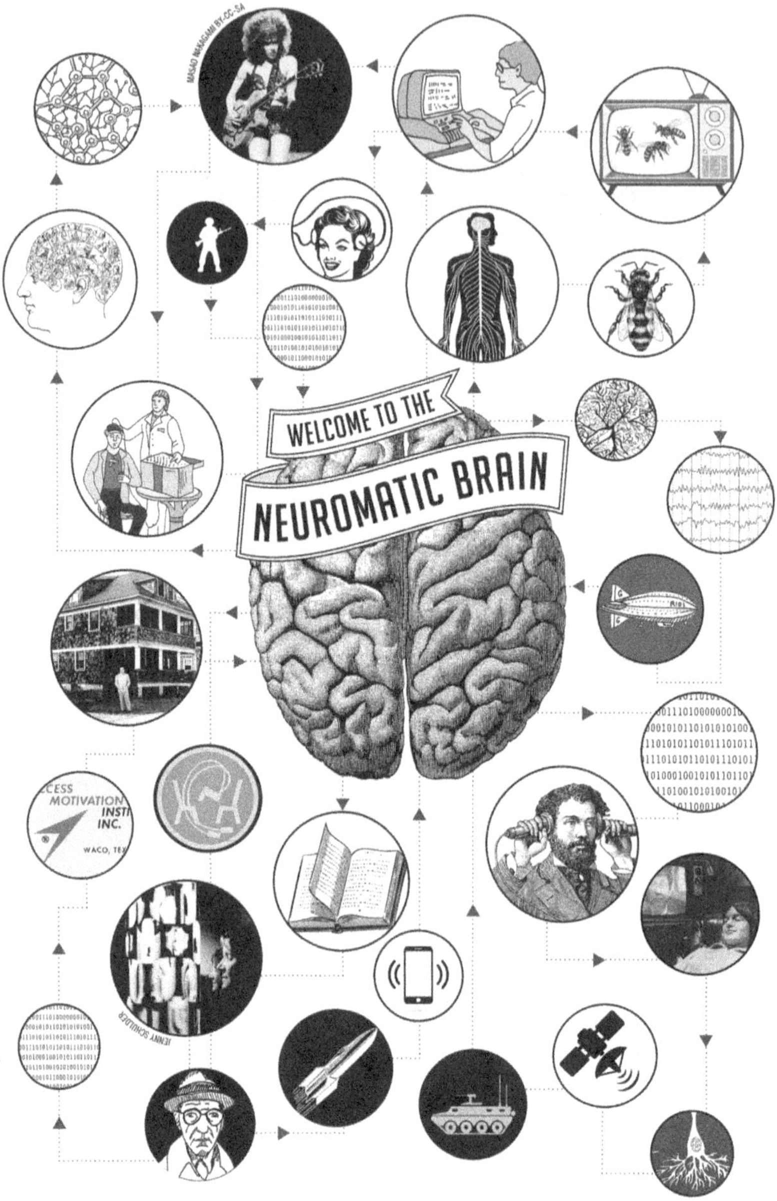

Welcome to the Neuromatic Brain. By Libby Modern.

PROLOGUE

Already Gone

Leave my head demagnetized
Tell me where the trauma lies
In the scan of pathogen
Or the shadow of my sin

CHARLOTTE GAINSBOURG, "IRM" (2009)

IT WAS A COOL AND RAINY October day as I made my way to Dr. Patrick McNamara's neuroscience laboratory. In through the glass doors, past the PatriotMart Cafeteria and convenience store, the ground floor lobby at the Jamaica Plain VA Hospital in Boston was full of aging soldiers in various states of disrepair. Memories of service and discipline and the blood traumas of war. Folding tables lined the entrance hall with Silly Bandz® bracelets. Hand-drawn placards issued a vaguely therapeutic invitation to veterans to stop, enjoy, and take in the effervescence of youth. The display of intricately threaded and multicolored wristbands, however, was falling flat.

This labor of elementary school children—congealed in bracelets whose components were manufactured by Brainchild Products of Toledo, Ohio—was failing to register with veterans and their caregivers as they walked or strolled past the folding tables. Some people had their route down. Others were actively searching for where to go, asking staff for directions, or else content sitting against the wall, seemingly waiting, as if their stilled presence would somehow make it all better. Walkers and oxygen tanks and wheelchair chatter leading the way. The whiff of sterile bureaucracy enveloping claims made by those who had been subject to the vicissitudes of generals and stray bullets.

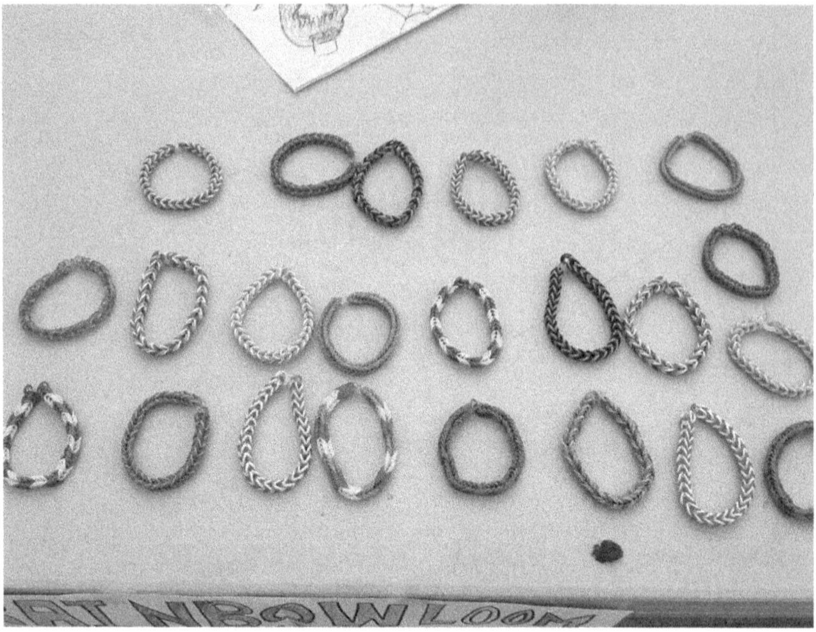

FIGURE 1. Silly Bandz® bracelets. Photo by author.

Dr. McNamara, the founding director of the Institute for the Bio-Cultural Study of Religion, and his team of researchers at Boston University have begun to discover fascinating correlations between left-onset Parkinson's disease and two things: 1) the erosion of religious interest and 2) increased difficulty in accessing religious concepts. By subjecting the brains of those who have been diagnosed with left-onset Parkinson's disease to intense forms of measurement, including magnetic resonance imaging, Dr. McNamara hopes to glimpse, in this particular form of neural degeneration, the fleeting fundament of human religiosity.[1]

These investigations into left-onset Parkinson's are part of a larger research agenda, one distilled in the inaugural issue of *Religion, Brain, and Behavior*, a journal cofounded by McNamara in 2011. And I quote:

1. Paul M. Butler, Patrick McNamara et al., "Disease-associated Differences in Religious Cognition in Patients with Parkinson's Disease," *Journal of Clinical and Experimental Neuropsychology* 33, no. 8 (2011): 917–28 and Paul M. Butler, Patrick McNamara et al., "Side of Onset in Parkinson's Disease and Alterations in Religiosity: Novel Behavioral Phenotypes," *Behavioural Neurology* 24, no. 2 (2011): 133–41; Erica Harris, Patrick McNamara, and Raymon Durso, "Possible Selves in Patients with Right- versus Left-onset Parkinson's Disease," *Aging, Neuropsychology, and Cognition* 24, no. 2 (2017): 198–215.

> Over the past two decades scholars in the historical, cultural, and psychological study of religion have been joined by a new breed of investigators: scientists interested in the biological foundations of religion and the links among its biological, behavioral, and cultural aspects. These scientists include cognitive psychologists, evolutionary anthropologists, cognitive neuroscientists, behavioral economists, and many others. Their aim is more or less the same as the religious studies scholars: to understand religion as a human phenomenon in a theologically neutral way. But these scientific approaches deploy techniques unfamiliar in older forms of the academic study of religion.[2]

Dr. McNamara has done much to advance neuroscientific investigations into religion over the past two decades.[3] He is agnostic about what his research will, in the end, discover but remains open to the possibility that religion can be explained. "We have the potential to figure religion out," he told me, "just like REM sleep. Not yet but perhaps soon given a patient, experimental, neurological approach to religion and cognition."[4]

In general, the goal of neuroscience has been to show how multiple levels of organization in the brain operate, interact, and correspond with one another. In doing so, neuroscientists are increasingly hopeful "that some of the major mysteries of mind-brain function will finally be explained." Such desire, they insist, "has become less pie-in-the-sky-romanticism and more a palpable project that can be actively and experimentally pursued."[5]

2. Patrick McNamara, Richard Sosis, and Wesley J. Wildman, "Editorial," *Religion, Brain and Behavior* 1, no. 1 (2011): 1.

3. A sampling of his contributions include: Patrick McNamara et al., "Supernatural Agent Cognitions in Dreams," *Journal of Cognition and Culture* 18, no. 3–4 (2018): 428–50; Patrick McNamara, "The Science and Theology of Dreams," *Theology and Science* 16, no. 4 (2018): 484–97; Patrick McNamara et al., "How Does Religiousness Protect against Risky Health Behaviors?," *Psychology of Religion and Spirituality* 2, no. 1 (2010): 30–34; Patrick McNamara, *The Neuroscience of Religious Experience* (Cambridge: Cambridge University Press, 2009); Patrick McNamara, ed., *Where God and Science Meet: How Brain and Evolutionary Studies Alter Our Understanding of Religion*, 3 vols. (Westport, CT: Praeger Publishers, 2006); Patrick McNamara, "The Motivational Origins of Religious Practices," *Zygon: A Journal of Religion and Science* 37, no. 1 (2002): 143–60.

4. Interview with Patrick McNamara, Jamaica Plain Veterans Administration Hospital, Boston, October 2013.

5. Patricia S. Churchland, Christof Koch, and Terrance J. Sejnowski, "What Is Computational Neuroscience?," in *Computational Neuroscience*, ed. Eric L. Schwartz (Cambridge, MA: MIT Press, 1993), 54.

Since its self-conscious inception in the 1980s, computational neuroscience has considered itself to be this cutting edge. Neuroscientists now imagine the end of psychology as a human, and therefore imprecise, science. As they condescendingly suggest, once the mathematical correspondence between brain and mind has been figured out, once the relationship between neurons and consciousness has been explained with a degree of replicable precision, psychology will have fulfilled its duty. According to neuroscientist Michael Gazzaniga (who founded the *Journal of Cognitive Neuroscience*), a proper and persuasive understanding of human behavior "will require neuroscience to think about how the rules and algorithms that govern all of the separate and distributed modules work together to yield the human condition."[6]

Neuroscience arrives, in other words, accompanied by a deep commitment to a thesis of progressive secularization. As the story goes, precise knowledge of the brain promises to liberate us from the "shackles of superstition" and the "concept of the soul."[7] The theological tendencies and/or religious biases of earlier inquiries into the brain will surely dissipate in the face of an overwhelming mathematics. With the increased rigor of experimental design and the refinement of techniques of measurement, the "module" of religion will be identified and subjected to sustained scrutiny. For as it turns out, there is "no ghost in the machine" after all. As one neuroscientist declares, "There is no scientific evidence for the existence of an immortal soul, in either our own species or any other species. There is, on the other hand, a growing body of scientific data which indicates that all animals, including ourselves, can for most, and perhaps even all, purposes be regarded as organic machines, devoid of anything mystical." Severe is the certainty, here, that computational neuroscience is a secularizing trend. For as "long as there is a systematic way (a nonmagical way) that a network" like the brain "performs its tasks, then in principle that way can be captured by an algorithm that formally specifies the relations between input and output."[8]

6. Michael S. Gazzaniga, *Who's in Charge?: Free Will and the Science of the Brain* (New York: Ecco, 2012), 218.

7. Nikolas Rose and Joelle M. Abi-Rached, *Neuro: The New Brain Sciences and the Management of Mind* (Princeton, NJ: Princeton University Press, 2013), 5, 30; Patricia S. Churchland, *Touching a Nerve: The Self as Brain* (New York: W. W. Norton, 2013), 63.

8. Rodney Cotterill, *No Ghost in the Machine: Modern Science and the Brain, the Mind and the Soul* (London: Heinemann, 1989), 7; Churchland, Koch, and Sejnowski, "What Is Computational Neuroscience?," 50.

Over the last few centuries, despite encomia to secularity, insights from the brain sciences have been revelatory. They have also been generative, aided and abetted by increasingly authoritative neuroimaging technologies—from the draftsman's hand to calipers and the silver staining of nervous tissue to electroencephalography and oscillators to MRI and neuro implants. Recent fanfare over such technologies has gilded the encroachments of neuroscience into existing scientific and humanistic conversations with the invention of new disciplinary spaces.[9] These include neuropsychiatry,[10] neurolaw,[11] neuroethics, neurogenetics, neuroeconomics,[12] neuromarketing, neuroaesthetics,[13] neuroergonomics,[14] neuroanthropology,[15] neurosociology,[16]

9. Melissa M. Littlefield and Jenell M. Johnson, "Introduction: Theorizing the Neuroscientific Turn—Critical Perspectives on a Translational Discipline," in *The Neuroscientific Turn: Transdisciplinarity in the Age of the Brain*, ed. Melissa M. Littlefield and Jenell M. Johnson (Ann Arbor: University of Michigan Press, 2012), 1–25; Tara H. Abraham, "Transcending Disciplines: Scientific Styles in Studies of the Brain in Mid-twentieth Century America," *Studies in History and Philosophy of Science*, Part C, 43, no. 2 (2012): 552–68.

10. J. B. Martin, "The Integration of Neurology, Psychiatry, and Neuroscience in the 21st Century," *American Journal of Psychiatry* 159, no. 5 (2002): 695–704.

11. Morris B. Hoffman, "Nine Neurolaw Predictions," *New Criminal Law Review: An International and Interdisciplinary Journal* 21, no. 2 (2018): 212–46.

12. Elana Ortiz-Teran et al., "Neural Implications of Investment Banking Experience in Decision-making under Risk and Ambiguity," *Journal of Neuroscience, Psychology, and Economics* 12, no. 1 (2019): 34–44; Paul W. Glimcher and Ernst Fehr, eds., *Neuroeconomics: Decision Making and the Brain*, 2nd ed. (London: Academic Press, 2014).

13. Marcus T. Pearce et al., "Neuroaesthetics: The Cognitive Neuroscience of Aesthetic Experience," *Perspectives on Psychological Science* 11 (2016): 265–79.

14. Ranjana K. Mehta and Raja Parasuraman, "Neuroergonomics: A Review of Applications to Physical and Cognitive Work," *Frontiers of Human Neuroscience* 7 (2013): 1–10.

15. Robert Turner et al., "The Brain in Culture and Culture in the Brain: A Review of Core Issues in Neuroanthropology," in *Progress in Brain Research* 178 (Special issue: Cultural Neuroscience: Cultural Influences on Brain Function, ed. J. Y. Chiao) (The Netherlands: Elsevier, 2009): 43–66; Victor Turner, "The New Neurosociology," in *On the Edge of the Bush: Anthropology as Experience*, ed. Edith L. Turner (Tucson: University of Arizona Press, 1985): 283–86.

16. Will Kalkhoff, Shane R. Thye, and Joshua Pollock, "Developments in Neurosociology," *Sociology Compass* 10, no. 3 (2016): 242–58; W. D. TenHouten, "A Neurosociological Model of Weberian, Instrumental Rationality: Its Cognitive, Conative, and Neurobiological Foundations," in *Handbook of Neurosociology*, ed. David D. Franks and Jonathan H. Turner (Dordrecht: Springer, 2013).

neurodesign,[17] neurocriminology,[18] and neurotheology.[19] These disciplinary formations, moreover, arrive under the sign of cognitive capitalism in which the latest discoveries by neuroscientists are mediated by what critical theorist Tony D. Sampson densely describes as "an economic regime that strives to situate an increasingly docile consumer-subject managed according to channeled attention, primed emotional engagement, and visceral affective stirrings."[20] In light of trends in consumer neuroscience, not to mention the more ominous applications of neuroscience by Facebook, Google, and Amazon for purposes of monitoring and predicting user behavior, the brain has become not only a money-making proposition of unprecedented potential but a synecdoche of one's human being.[21]

As a control subject in Dr. McNamara's experiment on cognition and religious coping strategies in Parkinson's Disease—funded by the John Templeton Foundation and the Department of Veterans Affairs—I became anonymized patient #41672 or, less precisely, an object of measurement and subject

17. Sanford Kwinter, "Neuroecology: Notes Toward a Synthesis," in *The Psychopathologies of Cognitive Capitalism*, pt. 2, ed. Warren Neidich (Berlin: Archive Books, 2014), 313–33.

18. Diana Concannon, *Neurocriminology: Forensic and Legal Applications, Public Policy Implications* (Boca Raton, FL: CRC Press, 2018).

19. Andrew B. Newberg, *Principles of Neurotheology* (Burlington, VT: Ashgate, 2013); Mario Beauregard and Denyse O'Leary, *The Spiritual Brain: A Neuroscientist's Case for the Existence of the Soul* (New York: HarperCollins, 2007); Justin L. Barrett, *Cognitive Science, Religion, and Theology: From Human Minds to Divine Minds* (West Conshohocken, PA: Templeton Press, 2011); Christopher Beem, *Democratic Humility: Reinhold Niebuhr, Neuroscience, and America's Political Crisis* (Lanham, MD: Lexington Books, 2015).

20. Arne De Boever and Warren Neidich, eds., *The Psychopathologies of Cognitive Capitalism*, pt. 1 (Berlin: Archive Books, 2013); Tony D. Sampson, *The Assemblage Brain: Sense Making in Neuroculture* (Minneapolis: University of Minnesota Press, 2016), xi.

21. Uma R. Karmarkar and Hilke Plassmann, "Consumer Neuroscience: Past, Present, and Future," *Organizational Research Methods* 22, no. 1 (January 2019): 174–95; Ale Smidts et al., "Advancing Consumer Neuroscience," *Marketing Letters* 25, no. 3 (2014): 257–67; Hilke Plassmann et al., "Marketing Actions Can Modulate Neural Representations of Experienced Pleasantness," *Proceedings of the National Academy of Sciences* 105, no. 3 (2008): 1050–54. See also Facebook's recently established Center for Marketing Science Innovation. Marty Swant, "Facebook Is Building Its Own Neuroscience Center to Study Marketing," *AdWeek* (May 23, 2017).

to new scientific techniques that have promised to transform the study of religion into a proper science.

On the first of two days in Dr. McNamara's laboratory I completed the required consent forms. I then began to fill out survey after survey that sought to establish a baseline of religious and/or secular proclivity. First up, the Magical-Ideation Scale and its list of statements and instruction to circle either true or false as they apply "to your beliefs":

Can some people make you aware of them by thinking about you?
<*true*>
Do things sometimes seem to be in different places when you get home, even though no one has been there?
<*false*>
Have you had the momentary feeling that you might not be human?
<*false; and this failure on my part is of continuing concern*>
Does the government refuse to tell us the truth about flying saucers?
<*well, I wouldn't necessarily use the word "refuse"*>
Do you sometimes have a feeling of gaining or losing energy when certain people look at you or touch you?
<*all the time*>
Have you noticed sounds on your records that are not there at other times?
<*I am not even going to dignify that with an answer*>

When I turned my attention to the Post-Critical Belief Scale—in which I registered the intensity of my agreement and disagreement, on a 1-to-7 scale, to a series of statements—I was struck by its thick historicity. These statements were not only precedented but very much part of a history whose measurable presence in this laboratory was not of primary concern. The first two statements from the Post-Critical Belief Scale, for example, weighed heavily on the historian of religion in me—but perhaps more heavily on those who have long struggled to define themselves as religious vis-à-vis the public potency of science and on those who sought to distinguish themselves from secular observers who measured their faith in terms they did not recognize.

The Bible holds a deeper truth that can only be revealed by personal reflection.
<*somewhat agree*>

> **If you want to understand the meaning of the miracle stories from the Bible, you should always place them in their historical context.**
> *<somewhat agree>*

The majority of statements in the Post-Critical Belief Scale referred to Jesus, to the Bible and attitudes toward it, as well as to a monotheistic God and degrees of belief in him. Other statements were clearly directed at assessing my secularity:

> **The scientific understanding of human life and the world makes a religious understanding obsolete.**
> *<strongly disagree>*

Despite the working assumption of Dr. McNamara and his team that there existed "a positive relationship between religiosity and self-control,"[22] I began to suspect that a discipline was being enacted in this laboratory. For as I was probed at ever-deeper levels, religion, conceived of as an internal affair revolving around cognition and choice, was being extracted from me.

During the next session I sat behind a computer and answered a series of questions that measured my impulsivity and capacity for self-control. Before each question appeared, I was primed with either religious or secular cues. I picked up on them only about halfway through the exercise. Civic or mundane primes flashed almost imperceptibly across the computer screen—Hospital, Axe, Civil, Jolt, Jungle, Blink, Chisel, Hut, Lodge, Block, Pepper. Religious primes flitted by at the edges of my vision—words like Bible, Nun, Prayer, Monk, Temple, Saint, Angel, Heaven, Sin, Hell, Demon, Devil. After each prime, I was immediately asked whether I would like to have, say, $28 today or $30 in 110 days, or $67 today or $85 in 35 days. Over and over again, the primes and the wagers changed and sometimes repeated,

22. Jonathan Morgan, Dustin Clark, Yorghos Tripodis, Christopher S. Halloran, April Minsky, Wesley J. Wildman, Raymon Durso, and Patrick McNamara, "Impacts of Religious Semantic Priming on an Intertemporal Discounting Task: Response Time Effects and Neural Correlates," *Neuropsychologia* 89 (2016): 403–13. On piety as an enhancer of self-control, see Michael E. McCullough and Brian L. B. Willoughby, "Religion, Self-regulation, and Self-control: Associations, Explanations, and Implications," *Psychological Bulletin* 135, no. 1 (2009): 69–93.

catalyzing anxiety over what I had just seen and whether I was being consistent in my own assessment of my desire for financial reward.[23]

I then filled out the Religion/Spirituality Questionnaire—the slash marking, perhaps, the most common sense groove of American religious history in which private is demarcated from public, self from institutions, agency from discourse. And as I would come to learn, the natural distinction between religion and spirituality was a working hypothesis of the entire project—specifically, that neural network connectivity between different parts of the brain corresponded to different capacities for experiential and doctrinal religious knowledge.[24]

As someone who has spent more than a little time in the nineteenth-century archive, this 1-to 7 continuum between experiential and doctrinal knowledge and the questions that stemmed from it were familiar—questions about the vagueness of divinity, inner peace, and the beauty of creation as well as questions about feelings of unity and personal actions outside an institutional frame ("How often do you pray privately other than at a church or synagogue?"). The last two questions of the survey put an exclamation point on this pixelated formation of American religious history:

> To what extent do you consider yourself a religious person?
> To what extent do you consider yourself a spiritual person?

My spiritual life history was then taken as a graduate student (whom I will call Steve) interviewed me over the course of two hours. Over and over

23. McNamara and his team achieved results that "suggest that religious primes influence discounting behavior via dopaminergic meso-limbic and right dorsolateral prefrontal supporting cognitive valuation and prospection processes" (Morgan et al., "Impacts of Religious Semantic Priming," 411). NB: because religious priming did not have a huge effect on my discount rate, I was informed that I had exhibited low impulsivity and a high degree of self-control.

24. On the differential of experiential/doctrinal knowledge and the turn to how people categorize religion, see Dimitrios Kapogiannis et al., "Cognitive and Neural Foundations of Religious Belief," *Proceedings of the National Academy of Sciences* 106, no. 12 (2009): 4876–81. This epistemic distinction is taken to a logical, if not ridiculous, conclusion in studies that conclude that there are physiological differences between religion and spirituality. See, e.g., Szabolcs Kéri and Oguz Kelemen, "Faith Unchanged: Spirituality, but Not Christian Beliefs and Attitudes, Is Altered in Newly Diagnosed Parkinson's Disease," *Religions* 7, no. 73 (2016): 1–9.

again I was asked about my personal takes on "ultimate reality," "ultimate meaning," and "fundamental beliefs." I was encouraged to imagine my spiritual life as a book with as many chapters as I wanted (although it was recommended that my book have between two and seven). I talked about my baptism at age eight, the presence of an absent father in my life, and about my use of recreational drugs as a form of psychic and analytical training. I talked about family trauma and the time that the devil followed me home after a Halloween party in Santa Barbara and pulled up to my driveway in a black pickup truck at 2 a.m. blasting the Eagles' "Already Gone." I related, in sentimental detail, my grandmother's warning before I set off to college in 1989 that I never, ever take a class in religious studies because those professors, she said, well, those professors don't believe in God.

Steve's questions elicited honesty, confession, truth, a hyperawareness of my individualism, or rather, my desire to at least be interesting to others because of some difference within me. In narrating my spiritual biography, I became a little paranoid—self-conscious, suspicious of my own motives as much as what might lie behind the questions. Was I filtering my responses, saying what some part of me wanted to hear, what I imagined Steve would want to hear, and at some level, uttering, in a vaguely strategic way, key words and phrases that might become data in the program, scanned for by the search algorithm?

The transcript of my spiritual life history—full of self-codings, confessional details, and verbal tics—was then scanned into NVivo. NVivo, whose original name was NUD*IST—for Non-numerical Unstructured Data Indexing Searching and Theorizing—is a qualitative data analysis program developed by QSR International in Melbourne, Australia. NVivo is a machine that goes where the human analyst cannot—into the organization and analysis of nonnumerical or unstructured data. The algorithm separates, sorts, arranges, classifies, and categorizes significant data points. It examines the relationships between patterns. And so on and so forth, linking, searching, modeling, linking. NVivo enables the scientist to test theories and gut feelings with the push of a search button. Observation after observation, NVivo helps gather evidence and build an argument through the analysis of pattern. Meaning is necessarily abstracted from the very beginning.

NVivo searched for patterns in my speech that hinted at religious proclivities or, inversely, secular leanings. This was its bread and butter. For as its makers proclaim,

FIGURE 2. NVivo banner, from http://www.qsrinternational.com/products_nvivo.aspx. "Connecting the dots in your data is faster, easier and more efficient with NVivo. More than just a tool for organizing and managing data, NVivo helps you think differently about your research, uncover more and back it all up with rigorous evidence."

> An understanding of human language can be especially powerful when applied to extract information and reveal meaning or sentiment in large amounts of text-based content (or unstructured information), especially the types of content that has typically been manually examined by people. Analysis that accurately understands the subtleties of language, for example, the choice of words, or the tone used, can provide useful knowledge and insight. NLP will play an important part in the continued development of tools that assist with the classification and analysis of data, with accuracy only improving as technology evolves.[25]

All of these tests and surveys—the "extraction of information"—were innocent enough but my anxiety increased as I went through the MRI prescreening for the third time in two days. First the graduate student, then the neuroscientist, and finally the clinician in the MRI suite. Clipboard out, pen in hand, she asked me once again:

25. https://www.qsrinternational.com/nvivo/research-and-data-and-ethics/natural-language-processing. Natural Language Philosophy (NLP) is one of the more subtle yet consequential legacies of the history I am narrating in this book. With roots in the problem of machine translation of human expression *as data*, the field of NLP is largely one of computer science application and algorithms that process representations *as a neural network* would do. Maite Giménez et al. "Semantic-based Padding in Convolutional Neural Networks for Improving the Performance in Natural Language Processing: A Case of Study in Sentiment Analysis," *Neurocomputing* 378 (February 2020): 315–23. In addition to NVivo, see the Google Cloud subscription service for developing "your own high-quality machine learning custom models to classify, extract, and detect sentiment with minimum effort"; https://cloud.google.com/natural-language/.

"Are you now or have you ever worked with metal or the cutting, grinding, or welding of metal? Have you ever been shot? Do you have any shrapnel in your body?"

"Yes, maybe," I said sheepishly. "When I was little my grandma used to break open thermometers and let me play with the mercury in a bowl."

"Will *that* still have an effect?" I asked.

"No, not on the MRI but mercury will erode the brain/blood barrier over time, which is not a good thing."

I then went off on a tangent. It was 1983. Early MTV. A few years before the MRI was approved by the FDA for clinical use. A broken-off thermometer atop a *TV Guide* on the coffee table. Mercury pulsating in a green glass bowl, eroding my brain/blood barrier over time. My grandmother, her intense evangelicalism, her struggles with mental illness, and her words of premillennial wisdom. "John Howard," she would say—Howard being my middle name and the name of *her* late husband who had died when my own mother was only two years old—"John Howard," she would say, "we are privileged to be living in the end times."

The nurse went on with her questions, unphased.

"Do you have any implants, clips, valves, stimulating devices, or dermal patches?"

"No," I said.

"Do you have a cardiac pacemaker, a neural stimulator, an implanted cardiac defibrillator, a cochlear implant, a dental implant, pacing wires, a hearing aid, an implanted insulin pump? Do you have a Swanz-Ganz catheter, any type of intravascular coil, filter, or stent, a heart valve, a penile prosthesis, a surgical clip or staple, dentures, a wire mesh, body piercings or a tattoo?"

These questions and this routine were for my own benefit, of course. For anything metal in me that was not embedded resolutely in bone could fly out of my skin at a very high speed once the magnets were turned on.

The MRI, then, is an oddly menacing device in that it can only, *safely*, measure flesh uncorrupted by the materiality of metal. For in the MRI there is a mortal threat for those who have taken even the most modest steps toward mechanical security and inscription—death simmering in a simple stent or heating up the metal bits in particular types of tattoo ink.

I got up on the exam table and laid down. My head was fitted into position and my left hand placed near a rubber ball that I could squeeze to signal that I was *not* doing fine for whatever reason, communicating with the technicians outside of the exam room, safe behind protective leaded

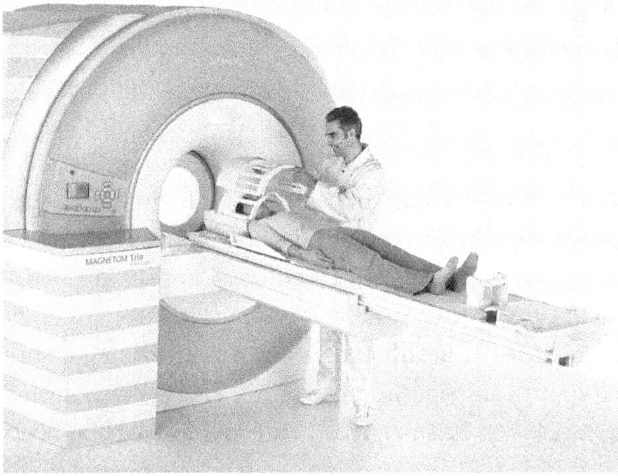

FIGURE 3. Magnetom 3T Trio, from https://www.siemens-healthineers.com/de-ch/magnetic-resonance-imaging/for-installed-base-business-only-do-not-publish/magnetom-trio-tim. Courtesy of Siemens Healthineers.

glass.[26] Everything, it seemed, was in working order. I stared up at the white ceiling. Someone pushed a button and my table glided into the bore. More adjustments were made and the technician informed me that she was now going to leave the room.

The magnetic susceptibilities of my hydrogen nuclei were about to be manipulated in a massive way. Each was about to assume a slightly different charge, in unison, so that scientists outside could then measure the difference between the oxygen rich and oxygen-poor blood running through my brain. Once my hydrogen protons were uniformly magnetized and their axes aligned, scientists could distinguish between these two blood states given their different magnetic properties. And once the computer had tracked the systemic flow of oxygen-rich blood in my brain, the scientists could chart real-time changes in neural activity due to the strange phenomenon that such activity causes the blood around it to become more oxygenated (known as BOLD, the blood oxygenation level dependent effect).[27]

26. Melvyn B. Ooi et al., "Prospective Real-time Correction for Arbitrary Head Motion Using Active Markers," *Magnetic Resonance in Medicine* 62, no. 4 (2009): 943–54.

27. The mechanics of the MRI are, of course, more complex than I am able to account for. As the MR-Technology Information Portal explains, "Protons are capable of absorbing energy if exposed to short radio wave pulses (electromagnetic energy) at their resonance

According to Robert Turner, physicist, MRI pioneer, former director of the Max Planck Institute for Human Cognitive and Brain Sciences, and son of anthropologists Edith and Victor Turner, such technological advances are leading this secular age. "The improved data quality now available," writes Turner, "allows the more unrealistic assumptions to be discarded, opening a way forward to far more realistic methods for brain functional analysis."[28] The improvement of data quality is dependent on a higher signal-to-noise ratio and image contrast generated by more powerful magnetic fields. The magnetic fields generated by the MRI are, under proper conditions, safe. In addition to issues of health and safety, it is a working assumption that there are no statistically significant cognitive effects precipitated by an MRI scan.[29] The possibility that magnetic fields could fuel a blip of spontaneous neuronal activity[30] or induce vertigo[31] or nausea or cause a DNA lesion[32] is consistently acknowledged but, lacking evidence, remains a peripheral concern.

---

frequency. After the absorption of this energy, the nuclei release this energy so that they return to their initial state of equilibrium. This transmission of energy by the nuclei as they return to their initial state is what is observed as the MRI signal. The subtle differing characteristic of that signal from different tissues combined with complex mathematical formulas analyzed on modern computers is what enables MRI imaging to distinguish between various organs"; https://www.mr-tip.com/.

28. Robert Turner, "Where Matters: New Approaches to Brain Analysis," in *Microstructural Parcellation of the Human Cerebral Cortex*, ed. Stefan Geyer and Robert Turner (Berlin: Springer-Verlag, 2013), 179.

29. Angela Heinrich et al., "Cognition and Sensation in Very High Static Magnetic Fields: A Randomized Case-crossover Study with Different Field Strengths," *Radiology* 266, no. 1 (2013): 236–45; Jöran Lepsien et al., "Investigation of Higher-order Cognitive Functions During Exposure to a High Static Magnetic Field," *Journal of Magnetic Resonance Imaging* 36 (2012): 835–40; Marc Schlamann et al., "Exposure to High-Field MRI Does Not Affect Cognitive Function," *Journal of Magnetic Resonance Imaging* 31, no. 5 (2010): 1061–66.

30. Marta Bianciardi et al., "Sources of Functional Magnetic Resonance Imaging Signal Fluctuations in the Human Brain at Rest: A 7 T Study," *Magnetic Resonance Imaging* 27, no. 8 (2009): 1019–29.

31. Paul M. Glover et al., "Magnetic-field-induced Vertigo: A Theoretical and Experimental Investigation," *Bioelectromagnetics* 28, no. 5 (2007): 349–61.

32. M. A. Hill et al., "Comments on Potential Health Effects of MRI-induced DNA Lesions: Quality Is More Important to Consider than Quantity," *European Heart Journal—Cardiovascular Imaging* 17, no. 11 (November 2016): 1230–38; Masha Fatahi and Oliver Speck, "Magnetic Resonance Imaging (MRI): A Review of Genetic Damage Investigations," Mutation Research/Reviews in *Mutation Research* 764 (2015): 51–63.

In study after study, the patient's experience of the MRI is also assumed to have little effect on what is being measured. As Turner and others surmise, experiences may be intense but rarely interfere with data capture or confound the brain's work-a-day processing of information.[33] While lacking a credentialed foundation for dissent I find this claim hard to believe.[34] For in this Magnetom 3T Trio there is utter abjection. There is the impulse to run screaming from the shaking table and clanking gears. There are the high-pitched, otherworldly blips cascading—sounds whose acoustic sheen is precision incarnate. There is the sense that the plane is really, truly going down this time. There is the terror and there is the management of terror.[35] There is the pounding of your flesh with sound and energy but nothing quite tangible. There is more terror.[36] And there is panic but there is the yogic

33. Jaane Rauschenberg, Robert Turner et al., "Multicenter Study of Subjective Acceptance during Magnetic Resonance Imaging at 7 and 9.4 T," *Investigative Radiology* 49, no. 5 (2014): 249–59. See also Valentina Hartwig et al., "Biological Effects and Safety in Magnetic Resonance Imaging: A Review," *International Journal of Environmental Research and Public Health* 6, no. 6 (2009): 1778–98.

34. Since my experience in the MRI there has been an uptick in interest in "multivariate pattern analyses and especially decoding analyses" that tackle the issue of confounds more directly. See, for example, Lukas Snoek, et al., "How to Control for Confounds in Decoding Analyses of Neuroimaging Data," *NeuroImage* 184 (2019): 741–760; Anil Rao, et al., "Predictive Modelling Using Neuroimaging Data in the Presence of Confounds," *NeuroImage* 150 (2017): 23–49.

35. In the MRI I was reminded of the claim that "religious worldviews provide a uniquely powerful form of existential security. Indeed, there may be no antidote to the human fear of death quite like religion." Kenneth E. Vail et. al., "A Terror Management Analysis of the Psychological Functions of Religion," *Personality and Social Psychology Review* 14 (2010): 85. Within cognitive studies of religion, Terror Management Theory (TMT) has considered religion to be "the paradigmatic means by which to manage our terror of death." Robert B. Arrowood et al., "On the Importance of Integrating Terror Management and Psychology of Religion," *Religion, Brain and Behavior* 8 (2018): 1.

36. While "neuroimaging techniques have contributed greatly to the identification of the structural and functional neuroanatomy of anxiety disorders," there has been little consideration of the MRI effect on the so-called "'fear network,' comprising the amygdala, insula, and anterior cingulate cortex." Kathrin Holzschneider and Christoph Mulert, "Neuroimaging in Anxiety Disorders," *Dialogues in Clinical Neuroscience* 13, no. 4 (2011): 453–61. According to contemporary neuroscientists, whatever transient emotions may inevitably arise in the MRI, rest assured, they can be mitigated through transparent instruction and disclosure forms. Niall W. Duncan and Georg Northoff, "Overview of Potential Procedural and Participant-related Confounds for Neuroimaging of the Resting State," *Journal of Psychiatry and Neuroscience* 38, no. 2 (2013): 84–96.

ease of breathing deeply, not swallowing, not moving, focusing on the most minimal tasks of breathing deeply, not swallowing, not moving.

To make a long story very short—one might say that in the MRI I encountered, on a visceral level, the abstractions of information, feedback, self-organization that fuel our contemporary moment. I was enveloped, albeit incompletely, by systems—technological[37] as well as social, political, economic, and historical.[38]

This particular MRI at the Boston VA is what anthropologist Talal Asad might call a formation of the secular.[39] In this MRI, religion had become something in need of measured explanation. For in order to figure out religion, it must be located and pinned down as a matter of "information processing in the brain related to one's beliefs about supernatural agents, god-concepts, and the relative interaction between these perceived metaphysical entities and one's self."[40] Indeed, one of the key epistemic and institutional logics of the secular is its commitment to religion as a matter of privately held and internally processed belief. Belief, in this convenient formulation, is a matter of choice, for better or for worse. The prospect of choice and the agency that it entails, then, is bolstered in the MRI by enthusiastic calculations of what lies behind the eyes and within the skull. Moreover, within this secular shell, there is a distinct difference between religious cognition and scientific cognition, the latter being called on to explain, publicly and

37. Ranajay Mandal et al., "Adaptive and Wireless Recordings of Electrophysiological Signals during Concurrent Magnetic Resonance Imaging," in *IEEE Transactions on Biomedical Engineering* 66, no. 6 (June 2019): 1649–57.

38. This affective state of relationality was more claustrophobic than the one chronicled by Patrick Jagoda in his *Network Aesthetics* (Chicago: University of Chicago Press, 2016). Whereas Jagoda is interested in an ordinary "form" of experience, my aim throughout this book will be on more intense and less neutral recognitions of one's self as part of a network. The experience of connectivity that I seek to conjure is not so much the background noise of modernity as much as it is signal evidence for what I will soon be calling the neuromatic. The brain, I argue, is what gives the aesthetics of the network its biopolitical edge. Indeed, my focus on the historical relation between cosmology and the brain has led me to conclude that the notion of the network—understood as flexible, expansive, interdependent, and at some fundamental level, self-organizing—has long assumed its power, reach, affect, and epistemic advantage by way of the brain.

39. Talal Asad, *Formations of the Secular: Christianity, Islam, Modernity* (Palo Alto, CA: Stanford University Press, 2003).

40. Butler, McNamara, et al., "Disease-associated Differences," 917.

with certainty, the former as part of an exercise in public reason and liberal governance. Indeed, the distinction between the religious and the secular, as a matter of common sense, is what makes legible my scan but also the brain scans of meditative monks, Franciscan nuns in the slow throes of mystical encounter, atheists and lay persons praying or under the influence of LSD, deep in trance, practicing extrasensory perception (ESP), speaking in tongues, and cultivating their spirituality.[41]

The data points generated in the MRI are real, in the ordinary sense. Yet the effects of measurement assume an extraordinary force in creating a culture that depends on particular readings of those data points. The representation of the brain on the screen, and according to the numbers, has its truth. There are patterns, repeated and repeatable, that reveal structure, flow, chemistry, mechanics. The measurements of those patterns, however, are made possible by desire, by history, by how existing measurements have been put to use and why. This is to say that the measurements of religion lack neither substance nor precision. But they do not account for the condi-

41. Richard J. Davidson and Antoine Lutz, "Buddha's Brain: Neuroplasticity and Meditation," *IEEE Signal Processing Magazine* 25, no. 1 (2008): 176–74; Andrew B. Newberg et al., "Cerebral Blood Flow during Meditative Prayer: Preliminary Findings and Methodological Issues," *Perceptual and Motor Skills* 97, no. 2 (October 2003): 625–30; P. K. Douglas, Sam Harris et al., "Performance Comparison of Machine Learning Algorithms and Number of Independent Components used in fMRI Decoding of Belief vs. Disbelief," *NeuroImage* 56, no. 2, (May 2011): 544–53; Dimitrios Kapogiannis et al., "Brain Networks Shaping Religious Belief," *Brain Connectivity* 4, no. 1 (2014): 70–79; Raymond L. Neubauer, "Prayer as an Interpersonal Relationship: A Neuroimaging Study," *Religion, Brain and Behavior* 4, no. 2 (2014): 92–103; Andrew B. Newberg, Nancy A. Wintering et al., "A Case Series Study of the Neurophysiological Effects of Altered States of Mind during Intense Islamic Prayer," *Journal of Physiology—Paris* 109 (2015): 214–20; Katrin H. Preller et al., "Effective Connectivity Changes in LSD-induced Altered States of Consciousness in Humans," *Proceedings of the National Academy of Sciences* 116, no. 7 (February 2019): 2743–48; Julio Fernando Peres et al., "Neuroimaging during Trance State: A Contribution to the Study of Dissociation," *PloS One* (2012): e49360; Bryan J. Williams, "Extrasensory Perception and the Brain Hemispheres: Where Does the Issue Stand Now?," *NeuroQuantology* 10, no. 2 (September 2012): 350–73; Andrew B. Newberg et al., "The Measurement of Regional Cerebral Blood Flow during Glossolalia: A Preliminary SPECT Study," *Psychiatry Research: Neuroimaging* 148 (2006): 67–71; Lisa Miller, "Neuroanatomical Correlates of Religiosity and Spirituality: A Study in Adults at High and Low Familial Risk for Depression," *JAMA Psychiatry* 71, no. 2 (2014): 128–35.

tions of their own possibility and therefore call into question the coherence of their powerful certainties.[42]

For in the MRI I sensed the presence of an agency, or, rather, I became hyperaware of my imagination of agency, a pattern generated by my fantasy impulse, or, better yet, through my desire to name that or who or what had gained full access to something so intimate that I did not even know of its existence. I was dependent on *this* machine for *this* truth. In the MRI I felt contingent practices of calculation achieve their mechanism within. Actions fueling concepts. Concepts fueling investigations. Investigations fueling imagination. Imaginings fueling actions.

In the MRI I experienced the invisible tendrils of discourse, which is to say that in the MRI I appreciated anew all those works of cultural and visual studies and critical ethnographies that had considered the MRI to be a particularly powerful construction of the social.[43] For in the MRI I became aware of my paranoia, mindful of how theories of information, feedback, and self-organization at midcentury continue to frame the social as comprised of discrete individuals in a giant communication machine.[44] And this, I must say, was a revelation. For in the MRI I began to consider what had brought me to it, lying on this particular examination table, thinking particular thoughts about all kinds of things that would eventually become the substance of this particular history of the brain. I began to consider all

42. See, for example, neurophilosopher Michael L. Anderson's strong statement of caution when it comes to the evidentiary authority of the MRI. "We must be ever mindful," writes Anderson, "that, although our scientific devices cost many millions of dollars more than does a set of calipers, measuring the change in local brain blood oxygenation puts us little closer to fundamental function than does measuring cranial bumps. . . . What we are typically doing in cognitive neuroimaging is measuring the response tendencies of local neural assemblies, and from this we are inferring computational functions." *After Phrenology: Neural Reuse and the Interactive Brain* (Cambridge, MA: MIT Press, 2014), xix.

43. Andreas Roepstorff, "Transforming Subjects into Objectivity—An 'Ethnography of Knowledge' in a Brain Imaging Laboratory," *Folk: Journal of the Danish Ethnographic Society* 44 (2002): 145–70; Joseph Dumit, *Picturing Personhood: Brain Scans and Biomedical Identity* (Princeton, NJ: Princeton University Press, 2004); Kelly A. Joyce, *Magnetic Appeal: MRI and the Myth of Transparency* (Ithaca, NY: Cornell University Press, 2008); Morana Alac, "Working with Brain Scans: Digital Images and Gestural Interaction in fMRI Laboratory," *Social Studies of Science* 38, no. 4 (2008): 483–508; Amit Prasad, *Imperial Technoscience: Transnational Histories of MRI in the United States, Britain, and India* (Cambridge, MA: MIT Press, 2014).

44. Michel Foucault, "Croître et multiplier," *Le Monde* 8037 (November 1970): 15–16.

those things that remain, at present, active yet epiphenomenal to the modifications of magnetic charge in oxygen-saturated blood as it is conscripted by active neurons.

Here, then, finally, is a picture of my brain on that cold and rainy afternoon.

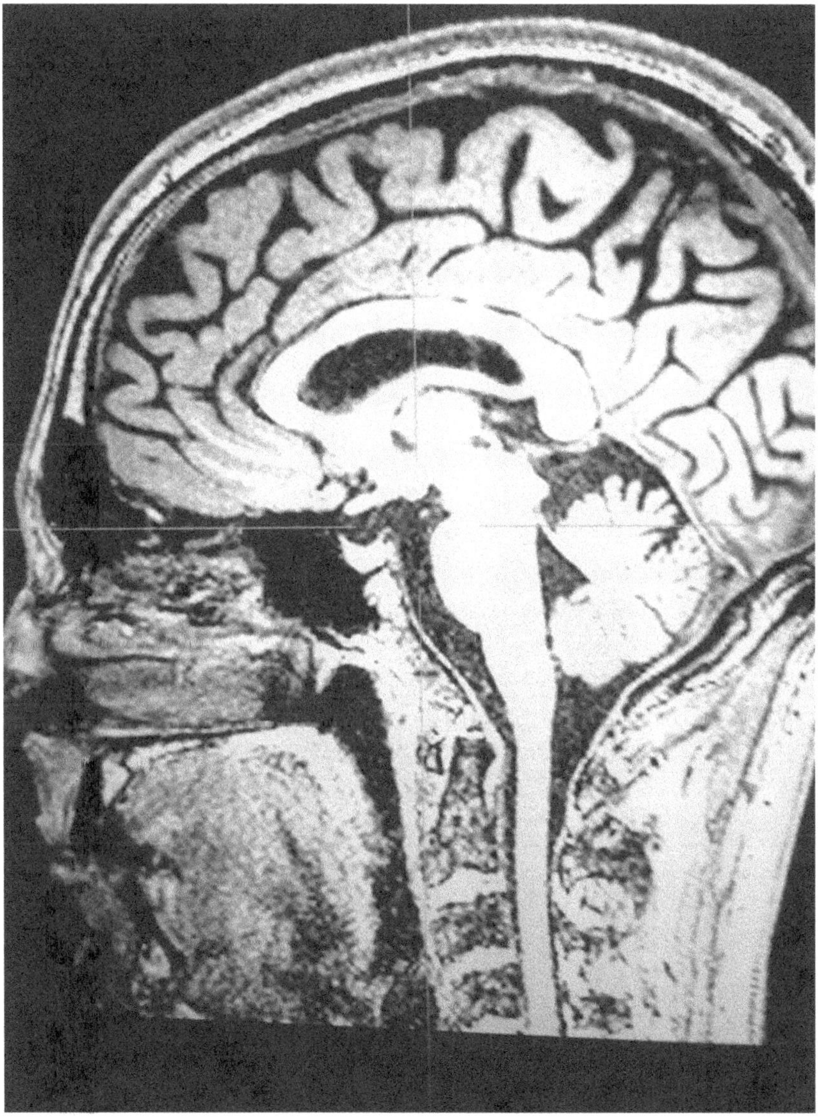

FIGURE 4. Author's MRI scan.

This picture captures my brain in the process of considering the history of the brain, in general. In addition to capturing the neural substrate of my brain, this picture offers a glimpse of what is at stake, analytically speaking, in the book now before you. For in this picture you can see the imprint of my mind upon the machine. But if you look closely, you can also see something else: namely, the imprint of some other kind of machine upon my mind and, perhaps, yours, too.

INTRODUCTION

The Brain—is wider than the Sky—
For—put them side by side—
The one the other will contain
With ease—and you—beside—

EMILY DICKINSON (ca. 1862)

The best model of the behavior of the brain is the behavior of the brain.

WARREN MCCULLOCH, *Transactions from the Ninth Conference on Cybernetics* (1952)

1. SATURATION

The brain beckons, inviting us to become a better, stronger, and faster version of ourselves—more productive, more empathetic, more successful, healthier, happier, wealthier, more sexy and serene.[1] For individuals, the optimization of and care for the brain have become moral obligations—

1. Rick Hanson, *Hardwiring Happiness: The New Brain Science of Contentment, Calm, and Confidence* (New York: Harmony Books, 2013); Richard O'Connor, *Rewire: Change Your Brain To Break Bad Habits, Overcome Addiction, and Conquer Self-Destructive Behavior* (New York: Penguin Books, 2015); Emily Nagoski, *Come As You Are: The Surprising New Science That Will Transform Your Sex Life* (New York: Simon and Schuster, 2015); Mark Robert Waldman and Chris Manning, *Neurowisdom: The New Brain Science of Money, Happiness, and Success* (New York: Diversion Books, 2017).

things that we should and must do.[2] *I imagine myself as a brain therefore I am. I can change myself and the world around me by changing my brain.* Present[3] manifestations of brain plasticity assume a range of pedagogical impulses, from the geriatric[4] to the explicitly pious[5] to the popular and pedestrian.[6] In every example, the potential to affect the neural fundament of cognition makes possible whatever freedoms may be achieved by the brain and the human that is attached to it. Such freedoms, in other words, are effects of brain activity; or, more precisely, they are effects of the brain in proper communication with itself and with the environment.[7] Neurons firing. Signals passed along through axons. Signals received by the dendrites of other neurons. Over 100 trillion synapses forming the "basic functional structure for information processing between neurons in the central nervous system, required for understanding . . . the functional properties of neural circuits and brain functions, and even the consciousness that emerges from them."[8] A precise mapping of the brain's neural architecture and patterned routes of processing. A faith in general trends, in "how movement toward more

2. Nikolas Rose and Joelle M. Abi-Rached, *Neuro: The New Brain Sciences and the Management of Mind* (Princeton, NJ: Princeton University Press, 2013), 23.

3. On "practices of cerebral self-help" that go back centuries in Anglo-European contexts, see Francisco Ortega, "Toward a Genealogy of Neuroascesis," in *Neurocultures: Glimpses into an Expanding Universe*, ed. Fernando Vidal and Francisco Ortega (Frankfurt: Peter Lang, 2011), 31. Jerzy Konorski coined the term "plasticity" to refer to the changes induced by neural excitability; *Conditioned Reflexes and Neuron Organization* (New York: Cambridge University Press, 1948), 80, 86.

4. Robert M. G. Reinhart and John A. Nguyen, "Working Memory Revived in Older Adults by Synchronizing Rhythmic Brain Circuits," *Nature Neuroscience* 22 (2019): 820–27.

5. Warren S. Brown and Brad D. Strawn, *The Physical Nature of Christian Life: Neuroscience, Psychology, and the Church* (Cambridge: Cambridge University Press, 2012). See also Andrew Newberg and Mark Robert Waldman, *How God Changes Your Brain: Breakthrough Findings from a Leading Neuroscientist* (New York: Ballantine Books, 2010).

6. Sharon Begley, *Train Your Mind, Change Your Brain: How a New Science Reveals Our Extraordinary Potential to Transform Ourselves* (with a foreword by the Dalai Lama) (New York: Ballantine, 2007); Dana Wilde, *Train Your Brain: How to Build a Million Dollar Business in Record Time* (Bloomington, IN: Balboa Press, 2013); Daniel G. Amen, *Feel Better Fast and Make It Last: Unlock Your Brain's Healing Potential to Overcome Negativity, Anxiety, Anger, Stress, and Trauma* (Carol Stream, IL: Tyndale House Publishers, 2018).

7. On theories of synaptogenesis and neurogenesis, see Tom G. Bowlig, "How Does Electroconvulsive Therapy Work? Theories on Its Mechanism," *Canadian Journal of Psychiatry* 56, no. 1 (January 2011): 13–18.

8. Hiroshi Kojima, "Information Processing in Synapses," in *Springer Handbook of Bio-/Neuroinformatics*, ed. Nikola Kasabov (Berlin: Springer Handbooks, 2014), 587–624.

realistic models of biological neurons might advance AI as we currently know it."[9]

The brain, in addition to the promises made on its behalf, is an idea whose materializations have been severe—not least for those white-collar workers whose brains have been fried by the allure of neuroenhancers; not least for those at-risk children being managed according to cognitive pedagogies or else prescribed Adderall regardless of whether they are symptomatic or not.[10] The biological turns in psychology[11] and psychiatry,[12] for example, have been taken up by the military, industry, and biomedical entrepreneurs. Projects now proliferate on neurogenetic engineering, transcranial magnetic stimulation, neuroprosthetics, and deep brain stimulation.[13] The brain has fueled all manner of institutional investments that promise a transformation of mind, soul, politics, higher education,[14] scholarly working conditions, and, if you were to grant figures like Silicon Valley's Ray Kurzweil even a

9. Matthew Roos, "Deep Learning Neurons versus Biological Neurons: Floating-point, Numbers, Spikes, and Neurotransmitters," *Towards Data Science* (March 14, 2019), https://towardsdatascience.com/deep-learning-versus-biological-neurons-floating-point-numbers-spikes-and-neurotransmitters-6eebfa3390e9.

10. Alan Schwarz, "Disorder or Not, Pills to Help in School," *New York Times* (October 9, 2012); Bethany Moreton, "S'More Inequality: The Neoliberal Marshmallow and the Corporate Reform of Education," *Social Text* 32, no. 3 (September 2014): 29–48; Margaret Talbot, "Brain Gain: The Underground World of 'Neuroenhancing' Drugs," *New Yorker* (April 20, 2009).

11. George Mandler, "Origins of the Cognitive (R)evolution," *Journal of History of the Behavioral Sciences* 38:4 (Fall 2002): 339–53.

12. Derek Richter, *Perspectives in Neuropsychiatry* (London: H. K. Lewis, 1950); W. R. Ashby, "Cybernetics," in *Recent Progress in Psychiatry*, ed. G. W. T. H. Fleming, 2nd ed. (London: J. and A. Churchill, 1950), 94–109. See also Anne Harrington, *Mind Fixers: Psychiatry's Troubled Search for the Biology of Mental Illness* (New York: W. W. Norton, 2019).

13. Robbin A. Miranda et al., "DARPA-Funded Efforts in the Development of Novel Brain-Computer Interface Technologies," *Journal of Neuroscience Methods* 244 (2015): 52–67; Ganesh R. Naik and Yina Guo, eds., *Emerging Theory and Practice in Neuroprosthetics* (Hershey, PA: IG Global, 2014). See also the neurotechnology company, Neuralink Corporation, founded by Elon Musk in 2016 that seeks to capture the information being processed by neurons and to use computers to store the results, analyze them, and to intervene. John Markoff, "Baby Steps to Linking Mind and Machine," *New York Times* (July 17, 2019): B1, B8. See also Adrian Curtin et al., "A Systematic Review of Integrated Functional Near-infrared Spectroscopy (fNIRS) and Transcranial Magnetic Stimulation (TMS) Studies," *Frontiers in Neuroscience* 13 (2019), https://doi.org/10.3389/fnins.2019.00084.

14. Josh Burk et al., "Neurodiversity: Creating an Inclusive College Classroom," Neurodiversity Initiative, William and Mary College, https://www.wm.edu/sites/neurodiversity/.

hint of legitimacy, immortality.[15] In addition to institutions, industries, and structures of feeling, materializations of the brain include strategies of, and new machines for, measuring brain activity. Such materializations, not surprisingly, occur alongside incredible feats of number crunching that have imparted a felt sense that the brain—mine and yours, too, dear reader—is an information processing device.[16]

Look around. Close your eyes. Listen. Cognitive models of neural processing have become abundantly present—in politics, business, and the academy, in pharmaceutical research and electroconvulsive treatment centers, in medical imaging suites, on children's television, in the percolating chatter of talk therapy and smartphone notifications. The brain has been taken up by preachers and magicians, by scientists, artists, and entrepreneurs, by engineers, athletes, and computer programmers, by psychologists and the CIA, by politicians, presidents, and policy makers, by criminologists, drug makers, and charter school administrators. The grammar of neuroscience is present in the TED talks of celebrity scientists and wannabe tech moguls, in brain-training apps such as www.lumosity.com, in the mundane tasks of swipe-laden socializing, in the kind of self that beams in and out of countless coordinated screens and digital devices distributed across the globe. These screens and these devices operate according to paradigmatic theories of neural networks and information processing that remain fun-

15. Kurzweil is the director of engineering for Google, cofounder and chancellor of Singularity University and chief executive officer of Kurzweil Technologies, Inc. Having long been involved in Silicon Valley projects, Kurzweil embodies both the mythic scope, ambition, and absurdity of the cognitive revolution. As a self-styled secular prophet of our information age (a group that includes many tech- and media-savvy futurists like Max Tegmark, Elon Musk, and your garden-variety *Wired* columnist), Kurzweil is, perhaps, best known for his predictions (and periodic updates) about when and how humans will inevitably transcend their biology and converge with machines. Everyone, according to Kurzweil, may soon become, at last, a pure distillation of their neuromatic self, living in a virtual world for eternity (or until the degradation of one's data or the catastrophic failure of server farms). See, e.g., *The Singularity Is Near: When Humans Transcend Biology* (New York: Penguin, 2005) and, with Terry Grossman, *Transcend: Nine Steps to Living Well Forever* (New York: Rodale, 2010).

16. Gopala K. Anumanchipalli et al., "Speech Synthesis from Neural Decoding of Spoken Sentences," *Nature* 568 (2019): 493–98; Haiguang Wen et al., "Neural Encoding and Decoding with Deep Learning for Dynamic Natural Vision," *Cerebral Cortex* 28, no. 12 (December 2018): 4136–60; Tyson Aflalo et al., "Decoding Motor Imagery from the Posterior Parietal Cortex of a Tetraplegic Human," *Science* 348 (May 22, 2015): 906–10.

damental to contemporary models of cognition. These screens and these devices, in other words, assume the human to be a complex, self-organizing system. This human processes information. This human is, at the end of the day, primarily and paradigmatically, a neuromolecular phenomenon. Society, in turn, is nothing more, and nothing less, than an aggregate of interactive brains.[17]

The brain, they say, is who we are, who we have always been, and who we will soon become. For we are about to assume our proper place in a cosmos saturated with self-organizing systems understood to operate according to the same transcendental logic as the brain.[18] Consequently, they say, the brain contains within it truths that are universal.[19] For "when each minute brain component has been located, its function identified and its interactions with each other component made clear—the resulting description will contain all there is to know about human nature and experience."[20]

17. Emily Martin, "Talking Back to Neuroreductionism," in *Cultural Bodies: Ethnography and Theory*, ed. Helen Thomas and Jamilah Ahmed (Oxford: Blackwell, 2004), 190–212.

18. Mohammad Zoqi Sarwani et al., "Personality Classification through Social Media Using Probabilistic Neural Network Algorithms," *International Journal of Artificial Intelligence and Robotics* (IJAIR) 1, no. 1 (2019): 9–15; Ahmed Sulaiman M. Alharbi and Elise de Doncker, "Twitter Sentiment Analysis with a Deep Neural Network: An Enhanced Approach Using User Behavioral Information," *Cognitive Systems Research* 54 (2019): 50–61.

19. There has been a steady increase in books that seek interpretive leverage on the seemingly endless decade of the brain in which we now find ourselves. Such books range from pointillist intellectual histories (Fernando Vidal, *The Sciences of the Soul: The Early Modern Origins of Psychology*, trans. Saskia Brown [Chicago: University of Chicago Press, 2006]) and pressing anthropological surveys of the scene (Joseph Dumit, *Picturing Personhood: Brain Scans and Biomedical Identity* [Princeton, NJ: Princeton University Press, 2003] and Nicolas Langlitz, *Neuropsychedelica: The Revival of Hallucinogenic Research Since the Decade of the Brain* [Berkeley: University of California Press, 2013]) to breezy and reactionary accounts (Paolo Legrenzi and Carlo Umilta, *Neuromania: On the Limits of Brain Science*, trans. Frances Anderson [New York: Oxford University, 2011]). See also Davi Johnson Thornton, *Brain Culture: Neuroscience and Popular Media* (New Brunswick, NJ: Rutgers University Press, 2011); Fernando Vidal and Francisco Ortega, *Being Brains: Making the Cerebral Subject* (New York: Fordham University Press, 2017); and Melissa M. Littlefield, *Instrumental Intimacy: EEG Wearables and Neuroscientific Control* (Baltimore: Johns Hopkins University Press, 2018).

20. Rita Carter, *Mapping the Mind* (Berkeley: University of California Press, 1999), 8.

2. APPROACHING THE NEUROMATIC (WITH A SHORT ENGINEERING ASIDE)

When the historian Michel Foucault flippantly writes that "Modernity begins when the human being begins to exist within his organism, inside the shell of his head," he is not to be taken literally.[21] He is suggesting, on the contrary, that the mind and its neural architecture have long served as the basis for a massive biopolitical project.[22] Ideas, according to Foucault, assume material force in the world—ideas fueling actions fueling concepts fueling investigations fueling imaginings fueling ideas fueling actions. And so on and so forth.[23] No beginning and no end. Which is to say that human beings do not, strictly speaking, exist solely inside their skulls. But the fact that they increasingly believe, act, and dream that they do adds no small bit of ballast to Foucauldian analyses of power and to this book which is, among

21. Michel Foucault, *The Order of Things: An Archeology of the Human Sciences* (New York: Vintage, 1994), 318.

22. A notorious but representative example being Karl W. Deutsch, *The Nerves of Government: Models of Political Communication and Control* (New York: Free Press, 1963). As Deutsch speculates, "If we think of an ethnic or cultural community as a network of communication channels, and of a state or a political system as a network of such channels and chains of command, we can measure the 'integration' of individuals in a people by their ability to receive and transmit information on wide ranges of different topics with relatively little delay or loss of relevant detail" (150).

23. I am inspired here by scholars who have attempted to think through issues of recursivity and cultural production. See, for example, Fernando Vidal's emphasis on the circular reinforcement of "brainhood." An anthropology of brainhood, argues Vidal, "predated reliable neuroscientific discoveries, and constituted a motivating factor of the research that, in turn, legitimized it" ("Brainhood, Anthropological Figure of Modernity," *History of the Human Sciences* 22, no. 1 [2009]: 14). Similarly, Kélina Gotman writes that the seductiveness of the neural metaphor is premised on its convenience and looping capaciousness in how it "shapes the biological and sociopolitical models it seeks to describe as much as it is shaped in turn by them, in a rhetorical to-and-fro between science and culture, scientific modeling and philosophical analysis" ("The Neural Metaphor," in *The Neuroscientific Turn: Transdisciplinarity in the Age of the Brain*, ed. Melissa M. Littlefield and Jenell M. Johnson [Ann Arbor: University of Michigan Press, 2012], 72). Rose and Abi-Rached have written of "the reciprocal relationship between concepts and technologies" (*Neuro*, 159). For a foundational statement on the "reciprocal production of science, technology, and society," see Andrew Pickering, *The Mangle of Practice: Time, Agency, and Science* (Chicago: University of Chicago Press, 1995), 1.

other things, a particular history of the brain as an "especially dense transfer point for relations of power" within this so-called secular modernity.[24]

For it is my hunch, shared by others, that to think about modernity is to think about the brain in terms of ideological density, epistemic reach, and the phenomenological swath that it cuts. Building on recent works that have noted the deep and abiding cognitive frame of modernity, an ever-encroaching neuro-ontology, and deployments of the brain in promiscuous projects of description, I want to think through how and why the brain has become so widely and intensely orientational. What to make of this model of a new kind of human who does not simply *have* a brain but whose very condition of existence *is* the brain? What to make of breathless copy that insists on a "synaptic self" and looping narratives of "how our brains become who we are"?[25] What lies behind this making of a brain-centered self that is subject to ethereal flows, a self so porous to the universe as to bleed into and, eventually, encompass it?[26]

A *neuromatic* brain has been integral to what one might call the metaphysics of this modernity.[27] Different parts and particles of the brain are identified and traced as part of a system, different locations are considered

24. Michel Foucault, *History of Sexuality* (New York: Vintage, 1990), 103.

25. See, e.g., the front matter of Joseph LeDoux, *Synaptic Self: How Our Brains Become Who We Are* (New York: Viking, 2002): "Synapses, the spaces between neurons, are the channels through which we think, act, imagine, feel, and remember, and also the means by which our most fundamental traits, preferences, and beliefs are encoded." On the capacity of the brain to detect and repair faulty synapses, see John Wade et al., "Self-repair in a Bidirectionally Coupled Astrocyte-neuron (AN) System Based on Retrograde Signaling," *Frontiers in Computational Neuroscience* 6, no. 76 (2012): 1–12.

26. According to Norbert Wiener (perhaps the most visible and self-conscious of cyberneticians), the human self was pure flow and the brain the agent of maintaining the proper fluidity. "The physical identity of an individual does not consist in the matter of which it is made," wrote Wiener. "We are but whirlpools in a river of ever-flowing water. We are not stuff that abides, but patterns that perpetuate themselves." Within this dynamic scheme agency consisted of nothing more than the maintenance of pattern, a hollow conceit that assumed substance only after the equations had been solved. Norbert Wiener, *The Human Use of Human Beings: Cybernetics and Society* (1950; New York: Houghton Mifflin, 1954), 101–2, 96. For a critique of the racializing politics of cybernetic plasticity, see Fred Moten, "The Touring Machine (Flesh Thought Inside Out)," in *Plastic Materialities: Politics, Legality and Metamorphosis in the Work of Catherine Malabou*, ed. Brenna Bhandar and Jonathan Goldberg-Hiller (Durham, NC: Duke University Press, 2019): 265–86.

27. Kay and Edwards have both used the notion of "cognitivism" to signify the discursive power and ontological privilege the brain assumed in the postwar developments in neurophysiology, information theory, and computer science. Lily E. Kay, "Cybernetics,

in their collective action, as leverage for cultivating the self and explaining the wide wide world.[28] The neuromatic, then, signals a particular physiology—how we imagine the brain to be in essence and how that particularity authorizes who we are seeking to become.

The neuromatic brain is currently evidenced in such commonsense talk of neural network connectivity and synaptic selves. The grammar, itself, is grounded in the aggressive framing of brain function in terms of the information-processing properties of its structural makeup. That framing, moreover, has the winning quality of being able to address heretofore underappreciated "emergent" feedback circuits and control loops, that is, "*network* effects of *system* effects."[29] Consequently, the *neuromatic* brain is something

---

Information, Life: The Emergence of Scriptural Representations of Heredity," *Configurations* 5, no. 1 (1997): 23–91, and Paul N. Edwards, *The Closed World: Computers and the Politics of Discourse in Cold War America* (Cambridge, MA: MIT Press, 1997), 2.

28. Egidio D'Angelo, "Toward the Connectomic Era," *Functional Neurology* 27, no. 2 (2012): 77. In claiming to have put theories of localized brain function to rest, the current embrace of a more expansive vision of nervous circuits spanning the entire brain and crossing anatomical boundaries is not necessarily new nor as revolutionary as often advertised. As historian Katja Guenther has argued, connectionism has long been embedded in localization theories, a tension between two paradigms that has played out in the American and European scenes of cognitive inquiry; *Localization and Its Discontents: A Genealogy of Psychoanalysis and the Neuro Disciplines* (Chicago: University of Chicago Press, 2015). Which is to say that theories of brain localization have ever been shadowed by theories of connectionism. In phrenology, for example, that exemplary strawman of localization, the faculties of mind may have been notoriously isolated but, more importantly, it was their communication with one another that guaranteed their optimal functioning; "Debate of the Faculties," *American Phrenological Journal* 13 (1851): 49–50, 73–75. By midcentury, as localization theories were being integrated into models that emphasized the "threadlike expansions" of neurons and their "mechanism," the brain became irreducible to one region or another or even to the parameters of neural tissue per se. In 1950, for example, scientists were confident that they were on the verge of a "comprehensive understanding" of the motor cortex as a specific location in the brain (where the German phrenologist Franz Joseph Gall had once designated the organ of acquisitiveness). They were also confident that this location was part of a "teeming multitude of living units that we must depend [on] for sensation, motion, understanding, consciousness"; Wilder Penfield and Theodore Rasmussen, *The Cerebral Cortex of Man: A Clinical Study of the Localization of Function* (New York: Macmillan, 1950), 1–2, 13.

29. Patricia S. Churchland, Christof Koch, and Terrence J. Sejnowski, "What Is Computational Neuroscience?" in *Computational Neuroscience*, ed. Eric L. Schwartz (Cambridge, MA: MIT Press, 1993), 46. Edwards, *Closed World*, 179–80.

more than mere flesh. It is a matter of neural networks processing information on a microscopic level, the brain interacting with its environment through the transfer of information, ever corresponding with itself and the world around—"information [being] the name for the content of what is exchanged with the outer world as we adjust to it, and make our adjustments felt upon it."[30]

The neuromatic brain was officially codified in the integration of information theory and theories of neural connectivity at midcentury. The basic paradigm for thinking about the brain soon became utterly persuasive—a vast neural network, processing information and communicating with itself in order to sustain itself and move the body forward in the world. There was an eloquence to the integration of the "logical calculus" of neural nets and information theory. Both of these intellectual streams were driven by the desire to control some aspect of the material world.

Claude E. Shannon's "Mathematical Theory of Communication" (1948) was published in the *Bell Labs Technical Journal* and presented to other engineers as an aid in understanding the relationship between inputs and outputs in a general communication system. Every communication system was comprised of six elements: 1) an information source, 2) a transmitter that transforms the source pattern into a signal, 3) a channel, 4) a receiver that decodes the signal back into pattern, 5) a destination, and 6) noise, or the inevitable degradation of signal. As part of the group of mathematicians who contributed to quality control at the American Telephone and Telegraph Company (AT&T), Shannon's demarcation was to address uncertainty of transmission head on—accounting for it in order to measure it, measuring it in order to predict and control it. Shannon's research agenda, moreover, was motivated by AT&T's drive to maximize profit by using its existing infrastructure more efficiently.[31] The mathematics were both pristine and practical.

Shannon's mathematical modeling of communication, first and foremost, was read for its potential commercial applications. As a general and groundbreaking proposition, Shannon proved that perfect communication between the source and destination could be achieved when communication channels were imperfect. In addition to using binary digits, or bits,

30. Wiener, *Human Use of Human Beings*, 17.

31. C. E. Shannon, "A Mathematical Theory of Communication," *Bell System Technical Journal* 27 (July 1948): 379–423, (October 1948): 623–56; Jonathan Sterne, *MP3: The Meaning of a Format* (Durham, NC: Duke University Press, 2012), 80.

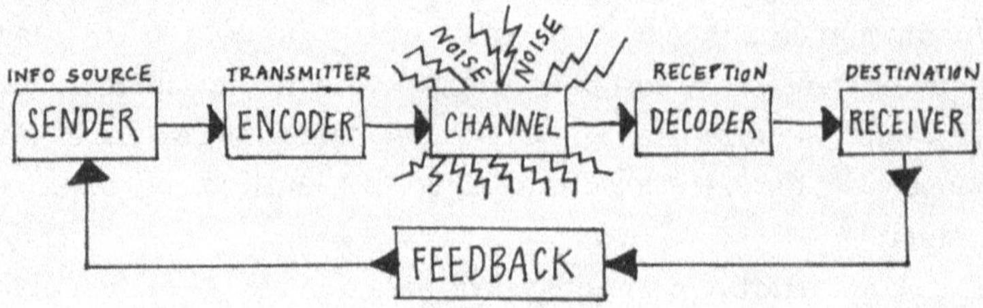

FIGURE 5. A mathematical theory of communication, as conceived by Claude E. Shannon in 1948. An approximation by Libby Modern.

to define any and all messages, Shannon's introduction of redundancy was revolutionary in that it suggested how a system could achieve error-free communication through channels riddled with noise. By way of repeating the transmission of any message enough times through a channel whose error rate was known, a communication system was able to not only guarantee accurate transmission of the message but also detect errors within the system as a whole and correct for them in future transmissions.[32]

In the decade following the publication of his groundbreaking article, Shannon witnessed the rapid uptake of information theory across numerous disciplines. As Shannon's friend at Bell Labs, J. R. Pierce, later commented, information theory "came as a bomb, and something of a delayed-action bomb."[33] Indeed, as it was read alongside emerging theories of automata and neural nets, information theory achieved much of its explosive charge.

In the development of digital computers, for example, information theory was essential in securing all but perfect communication between "organs" within a computer and between that organ network and its environment. Pioneering mathematicians such as John von Neumann relied on Shannon's theory to make real in the world the axiomatic propositions of Alan Turing. At the Institute for Advanced Study in Princeton, New Jersey, von Neumann began building what was essentially a Turing machine: a central processing

32. John von Neumann, *The Computer and the Brain*, 3rd ed., with foreword by Ray Kurzweil (1958; New Haven, CT: Yale University Press, 2012), xvi–xvii.

33. Pierce was Shannon's colleague at Bell Laboratories and best known for his work on satellite communication systems. Pierce was also a science fiction writer who published short stories under the pseudonym of J. J. Coupling in *Astounding Science Fiction*, the same magazine in which L. Ron Hubbard's theory of Dianetics first appeared in March of 1950.

organ where numbers were crunched and logical operations were carried out; a memory organ where the program and data that were generated were stored, a mass storage organ, and input and output channels.[34] As he theorized how to arrange those organs within a computer, von Neumann realized that Shannon's theory had made it possible to secure communication between these organs by way of parallel processing of information.

In addition to information theory, von Neumann's attempt to reverse engineer a human brain borrowed from the formal mathematics of neural processing worked out in "A Logical Calculus of the Ideas Immanent in Nervous Activity." In this 1943 article, the psychologist Warren McCulloch and mathematical prodigy Walter Pitts had provided a theory and framework for calculating the pathways through which neurons received, transmitted, and coordinated electrical signals. "A Logical Calculus" had been received with little fanfare among biologists and psychologists.[35] But it soon gained significant traction as it was read in dialogue with Shannon and featured prominently in Norbert Wiener's *Cybernetics* (1948) and in von Neumann's "The General and Logical Theory of Automata."[36]

For McCulloch and Pitts, the neuron was an ideal type, the truth of which was a matter of form rather than substance. According to McCulloch and Pitts, a neuron either fired or it did not and therefore could be represented as a proposition—on or off, a 1 or a 0. Each proposition related to

34. Von Neumann, *The Computer and the Brain*, xx.

35. Warren S. McCulloch and Walter Pitts, "A Logical Calculus of the Ideas Immanent in Nervous Activity," *Bulletin of Mathematical Biophysics* 5, no. 4 (1943): 115–33. "A Logical Calculus" owes much to the mathematical modeling of biological systems spearheaded by Walter Cannon at Harvard Medical School (of whom Wiener had been a colleague). Wilfrid Rall, "Some Historical Notes," in *Computational Neuroscience*, 3.

36. Such traction was already evident in Walter Pitts and Warren S. McCulloch, "How We Know Universals: The Perception of Auditory and Visual Forms," *Bulletin of Mathematical Biophysics* 9, no. 3 (1947): 127–47. The cross-fertilization of information theory and neural nets was vividly on display in articles collected in *Automata Studies*, ed. C. E. Shannon and J. McCarthy (Princeton, NJ: Princeton University Press, 1956). In the first essay of this volume (published earlier as Rand Research Memorandum 704 [1951]), the mathematician S. C. Kleene declared that the McCulloch-Pitts assumptions gave a nerve net the character of a "digital automaton," as opposed to an analog mechanism (5). See also Mary A. B. Brazier, "Neural Nets and the Integration of Behavior," in Derek Richter, *Perspectives in Neuropsychiatry* (London: H. K. Lewis, 1950): 35–45. The integration of information theory and neural networks was also enshrined in proximate articles in *Fortune* by Francis Bello; see, e.g., his "The Information Theory," *Fortune* (December 1953): 136–40ff. and "New Light on the Brain," *Fortune* (January 1955), 104–7.

other propositions. And so on and so forth. Just as Shannon had begun thinking about Boolean algebra in relation to the electromechanical relays of the Differential Analyzer at MIT in the late 1930s, McCulloch and Pitts had applied Boolean logic to the neural network. In figuring each neuron as an on/off valve for transmitting an electrical pulse,[37] McCulloch and Pitts were approaching the brain in terms of computational processing. For them, the network of neurons was an elegant problem of logic—the brain is made up of a network of neurons, each one firing or not depending on the number and (perhaps even patterns) of incoming signals across the axons that were connected to it. Each neuron "communicated" with the other and, in tandem, communicated with the outside world.[38]

As an initial foray into mapping the relationality of these propositions in the form of complex mathematical equations, McCulloch and Pitts approached the brain as a vast communication network bent on processing information.[39] They had succeeded in their goal "to construct theory enough to be able to state how a nervous system could do anything." In arguing that a network of ideal neurons could calculate all calculable problems, McCulloch and Pitts provided the neural equivalent of a Turing machine. Or, as von Neumann contended, McCulloch and Pitts had solved the "purely formalistic" problem of finding an "equivalent network" that would "explain" the computational function of the central nervous system.[40] McCulloch

37. C. E. Shannon, "A Symbolic Analysis of Relay and Switching Circuits," *American Institute of Electrical Engineers Transactions* 57 (1938): 713–723.

38. Von Neumann, *The Computer and the Brain*, 55.

39. Joseph Dumit, "Plastic Diagrams: Circuits in the Brain and How They Got There," in *Plasticity and Pathology: On the Formation of the Neural Subject*, ed. David Bates and Nima Bassiri (New York: Fordham University Press, 2016), 225. As Dumit points out, such an approach opened up the possibility of creating a map that coordinated the spatial and temporal registers of neural activity. In the wake of "A Logical Calculus," the neuropsychologist Donald Hebb wrote of how information was represented and stored in the brain by way of neurons working in groups. His was a call to replace behaviorism's emphasis on conditioning and response with a model of the brain as an information processing device. D. O. Hebb, *The Organization of Behavior: A Neuropsychological Theory* (New York: John Wiley and Sons, 1949), xiii–xiv.

40. Their "theory" was modeled on neuroanatomical discoveries of the previous decade but, as McCulloch later acknowledged, the explanatory purchase of their scheme was not dependent on the latest discoveries in neuroanatomy. It was, in other words, doubly disembodied—an abstraction of a logically informed approximation. Which is to say that McCulloch's science was wholly cybernetic in its desire for control rather than realist representation. John von Neumann, "A General and Logical Theory of Automata" and discus-

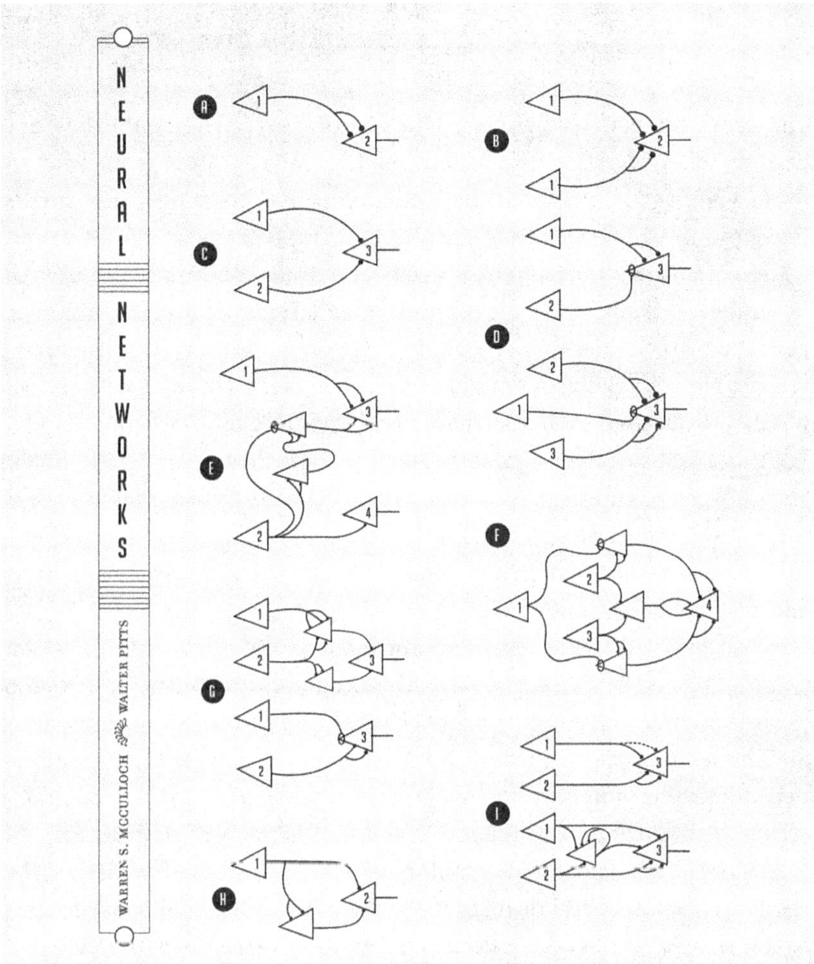

FIGURE 6. Neural nets schematic, from McCulloch and Pitts, "Logical Calculus," *Bulletin of Mathematical Biophysics* 5:4 (1943). An approximation by Libby Modern.

and Pitts, in other words, had eviscerated the brain, reducing it to a logical problem to be solved. The only thing left was the numbers.[41]

---

sion in *Cerebral Mechanisms in Behavior: The Hixon Symposium*, ed. Lloyd A. Jeffress (New York: Wiley, 1951), 32–34.

41. H. D. Landahl, Warren S. McCulloch, and Walter Pitts, "A Statistical Consequence of the Logical Calculus of Nervous Nets," *Bulletin of Mathematical Biophysics* 5, no. 4 (1943): 135–37.

The proposition that cognition is a neuromatic phenomenon has now become self-evident for neuro- and cognitive scientists, not to mention for those who live vis-à-vis brain-built environments—from the black mirrors they hold in their hands to the neuromodulating drugs they take, with or without a prescription. Here is a variety of holism whose authority is based in numerical truth[42]—circuit diagram upon circuit diagram, a computer-generated compilation of the brain in three dimensions. It codes incoming sensory signals—patterns of light or sound—in terms of information, tracing their impact on an initial chain of neurons and following the tremulations and loops of the initial signal as it spreads inward on parallel tracks and outward according to its own organized fashion.[43] Remarkable, indeed, is the system that corrects its own internal communications, that accounts for the noise within yet remains a black box, still in some respects unknown and at some level unknowable.

With over 10^{14} synapses, the brain remains the horizon of organized complexity—be it of self or society. This brain, one might say, is a fetish of the highest order, or, perhaps, an "enchanted loom" to borrow a phrase from Sir Charles Sherrington, whose work on neural communications within a network setting won him a Nobel Prize in 1932. Sherrington was central to seeing the brain as a mechanism—a "shifting harmony of subpatterns" and "relays of electrical potentials running along nerve-paths"—but insisted that the "step from electrical disturbance in the brain to the mental experience" would remain, at the end of the day, a "mystery," generative of much good in its ultimate insolvability.[44] The brain is enchanted, then, to offer a slightly more ominous spin than Sherrington, not simply because of its opaque complexity. On the contrary, the brain is enchanted because of the fierce wishfulness that it has catalyzed, the promises that it has precipitated, and the

42. Anne Harrington has explored connectionism's roots in holism before it broke bad and became fodder for eugenicists; *A Reenchanted Science: Holism in German Culture from Wilhelm II to Hitler* (Princeton, NJ: Princeton University Press, 1999).

43. R. Cameron Craddock et al., "Connectomics and New Approaches for Analyzing Human Brain Functional Connectivity," *Gigascience* 4, no. 13 (2015), 1–12; Olaf Sporns and Danielle S. Bassett, "Editorial: New Trends in Connectomics," *Network Neuroscience* 2, no. 2 (2018): 125–27.

44. Sir Charles Sherrington, *Man on His Nature* (The Gifford Lectures, 1937–38) (Cambridge: Cambridge University Press, 1940), 225, 210, 112, 124.

disciplines that have accompanied its designs. Everywhere it is conjured, the neuromatic brain is accompanied by "transcendent expectation."[45]

3. BLURRED LINES

Neuromatic offers a fugitive history of a cognitive revolution that is vast in scope and long in the making.[46] Mine are stories that foreground the ideas, techniques, and practices that have gone into the assembly and present maintenance of the brain as the locus of all that is human and all that is not—from the magnetic utterances of the MRI to the claims of contemporary cognitive scientists of religion; from eighteenth-century brain anatomies to theologies built around them; from the spread of phrenological cabinets and mental pieties in the nineteenth century to the discovery of the motor cortex and the emergence of the brain wave as a measurable manifestation of cognition; from military-fueled research into neural networks and artificial intelligence to the founding of brain-centric religious organizations such as Scientology and Eileen Garrett's Parapsychology Foundation, Inc.; from the deployments of cognitive paradigms in electric shock treatment in the 1970s to the work of Barbara Brown, neurofeedback pioneer, whose goal was to "teach man to perceive and to control some of his brain functions."[47]

45. Hans Blumenberg, *The Legitimacy of the Modern Age* (Cambridge, MA: MIT Press, 1985), 31, 68. As physiologist Richard Andersen has recently written, neuroscience has been able "to tap . . . into high-level neuro-processing—the intent to initiate an action—and then convey . . . the relevant electrical signals to a robotic arm." The robot and the brain can eventually be made to speak the same language, to one another, back and forth, communicating over a seemingly unbridgeable gulf. A miracle of sorts. "I get goosebumps every time I see it," Andersen admits. Richard Andersen, "The Intention Machine: A New Generation of Brain-Machine Interface Can Deduce What a Person Wants," *Scientific American* 320, no. 4 (April 2019): 24–31.

46. I am using the term "cognitive revolution" in a much broader sense than scholars who focus mainly on the cognitive sciences. For examples of the latter, see George Mandler, *A History of Modern Experimental Psychology from James and Wundt to Cognitive Science* (Cambridge, MA: MIT Press, 2007); Bernard J. Baars, *The Cognitive Revolution in Psychology* (New York: Guilford Press, 1986); Howard Gardner, *The Mind's New Science: A History of the Cognitive Revolution* (New York: Basic Books, 1987).

47. Barbara B. Brown, *New Mind, New Body: Bio-feedback: New Directions for the Mind* (New York: Harper and Row, 1974), 57.

Narrated self-consciously as a history of the present, *Neuromatic* offers a wide-ranging, meticulous, and joyously bizarre history that undergirds our evolving and enthusiastic subservience to the brain. For the brain, or something like it, is now figured as the site of whatever humanity is and will be and, perhaps, what constitutes its proper object of worship. One might even go so far as to argue that the current scope and practical impact of universalizing the brain resembles nothing less than a religious revival, spreading out and in with every click and swipe and study and scheme of neural enhancement.[48] But that is not my argument, exactly. Rather than make the case for religious and/or secular passions about the brain, I seek instead to conjure the brain as a nexus of power relations, as a site of animated commitment to so-called secular *and* religious orders both.

Such commitments, not to mention the enthusiasm, trouble the so-called divide between the religious and scientific, faith and reason. Consequently, I dwell on the brain as an object of multifarious reverence, from the bent mechanistic philosophies of the eighteenth century in which the nervous system assumed properties of self-organization in relationship to its environment, to appeals to the ethereal emanations of consciousness in the nineteenth century, to the measurement of brain waves and mapping of neural networks in the twentieth century, to contemporary claims to have discovered the neural correlates of religion.

In moving back and forth between professional scientists and Protestant revivalists, philosophers, parapsychologists, and pietists, speed readers and spiritualists, mathematicians and cyberneticians, mesmerists, artists, and political activists, I take up the relationship between religion and science from an odd angle. Rather than assume an ontic distinction between religion and science from the beginning, I focus on the production of the religious-secular difference as it has inflected a range of subject positions, politics, and epistemologies, regardless of whether they call themselves religious or secular or something else altogether. From this perspective, the neuroscientific professions prove to be particularly powerful formations of the secular order—which is to say that the history, authority, and reach of these sciences are not merely bound up (sometimes antagonistically, sometimes not) in the

48. For a provocative take that moves against the self-satisfied narratives of secular progress that revolve around the brain and "the essentials of human nature," see Max Stadler, "The Neuromance of Cerebral History," in *Critical Neuroscience: A Handbook of the Social and Cultural Contexts of Neuroscience*, ed. Suparna Choudhury and Jan Slaby (Oxford: Blackwell Publishing, 2016), 151.

history of "religion" but have also produced particular forms of religion at every turn. For over the course of the past few hundred years, the "scientific" consideration of the brain vis-à-vis "religion" *and* the deliberate practice of piety as a brain-centered enterprise are part of the *same* discursive weave.

Indeed, it would not be an exaggeration to claim that the turn to the secular that marks certain strains of modernity has, from the beginning, been bound up in the explanatory allure of the brain. In the opening salvos of the Enlightenment, for example, the brain became a site of interpretive struggle as the politics of secular differentiation took hold. As the quest for the seat of the soul became ever more empirical, the very concept of soul was rearticulated in order to address new, pressing questions about the self and world. For Thomas Willis (1620–75) and others, the brain served what God once did, as a spur and resource for speaking of things true and everlasting. Yet the brain was not considered God per se but rather the center of a humane universalism. Ironically, such claims to timelessness persist and call into question the secularity of any secular age and the assumption that the neuro- and cognitive sciences are categorically (or for that matter, affectively) different from the religions they claim to study.

Consequently, I am not interested in tracking either religion's decline and/or stubborn persistence (a tracking that grounds and makes possible the so-called secularity of scientific investigations). *The religious is not something out there in the world or in the brain waiting to be discovered, taken up, and dissected or, for that matter, waiting to be transformed into the secular.* On the contrary, it is my contention that religion is both an end and means of incredible discursive investment. *Neuromatic* dwells within moments and stories of this investment—and on those who produced religion as a neural matter, who framed the brain as the epistemic and political horizon of a strictly human world or else designated the brain as a matter of the utmost religious importance: on *persons, whether they were self-consciously secular, religious, or somewhere in between.*[49] So rather than conform to a secular framing of the cognitive sciences as emerging from the religious or (which amounts to the same thing) rehearsing the history of piety impinging

49. For an effort to split the difference between naturalism and constructionism vis-à-vis religious experience, see Ann Taves, *Religious Experience Reconsidered: A Building-Block Approach to the Study of Religion and Other Special Things* (Princeton, NJ: Princeton University Press, 2009). My approach differs from Taves's in that I subject to genealogical scrutiny not only the category of religion but also the authority of scientists who claim to be studying religion and its experiential correlates.

on scientific practice, I seek to narrate a history of the brain itself without recourse to the comforts of secular critique.[50]

The blurred lines that I limn within and across chapters persistently return to cybernetics—that mid-twentieth-century coalition of mathematicians, military planners, neurophysiologists, electrical engineers, psychologists, social scientists, economists, corporate strategists, artists, writers, parapsychologists, do-it-yourself religionists, biofeedback practitioners, and computer programmers who considered the intimate relationality between biological and mechanical worlds. Within the cybernetic fold, the brain served as both material object and metaphysical horizon, a window on the cosmos whose view instructed, disciplined, and, quite literally, tied down individuals, as in the case of electric shock therapy.

In addition to addressing the oft-cited God-talk of cyberneticians, my aim is to gain leverage on the more pedestrian, pressing, *and* profound claim that *their* scientific method was "an essential part of nature."[51] I argue that the religious register of cybernetics does not simply lie in the so-called mystical leanings and investments of its practitioners but also in how their practices consolidated centuries of neural and metaphysical speculation and contributed to a pervasive understanding of self and world as essentially neuromatic. Each of the chapters addresses what might be called the long history of the neuromatic brain before, during, and after it assumed institutional authority within the cybernetic fold.

Chapter 1 examines the conceptual present of the cognitive science of religion (CSR). It tends, as any empirically minded, scientific approach would, to ecological confounds, cultural artifacts, and the conditions that make possible (and increasingly legible) cognitive investigations into religion. I argue that CSR is part and parcel of a contemporary secular imaginary not because of its scientific pose vis-à-vis the religious but rather because of the kind of human it assumes and instantiates with every pronouncement it makes about human religiosity—a human hardwired to believe but capable, at the end of the day, of overcoming this proclivity.[52]

50. Talal Asad, Wendy Brown, Judith Butler, and Saba Mahmood, *Is Critique Secular?: Blasphemy, Injury, and Free Speech* (New York: Fordham University Press, 2013).

51. L. J. Fogel, A. J. Owens, and M. J. Walsh, "Intelligent Decision-making Through a Simulation of Evolution," *Simulation* 5, no. 4 (1965): 267–79.

52. For with their increasingly precise analyses of religion within the brain, the vast majority of cognitive and neuroscientific studies reproduce and reify key differentials within the categorical bulwark of secular modernity. Indeed, the brain's power and status reside in its historical role in securing a logic that makes compatible all manner of

I am particularly interested in what cognitive scientists refer to as the "hyperactive agency detection device"—the bundle of cognitive processes that prime humans to scan for and believe in supernatural agents.[53] By moving across a number of sites integral to the making of the conceptual infrastructure of hyperactive agency detection, I situate CSR as a formation of secularism with a distinctive but flickering Protestant character. In doing so, I consider the degree to which the imagination of religion by neuro- and cognitive scientists in this secular age is as much, if not more so, about the intensities of psychic investment, epistemic certitude, and passionate sociality that these same scientists dismissively ascribe to religion.[54] Indeed, it is precisely the pathological primacy they ascribe to religion, or something like it, that signals their secular bona fides: primitive proclivities ever in need of colonization, containment, or, at the very least, mansplaining correction.

Which is to say that much of this secular age has emerged in and through differences that are assumed to be neurological and therefore universal. For within the historical performance of so-called secular sciences such as CSR,

---

differences—between the religious and the secular, men and women, black and white, self and other, reason and madness. Whether being weaponized or radically questioned, these are the differences that ground our world and animate our view of it. A network of associated differentials building up and out, threading their way across skin and synapse and weaving their way into consciousness. Differentials whose interconnected, interactive, and looping quality generates powerful institutions and petty habits, routine rituals and ideological regimes. Historically speaking, these differentials have long worked together in their discursive production of one another and the metaphysics of this modernity. For recent work on the fraught categorical production of secularism, see Saba Mahmood, *Religious Difference in a Secular Age: A Minority Report* (Princeton, NJ: Princeton University Press, 2015), and Peter Coviello, *Make Yourselves Gods: Mormons and the Unfinished Business of American Secularism* (Chicago: University of Chicago Press, 2019).

53. Stewart. E. Guthrie, "A Cognitive Theory of Religion," *Current Anthropology* 21, no. 2 (1980): 181–203; Justin L. Barrett, "Exploring the Natural Foundations of Religion," *Trends in Cog Sciences* 4 (2000): 29–34; Pascal Boyer, *Religion Explained: The Evolutionary Origins of Religious Thought* (New York: Basic Books, 2001); Ilkka Pyysiäinen, *Supernatural Agents: Why We Believe in Souls, Gods, and Buddhas* (New York: Oxford University Press, 2009). See also Uffe Schjoedt, Wesley J. Wildman, Richard Sosis, and Joseph Bulbulia, "Vikings, Virtual Reality, and Supernatural Agents in Predictive Minds," 1, and Marc Andersen, "Predictive Coding in Agency Detection," both in a special issue of *Religion, Brain and Behavior* 9, no. 1 (2019): 65–84.

54. Or, to put this another way, the intensities associated with religion-making are the through line of this secular modernity.

religion is more often than not associated with the feminine.[55] In the eighteenth century, for example, an ancient connection between nerves, rationality, and male sexual prowess was renewed as the brain inspired scientific, technological, medical, and philosophical advance. In the nineteenth century, American neurology emerged, in part, as a diagnostic response to the "mediomania" of spiritualist women. In the mid-twentieth century, cybernetic theories of automata and disembodied cognition were strategies to overcome "biological noise" and other feminized threats to communication.[56] Presently, the masculine fantasies coded into neuroscientific understanding are often noted but not sufficiently explained.

To be sure, much can be made of the overrepresentation of male scientists and female test subjects within the history of neuro- and cognitive sciences. But numbers alone do not address the discursive processes of neuromation.[57] For cognitive scientists are the most obvious and well-funded

55. Jonathan W. VanRyzin et al., "Microglial Phagocytosis of Newborn Cells is Induced by Endocannabinoids and Sculpts Sex Differences in Juvenile Rat Social Play," *Neuron* 102, no. 2 (2019): 435–49; Lary Cahill, "Denying the Neuroscience of Sex Differences," *Quillette* (March 29, 2019), https://quillette.com/2019/03/29/denying-the-neuroscience-of-sex-differences/. See also the policy of the NIH, adopted in 2016, that requires researchers to consider sex as a biological variable. For pushback against the more blithe biologizations of sexual difference, see Rebecca Jordan-Young and Raffaella I. Rumiati, "Hardwired for Sexism? Approaches to Sex/Gender in Neuroscience," *Neuroethics* 5, no. 3 (2012): 305–15; Cordelia Fine, *Delusions of Gender: How Our Minds, Society, and Neurosexism Create Difference* (New York: W. W. Norton, 2010); Robyn Bluhm, Anne Jaap Jacobson, and Heidi Lene Maibom, eds., *Neurofeminism and the Intersection of Feminist Theory and Cognitive Science* (New York: Palgrave MacMillan, 2012); Gina Rippon, *The Gendered Brain: The New Neuroscience That Shatters the Myth of the Female Brain* (New York: Random House, 2019). Cahill calls Rippon a form of bad science, i.e., bad religion. These "anti-sex difference writers," warns Cahilll, treat "brain plasticity" as "a magic talisman with no limitations that can explain away sex differences . . . and resurrecting 19th century [*sic*] arguments almost no neuroscientist knows of, or cares about."

56. Joan Wallach Scott, *Sex and Secularism* (Princeton, NJ: Princeton University Press, 2018), 64–65; K. S. Lion and D. F. Winter, "A Method for the Discrimination between Signal and Random Noise of Electrobiological Potentials," *Electroencephalography and Clinical Neurophysiology* 5 (1953): 109–11.

57. The data of overrepresentation are bound up in a longer history in which sexual difference is made intelligible (and politically generative) by cognitive inquiries. Daphne Joel and Luba Vikhanski, *Gender Mosaic: Beyond the Myth of the Male and Female Brain* (New York: Little, Brown Spark, 2019). *Gender Mosaic* is part of a spate of recent work that identifies sexism in terms of representation—the representation of female test subjects and the crass claims that male and female brains are fundamentally different in particular pro-

edge of religion-making in the present. And the questions they pose and the answers they provide are part of a longer, wider, and more promiscuous history of charging religion with cognitive mechanics and neural implication. In order to offer historical leverage on the depth, density, and diffusion of this charge, mine is a slanted albeit strategic response to Pascal Boyer, one of the most visible figures in the contemporary cognitive science of religion. Boyer is cocksure in his argument that most scholars of religion are "lackadaisical" in their "approach to explaining religious thought and behavior." Consequently, "the field has become . . . unresponsive to actual scientific proposals."[58] As an intervention into the cognitive turn within Religious Studies, the first chapter turns the tables on a discipline that presents itself as beyond critique.

Chapter 2 revisits the engineering imagination of Emanuel Swedenborg (1688–1772) and the history it anticipated. In 1718 the Swedish polymath wrote a treatise on brain vibrations as he supervised the transportation of Royal Swedish Navy ships during the Great Northern War. Swedenborg drew from Willis's gallery of images and descriptions of brain function in terms of fluids, juices, ropes, fibers, spirits, sugar cane stalks, and a church organ blowing air into pipes.[59] In studying brain "tremulations" alongside bomb trajectories, Swedenborg signified a lasting synergy between metaphysical speculation and military planning. He described tremulations in terms of a higher physics that transcended mere linearity, a movement that was both mechanistic *and* the hinge of divine flux.[60] As Swedenborg would

---

cessing registers. Joel and Vikhanski seek to subvert the strictures of this binary by pointing out how the brain and, by extension, one's biological identity, is a patchwork of "masculine" and "feminine" traits. This notion of gender as combinatory within the individual, however subtle its rendering, leaves in place difference as the organizing metaphor of sexuality.

58. Pascal Boyer, "Explaining Religious Concepts: Levi-Strauss the Brilliant and Problematic Ancestor," in *Mental Culture: Classical Social Theory and the Cognitive Science of Religion*, ed. Dimitris Xygalatas and William W. McCorkle Jr. (Durham, UK: Acumen Publishing, 2013), 164–75.

59. Wes Wallace, "The Vibrating Nerve Impulse in Newton, Willis and Gassendi: First Steps in a Mechanical Theory of Communication," *Brain and Cognition* 51, no. 1 (2003), 77.

60. As Jonathan Sheehan and Dror Wahrman have shown, Swedenborg was part of a larger epistemic shift within the early modern period—a secular stirring that precipitated all manner of materialisms "beyond mechanism"; *Invisible Hands: Self-Organization and the Eighteenth Century* (Chicago: University of Chicago Press, 2015), 26. Jessica Riskin has argued that a mechanistic approach to science left much room for theology despite its insistence to the contrary. "A material world lacking agency assumed, indeed required, a

later insist, "I have pursued this anatomy [of the brain] solely for the purpose of discovering the soul" and addressing topics of "too sublime a nature to be explained to the common understanding."[61] And while Swedenborg's conception of the brain as a self-organizing entity had little impact during his lifetime, it would have a substantial impact on the profusion of metaphysical pieties in nineteenth-century America. These pieties, in turn, would till the ground for neurophysiological consensus in the twentieth.[62]

In tracing the afterlives of Swedenborg through the nineteenth century into the twentieth, I contemplate what it means for the brain to have become both the source and spur of piety. I revisit a cast of characters whose interest in "active relations" between brain and environment offers a neural backstory of information theory—the shared mathematical grammar of the cybernetic sciences. For how in the world did information become the default frame of neural activity? How did information theory become the means by which to measure, explain and/or experience the logic of correspondence between the inside of the head and whatever might lie at the end of the universe? In addressing such questions, my "genealogy of information" moves from eighteenth-century visions of cerebral organization to the spread of phrenological cabinets in the nineteenth century to ethnographies

---

supernatural god," or, at the very least, theories of active mechanism in which the things of nature contain within them their own "sources of action"; *The Restless Clock: A History of the Centuries-Long Argument Over What Makes Living Things Tick* (Chicago: University of Chicago Press, 2016), 4, 7.

61. Emanuel Swedenborg, *The Soul, or Rational Psychology* (New York: New-Church Press, 1887), xxx; Emanuel Swedenborg, *The Brain, Considered Anatomically, Physiologically and Philosophically*, vol. 1, ed. and trans. R. L. Tafel (London: Speirs, 1882), 80.

62. To be sure, the mentalization of piety occurred within more mainline religious venues than explored herein. See, e.g., Rev. J. Bourne Jones, *The Mind: Its Faculties, and Their Culture, A Lecture* (London: Elliot Stock, 1877) and John R. Macduff, *Mind of Jesus* (New York: American Tract Society, ca. 1860). Even the best studies of religion, psychology, and/or medicine hew to denominational lines and church history pedigrees. See, e.g., Donald Meyer, *The Positive Thinkers: Religion as a Pop Psychology from Mary Baker Eddy to Oral Roberts* (New York: Pantheon Books, 1980); Ann Taves, *Fits, Traces, and Visions: Experiencing Religion and Explaining Experience from Wesley to James* (Princeton, NJ: Princeton University Press, 1999); Christopher White, *Unsettled Minds: Psychology and the American Search for Spiritual Assurance, 1830–1940* (Berkeley: University of California Press, 2008); Anne Harrington, *The Cure Within: A History of Mind-Body Medicine* (New York: W. W. Norton, 2008); Tanya M. Luhrmann, *When God Talks Back: Understanding the American Evangelical Relationship with God* (New York: Random House, 2013).

of the spirit land of the Diakka—racialized spirits who "psychologize your nervous and muscular systems" for their own amusement—to the discovery of the motor cortex in the late nineteenth century and the emergence of electroencephalography in the early part of the twentieth century to the discovery of neural nets and the displacement of behaviorist paradigms in the twentieth century. In tracing the blur of religious and scientific enthusiasms that culminates in the emergence of cybernetics in the Cold War years, I unapologetically include those who would not be recognized as sufficiently scientific in their practice.[63]

Chapter 3 explores the occult margins of cybernetics at midcentury and how the fantasies of cognition as neuromatic and disembodied play out in the arenas of parapsychology, Scientology, and high modernist magic. Here I focus on the relationship between four figures whose shared preoccupation with cybernetics resulted in bizarre but predictable applications beyond the transmission of electronic messages across a wire: the medium Eileen Garrett, who founded the Parapsychology Foundation in New York City in 1951; the artist Brion Gysin, whose first book on the ethereal "system" of white racism was published by Garrett; the experimental writer William S. Burroughs, whose "cut-up" experiments with Gysin involved slicing and cutting, folding and taping words (his own as well as others') together in order to reveal their deleterious effects on "psycho-sensory processing"; and L. Ron Hubbard, author of *Dianetics: The Modern Science of Mental Health* (1950) and the founder of Scientology. Garrett, Gysin, Burroughs, and Hubbard, I argue, each distilled a metaphysical directive undergirding neuromatic theories of communication by using them to make sense of supernatural orders of existence, variously construed.

In the shadows of the cybernetic mainstream, each of these figures sought knowledge about the relationship between human cognition and patterns that transcended the merely human realm. And each insisted that one must possess the technical means for addressing those patterns and what lies beyond them—patterns in language, patterns that controlled indi-

63. For those caught up in securing the secular status of their profession and distancing it from the mystifications of religion, phrenology often serves as a synecdoche of the pseudo and the entire history of errant neural inquiry. See, e.g., Rita Carter, *Mapping the Mind* (Berkeley: University of California Press, 1999), 41; Robert Restak, *The Naked Brain: How the Emerging Neurosociety Is Changing How We Live, Work, and Love* (New York: Harmony Books, 2006), 214. On phrenology as "a false start" that "belongs to the pre-history of neuroscience, rather than its early history," see Elkhonon Goldberg, *The New Executive Brain: Frontal Lobes in a Complex World* (New York: Oxford University Press, 2009), 37.

vidual lives, patterns that were woven into every brain and into the very fabric of the cosmos. Yet in their compulsion to recognize pattern, Hubbard, Garrett, Gysin, and Burroughs also strayed from the directives of disembodiment exhibited in the cybernetic mainstream. So even as they distilled the metaphysical impulse of a burgeoning cybernetic sprawl, they each forwarded a different and more capacious kind of neuromation in which eros, history, affect, and power played a significant role in imagining the ends and means of the brain. In other words, Hubbard, Garrett, Gysin, and Burroughs acknowledged that which more stringent scientific protocols did not—the desires of the flesh, to be sure, as well as *how* those desires are coded by the positivity of discourse.

Chapter 4 offers yet another tendril history by exploring the neuromatic character of electric shock therapy (ECT). The chapter tells the story of psychiatric reform as it played out at Napa State Mental Hospital in Napa, California, from the late nineteenth century to the late 1970s. Opening as Napa Asylum for the Insane in 1875, it proudly implemented protocols of "moral treatment" by targeting the cultivation of conscience in a gendered key. By the end of World War II, Napa State had become an enthusiastic adopter of ECT as the brain increasingly replaced the genitals as the source of sexual pathology and site of sexual differentiation.[64] I argue that as cybernetic strategies of organizational management took hold within mental asylums such as Napa State, ECT became an efficient and effective form of gender torture. For whatever might have been deviant, dangerous, or errant about one's sexuality or erotic comportment, ECT became the preferred means of addressing such "failure of secular embodiment."[65]

I interrogate this preference as it plays out at Napa State and other locales such as Lansing, Michigan, where Dr. H. C. Tien, an independent electroconvulsive therapist, ran the Michigan Institute of Psychosynthesis. Beginning in the late 1960s Tien also advocated a "systems approach" to psychiatry and insisted that Norbert Wiener's *Cybernetics* "should be required reading for every psychiatrist." Tien's innovation was ELT, which technically stood for electrolytic therapy or, as Tien was quick to point out, "electric love therapy." The majority of Tien's patients were white working-class women whose

64. As resident physician Dr. John Champlin remarked in 1976, the "deep humanism" of Norbert Wiener's conception of cybernetics and general systems theory were privileged levers of psychiatric practice at Napa State; "New Flowers of Kent," *Napa Quarterly* (Winter 1976).

65. Coviello, *Make Yourselves Gods*, 235–36.

families, relationships, and psyches had fallen on hard times. Tien's practice of ELT, which involved an elaborate closed-circuit television system and the promise of personality reprogramming, was a culmination of envisioning human cognition as a neuromatic proposition.[66] For, as the electrodes were applied to a shaved section of the scalp and 130 volts of current delivered for less than a second at a time, the brain *became* a machine and was not simply *like* a machine.[67]

At the end of the day, I insist that the neuromatic brain must be seen in light of its history, in light of its massively institutionalized capacities to transform a person into a socially productive concept. For this triumph of the neuromatic brain has signaled more than just the reduction of the brain to an information-processing device. It has also signaled the reduction of everything within its purview—and everything it could *imagine to be* within its purview—to systems that traffic in information. And once the human has been reduced, in theory, to a system, chock full of neurons processing information, the human is then animated by the same logic that animates the birds and the bees, ant colonies, weather systems, plant ecologies, thermostats, telephone wires, antiaircraft missiles, and all the machines that we *serve* and are *served by* and that fill our days with the joys of social mediation.[68] Forever and ever. Amen.

66. John M. Friedberg, "Shock Treatment, Brain Damage and Memory Loss: A Neurological Perspective," *American Journal of Psychiatry* 134, no. 9 (September 1977): 1010–13.

67. This overcoming of the brain-computer metaphor—acknowledging the substantive differences between brains and computers in order to insist on their formal similarities—is, perhaps, the lasting legacy of the neuromatic paradigm. Indeed, the literalism built into this cyborg ontology is still a talking point among cognitive scientists and their fellow travelers. Cognitive scientist Zenon Pylyshyn distilled the cybernetic complaint that "there is no reason why computation ought to be treated merely as a metaphor for cognition, as opposed to a hypothesis about the literal nature of cognition"; see his "Cognition and Computation: Issues in the Foundations of Cognitive Science," *Behavioral and Brain Science* 3, no. 1 (1980), 114. For a more recent effort to overcome the metaphorical gap between culturally laden categories and the truth of "nonordinary experiences," see Ann Taves et al., "What Counts as Religious Experience? The Inventory of Nonordinary Experiences as a Tool for Analysis across Cultures," forthcoming in *Archive for the Psychology of Religion*; preprint DOI: 10.31234/osf.io/ux28d.

68. The revolutionary decree of cybernetics was that the "human organism can be regarded as a complex information-processing system" that informed "the entire analysis"

4. CYBERNETICS AND THE QUESTION OF RELIGION

Within what critical theorists Eve Sedgwick and Adam Frank have called the "cybernetic fold," there was unwavering conviction in the efficacy of approaching the brain, systematically, as a system, in order to build systematically upon it.[69] With distinct roots in the eighteenth[70] and nineteenth centuries,[71] cybernetic styles of reasoning were marked by a mathematics of mystical ascent. Because of their faith in machines to take the measure of networks, those within the cybernetic fold had come to privilege systemic relationalities over singular entities. A vast interconnected universe was being imagined into being. Indeed, an elegant and expansive systematicity inhered to processes both microscopic and metaphysical—in the movement of an individual neuron, in genetic coding,[72] in the evolution of species,[73] in the market,[74]

---

of other dynamic systems. James G. March and Herbert A. Simon, *Organizations* (New York: John Wiley and Sons, 1958), 9. On the human as part of a scalable universe defined in terms of information, see Harold M. Schroder et al., *Human Information Processing: Individuals and Groups Functioning in Complex Social Situations* (New York: Holt, Rinehart and Winston, 1967).

69. Eve Kosofsky Sedgwick and Adam Frank, "Shame in the Cybernetic Fold: Reading Silvan Tomkins," *Critical Inquiry* 21, no. 2 (1995): 496–522.

70. Otto Mayr, *Authority, Liberty, and Automatic Machinery in Early Modern Europe* (Baltimore: Johns Hopkins University Press, 1989); Sheehan and Wahrman, *Invisible Hands*; Riskin, *Restless Clock*.

71. James R. Beniger, *The Control Revolution: Technological and Economic Origins of the Information Society* (Cambridge, MA: Harvard University Press, 1986); Anson Rabinbach, *The Human Motor: Energy, Fatigue, and the Origins of Modernity* (Berkeley: University of California Press, 1992); Lauren Otis, *Networking: Communicating with Bodies and Machines in the Nineteenth Century* (Ann Arbor: University of Michigan Press, 2001); David A. Mindell, *Between Human and Machine: Feedback, Control, and Computing before Cybernetics* (Baltimore: Johns Hopkins University Press, 2002); Evelyn Fox Keller, "Organisms, Machines, and Thunderstorms: A History of Self-Organization, Part One," *Historical Studies and the Natural Sciences* 38, no. 1 (2008): 45–75.

72. Lily E. Kay, *The Molecular Vision of Life: Caltech, the Rockefeller Foundation, and the Rise of the New Biology* (New York: Oxford University Press, 1996).

73. Donna J. Haraway, "The High Cost of Information in Post–World War II Evolutionary Biology: Ergonomics, Semiotics, and the Sociobiology of Communication Systems," *Philosophical Forum* 13, no. 2–3 (Winter–Spring 1981–82): 244–78.

74. Philip Mirowski, *Machine Dreams: Economics Becomes a Cyborg Science* (Cambridge: Cambridge University Press, 2002).

in the visual cortex of a frog,[75] in politics[76] and the practice of international relations,[77] in the demographic arcs of populations,[78] in art,[79] architecture[80] and design.[81] Consequently, cybernetics considered itself to be a holistic science "devoted to the study of system[s]" as they were found in every conceivable nook of existence.

Cybernetics came of age (and achieved official nomination) with the surprise success of Norbert Wiener's *Cybernetics: or, Control and Communication in the Animal and the Machine* (1948). Palpable, too, was the buzz generated by ten interdisciplinary conferences sponsored by the Josiah Macy Foundation between 1946 and 1953 on the topic "Circular Causal and Feedback Mechanisms in Biological and Social Systems."[82] As anthropologist Margaret Mead noted in 1968, cybernetics was "a form of cross-disciplinary thought which made it possible for members of many disciplines to communicate with each other easily in a language which all could understand."[83] Drawing from engineering advances in radar, code-breaking, and antiaircraft systems, figures like Wiener, Mead, Warren McCulloch, Walter

75. J. Y. Lettvin, H. R. Maturana, W. S. McCulloch, and W. H. Pitts, "What the Frog's Eye Tells the Frog's Brain," *Proceedings of the IRE* 47, no. 11 (November 1959): 1940–51.

76. Deutsch, *Nerves of Government*.

77. Nicolas Guilhot, "Cyborg Pantocrator: International Relations Theory from Decisionism to Rational Choice," *Journal of the Behavioral Sciences* 47, no. 3 (Summer 2011): 279–301; Jonathan Eberhart, "About the Systems System," *Science News* 91, no. 1 (1967).

78. Hudson Hoagland, "Cybernetics of Population Control," *Bulletin of the Atomic Scientists* 20, no. 2 (1964): 2–6; A. J. Coale, *The Growth and Structure of Human Population* (Princeton, NJ: Princeton University Press, 1973); Renato M. Capocelli and Luigi M. Ricciardi, "A Cybernetic Approach to Population Dynamics Modeling," *Cybernetics and System* 9, no. 3 (1979): 297–312.

79. Jack W. Burnham, "Systems Esthetics," *Artform* 7, no. 1 (September 1968): 30–35. See also Jasia Reichardt, ed., *Cybernetic Serendipity: The Computer and the Arts* (New York: Praeger, 1969).

80. Reinhold Martin, *The Organizational Complex: Architecture, Media, and Corporate Space* (Cambridge, MA: MIT Press, 2003).

81. Gyorgy Kepes, *The New Landscape in Art and Science* (Chicago: Paul Theobold, 1956); Peter Seitz, *Design Quarterly 63: A Clip-On Architecture* (Minneapolis: Walker Art Center, 1965).

82. Norbert Wiener, *Cybernetics: or, Control and Communication in the Animal and the Machine* (New York: Wiley, 1948); Claus Pias, ed., *Cybernetics: The Macy Conferences 1946–1953: The Complete Transactions* (Chicago: University of Chicago Press, 2016).

83. Margaret Mead, "Cybernetics of Cybernetics," in *Purposive Systems: Proceedings of the First Annual Symposium of the American Society for Cybernetics*, ed. Heinz von Foerster et al. (New York: Spartan Books, 1968), 2.

Pitts, Claude Shannon, Warren Weaver, Gregory Bateson, John von Neumann, Stafford Beer, Mary A. B. Brazier, and William Grey Walter[84] formed a self-conscious scientific community invested in thinking through issues of feedback control, self-organization, information theory, and neural networks.[85] As the specter of atomic annihilation hovered over places like MIT, the University of Illinois, and the Institute for Advanced Study, cybernetics acquired a certain allure and was taken up in popular media, science fiction, and psychotherapies both old and new age.[86] These scripts promised

84. William Grey Walter: henceforth referred to as Grey Walter or by his surname, Walter. Friends called him by his middle name, Grey, and after he had established his reputation, William was used less and less.

85. More and more pressingly good works have appeared that address cybernetics as a consequential congealing of scientists, institutional resources, and methodologies. Significant studies include Donna Haraway, "Signs of Dominance: From a Physiology to a Cybernetics of Primate Society, C. R. Carpenter, 1930–70," *Studies in History of Biology* 6 (1983): 129–219; Steve Joshua Heims, *The Cybernetics Group* (Cambridge, MA: MIT Press, 1991); Geof Bowker, "How to Be Universal: Some Cybernetic Strategies, 1943–70," *Social Studies of Science* 23, no. 1 (February 1993): 107–27; Peter Galison, "The Ontology of the Enemy: Norbert Wiener and the Cybernetic Vision," *Critical Inquiry* 21, no. 1 (Autumn 1994): 228–66; N. Katherine Hayles, *How We Became Posthuman: Virtual Bodies in Cybernetics, Literature, and Informatics* (Chicago: University of Chicago Press, 1999); Jean-Pierre Dupuy, *The Mechanization of the Mind: On the Origins of Cognitive Science*, trans. M. B. Debevoise (Princeton, NJ: Princeton University Press, 2000); Bruce Clark and Mark Hansen, eds., *Emergence and Embodiment: New Essays on Second-Order Systems Theory* (Durham, NC: Duke University Press, 2009); Eden Medina, *Cybernetic Revolutionaries: Technology and Politics in Allende's Chile* (Cambridge, MA: MIT Press, 2011); Elizabeth A. Wilson, *Affect and Artificial Intelligence* (Seattle: University of Washington Press, 2011); Andrew Pickering, *The Cybernetic Brain: Sketches of Another Future* (Chicago: University of Chicago Press, 2011); Orit Halpern, *Beautiful Data: A History of Vision and Reason since 1945* (Durham, NC: Duke University Press, 2014); Ronald R. Kline, *The Cybernetics Moment: Or Why We Call Our Age the Information Age* (Baltimore: Johns Hopkins University Press, 2015); Daniel Belgrad, *The Culture of Feedback: Ecological Thinking in Seventies America* (Chicago: University of Chicago Press, 2019).

86. The "elusive quality" and "romantic aura" of the word "cybernetics" was "useful" according to J. R. Pierce, Claude Shannon's colleague at Bell Labs. "Indeed," wrote Pierce, the word "cybernetics" could help "add a little glamour to a person, to a subject, or even a book." J. R. Pierce, *Symbols, Signals, and Noise: An Introduction to Information Theory* (1961; New York: Dover, 1980), 208, 228. As a reviewer of the second edition of Wiener's *Cybernetics* (1961) put it, cybernetics had "triggered an effort in many laboratories all over the world," covering "so wide a field that few symposia nowadays try to cover the whole of it"; F. L. H. M. Stumpers, "Review of Norbert Wiener's *Cybernetics*, 2nd edition," *IRE Transactions on Information Theory* 8 (July 1962): 332. With media accounts of the apocalyptic

to make "tremendous" contributions to "human welfare" in the form of military, medical, biological, emotional, political, ethical, and economic advance.[87] As the philosopher of science Jean-Pierre Dupuy summarizes:

> Cybernetics is responsible for introducing the logico-mathematical style of formalism and conceptualization to the sciences of the brain and the nervous system; for conceiving the design of information-process machines and laying the foundations of artificial intelligence; for producing the "metascience" of systems theory, which has left its mark on all the human sciences, from family therapy to cultural anthropology; for providing the major source of inspiration for conceptual innovations in economics, operations research, game theory, decision and rational choice theory, political science, sociology, and still other disciplines.[88]

Neural metaphors were pervasive among cyberneticians. They extended the notion of neural networks into previously uncharted territories—business, society, and mental health.[89] Indeed, the most significant legacy of cybernetics was its seeding of a cognitive revolution that spanned brain research and imaging, behavioral therapy, psychological counseling, digital computation, military planning, psychopharmaceuticals, and practices of neuromodulation such as ECT. Throughout the 1960s, for example, the field of operational research (OR) emerged as systems analysts offered their clients strategic plans to shore up the distribution of resources within large organizations. These plans were enthusiastically based on the neuromatic wager.[90] In one of those disturbing recursivities of history, such plans were soon adopted

---

possibilities of these new sciences, dire warnings about our robot children, and ready-made plots for science fiction writers, the initial legitimacy of cybernetics as a generalist term eroded under the weight of exaggeration, critique, and ridicule.

87. Warren Weaver, "Science and Complexity," *American Scientist* 36 (October 1948): 543–44. The spirit of social regulation infused cybernetics, a word first coined by the physicist André Ampère-Marie who used *cybernétique* to refer to the art of governing nations ("Essai sur la philosophie des sciences," 1838).

88. Dupuy, *Mechanization of Mind*, 43.

89. Dupuy, *Mechanization of Mind*, 82. On the cybernetic saturation within the humanities and human sciences, see the essays collected in a special issue of *History of the Human Sciences* 33, no. 1 (2020), edited by Stefanos Geroulanos and Leif Weatherby.

90. Agatha Hughes and Thomas Hughes, eds., *Systems, Experts, and Computers: The Systems Approach in Management and Engineering, World War II and After* (Cambridge, MA: MIT Press, 2011).

by mental health reformers, particularly in managing medical therapeutic resources, staffing levels, and patient queues within mental hospitals. By the 1970s the target of OR's cybernetic strategies was drifting ever inward, by way of medical institutions, back into the brain.

The notion of a brain that processes information at the level of the neuron was not simply a reflection of the plausibility of the computational metaphor. On the contrary, as the historian Elizabeth A. Wilson has argued, the relatively rapid transmission of this notion was also perpetuated by the masculine fantasy to achieve a state of disembodied cognition. For what is "presupposed about embodiment in this metaphor," writes Wilson, "fits with certain masculinist presumptions about psychological functioning." As Wilson shows, the cybernetic project of envisioning the brain without flesh—thinking without the disruptions of desire, thinking without political bias, thinking clearly about the problems at hand—was driven by an intellectual curiosity to discover a formal equivalence between human and machine. The scene of neuromatic discovery, then, idealized intimacy between men and machines, an idealization that came at the exclusion of nonnervous flesh, in general, and women in particular.[91] Indeed, intense forms of male bonding animated the first generation of cybernetic scientists—a homosociality that was part and parcel of their fierce wishfulness to develop a purely abstract understanding of cognition. Wresting control of the flesh from the flesh. All glitches, at the end of the day, anticipated.[92]

91. Wilson notes the Cartesian compulsion to secure the "autonomy of cognition" despite its "irreducible and perpetual debt to the body"—a frustration that fuels the "containment of the corporeal to the feminine"; see her "'Loving the Computer': Cognition, Embodiment, and the Influencing Machine," *Theory and Psychology* 6, no. 4 (1996): 579, 583–85. There are, of course, radical possibilities for reimagining sexual difference within the cybernetic fold—from the edges of Turing's playful yet imperfect dissolution of gender binaries in evaluating machine intelligence to Donna Haraway's "A Cyborg Manifesto: Science, Technology, and Socialist-Feminism in the Late Twentieth Century," *Socialist Review* 80 (1985): 65–108. But gender was often reinscribed after the first hint of dissolution—see, e.g., Tyler Curtain, "The 'Sinister Fruitiness' of Machines: Neuromancer, Internet Sexuality, and the Turing Test," in *Novel Gazing: Queer Readings in Fiction*, ed. Eve Kosofsky Sedgwick (Durham, NC: Duke University Press, 1997): 142–45.

92. Wilson, *Affect and Artificial Intelligence*, 123. In the mathematics department at Bell Labs and in the social circles of McCulloch and Pitts at the University of Chicago, for example, the motive to mechanize the human corresponded to a sublimated desire for physical contact. And "A Logical Calculus" arose within conversations at the University of Chicago about mathematical modeling of biological systems. Pitts attended weekly seminars on mathematical biology at the University of Chicago. These seminars revolved around

Given the mathematics of mystical ascent undergirding them, cybernetic claims often had more than a hint of theology about them. Cyberneticians, in general, strove to replace bodily ritual with mental mythos. Their proclivity for abstraction was intrinsic to their desire to translate physiology into the precision of physics. In 1966, operations researcher Stafford Beer speculated that

> our own responsibility, in cybernetic terminology, is to play the role of a tiny controlled sub-system of a total control system. . . . [Man's] sub-system is a microcosm of the total system. Any one cell in his body contains his whole genetic blueprint, coded in a molecule of DNA. As a whole person he contains a blueprint of the universe: the Kingdom of God is, in this sense too, within him.[93]

There have been a number of scholars who have provocatively pointed to such statements as evidence for the religious hue of cybernetics. Such studies have also shed light on how cybernetic investigators dabbled in spiritualism, parapsychology, and the attainment of "supernormal knowledge."[94] These studies have emphasized how questions of religion and theology stick to the grammar of cybernetics without necessarily considering the questions posed or the answers provided beyond philosophy, the curiosities of phrasing, or particularities of personal biography. Such studies, then, are helpful

---

the work of mathematical physicist Nicolas Rashevsky, and "A Logical Calculus" was subsequently published in Rashevsky's journal. Nicolas Rashevsky, "Mathematical Biophysics and Psychology," *Psychometrika* 1 (1936): 1–26; Jack D. Cowan, "Interview with James A. Anderson and Edward Rosenfeld," in *Talking Nets: An Oral History of Neural Networks*, ed. James A. Anderson and Edward Rosenfeld (Cambridge, MA: MIT Press, 1998): 104–5. See also Tara H. Abraham, "Nicolas Rashevsky's Mathematical Biophysics," *Journal of the History of Biology* 37 (2004): 333–85, and Tara H. Abraham, "(Physio)logical Circuits: The Intellectual Origins of the McCulloch-Pitts Neural Networks," *Journal of the History of the Behavioral Sciences* 38, no. 1 (2002): 3–25.

93. Stafford Beer, "Cybernetics and the Knowledge of God," *Month* 34 (1966): 302–3.

94. Pickering, *Cybernetic Brain*, 318. Pickering points to the "spirituality" of British cyberneticians such as W. Ross Ashby and Stafford Beer, highlighting their "faith in the agency of matter" (236) and suggesting the term "hylozoism" to name it, a blend of eastern and western concepts of a spirit-infused nature. Edwards writes that cybernetics approached the "problem of spiritual growth" by way of "mystical powers, animals, or other life forms" (Edwards, *Closed World*, 312). See also John Shiga, "Of Other Networks: Closed-World and Green-World Networks in the Work of John C. Lilly," *Amodern 2: Network Archaeology* (October 2013), http://amodern.net/article/of-other-networks/.

to the degree that cybernetic researchers were, indeed, prone to theological speculation and involved in occult investigations.[95] Yet with few exceptions[96] there is little work that rigorously situates the particular mix of engineering prowess and cosmic promise within a sustained history of religion or, for that matter, within the purview of religious studies.

In contrast to studies that bind their analytic of religion to the intentions of individual cyberneticians, I am not particularly interested in the creative, radically open, disunified, groovy, and/or liberatory dimensions of their projects.[97] Indeed, I want to resist reading the abstract freedoms promised by cybernetics as anything more than that. I want to resist reading the cybernetic conceits of unpredictability and randomness as anything more than wishful thinking. I want, instead, to focus on the more local constraints that such promises generated and the disciplinary air they maintained.[98] For once one begins to live life as though one were an information

95. See, e.g., John Cunningham Lilly, *Man and Dolphin* (New York: Doubleday, 1961) and Stewart Brand, *II Cybernetic Frontiers* (New York: Random House, 1974).

96. Lily E. Kay, "From Logical Neurons to Poetic Embodiments of Mind: Warren S. McCulloch's Project in Neuroscience," *Science in Context* 14, no. 4 (2002): 591–614; John R. Williams, "World Futures," *Critical Inquiry* 42, no. 3 (Spring 2016): 473–546. As Fred Turner has demonstrated, the popularization of cybernetics in the 1960s and 1970s, when divorced from the strictures of peer review, often cut toward the mystical and the utopian; see his *From Counterculture to Cyberculture: Stewart Brand, the Whole Earth Network, and the Rise of Digital Utopianism* (Chicago: University of Chicago Press, 2006).

97. Pickering, for example, commends the cybernetic impulse to "fuse . . . science and spirituality rather than setting them at each other's throats" (*Cybernetic Brain*, 302). How else, he asks, might we conceive of "another future . . . different from the ones that are more readily imagined." The recourse to a resolving utopianism that plagues some accounts of cybernetics (and conciliatory approaches to science and religion, in general) flattens the jagged ambiguity of the situation and occludes a sustained analysis of power/knowledge. Similarly, an emphasis on fissures within cybernetic debates can quickly become pointillist at the expense of considering discursive compatibilities. See, e.g., Ronald Kline, "How Disunity Matters to the History of Cybernetics in the Human Sciences in the United States, 1940–80," *History of the Human Sciences* 33, no. 1 (2020): 12–35.

98. Hayles, *How We Became Posthuman*, 285–86. Hayles calls reflexivity a 'threatening and subversive idea" that gives pause to overreading the determinism of technological advance (8). Mine, then, is an intervention of sorts, a counter to those critical studies that suggest that the turn from homeostasis to reflexivity in second-order cybernetic discourse was a moment of incredible possibility. Which is to say that the "religious" dimension of cybernetics does not necessarily suggest liberatory potential. Rather, whatever is religious about cybernetics signals the complicity of its concepts and practices in the making of our secular age.

processing device, and the more the environment mirrors and makes way for further refinement of that performance, the more the metaphorical distinction between mind and machine becomes rather hazy. So when religious aptitude, sexual difference, or racial identity is evidenced within the brain, might such measures of ionic currents across nerve cell membranes have as much to do with the impressibility of categorical binaries (male vs. female, religious vs. secular, black vs. white) in the culture at large?[99]

5. CYBERNETIC THESES OF SECULARIZATION

Cyberneticians, of course, were not the first to see the brain as a leverage of religious reform or to use the brain to integrate physiological inquiry and metaphysical speculation.[100] By the sixteenth century, as the Protestant Reformation was taking hold, new translations of Galen began to appear that emphasized the importance of anatomy in medical matters while retaining an original emphasis on the physiological conditions under which and through which the soul could exercise its functions. Martin Luther's friend and fellow reformer, Phillip Melanchthon (1497–1560) initiated a revision of Aristotle's *scientia de anima* in terms of what he called *psychologia*, a philosophical inquiry into the soul that would also serve as an extension of theology. By the end of the seventeenth century, ongoing investigations into the physics of the soul-body connection became subject to both an emergent mechanistic philosophy and more rigorous experimental criteria.[101]

99. See, e.g., Jordan R. Young and Rafaella I. Rumiati, "Hardwired for Sexism? Approaches to Sex/Gender in Neuroscience," *Neuroethics* 5, no. 3 (December 2012): 305–15. *Neuromatic* adheres to a critical reading of brain plasticity and what Kyla Schuller has read as the gendered and racialized histories of nervous impressibility; see her *The Biopolitics of Feeling: Race, Sex, and Science in the Nineteenth Century* (Durham, NC: Duke University Press, 2017). Or as Victoria Taylor-Pitts puts it in her feminist critique of neural plasticity, "brain knowledge is not simply shaping what we think brains are, but is informing practices that literally, materially shape them" (*The Brain's Body: Neuroscience and Corporeal Politics* (Durham, NC: Duke University Press, 2016), 7).

100. For medieval stirrings of neuromysticism, see Scott E. Hendrix and Christopher J. May, "Neuroscience and the Quest for God," in *The Neuroscientific Turn*, ed. Littlefield and Johnson, 105–19.

101. In *Treatise of Man* (1664), René Descartes posited the pineal gland as the hinge between body and soul where animal spirits passed through. In impressing themselves upon the pores of the brain, the brain also impressed itself upon these spirits in this deli-

Here began an uneven and fraught attempt to establish the brain at the center of a natural philosophy—an empirical and measured approach that would soon become self-consciously secular in its epistemic aspirations. For example, Thomas Willis—celebrity scientist of his time, tutor of John Locke at Oxford, and coiner of the term "neurologie"—offered detailed anatomical drawings of the brain in the service of a politicized theology. Such was the crassly human materiality that he termed, in a nod to his Christian anthropology, the "chapel of Deity."[102] Willis, for his part, was eager to use the brain as political leverage in protecting the authority of the Church of England. Willis's goal was to provide a neurophysiological rationale for what constituted good religion and proper political order. In his "Anglican optimism," Willis pursued the study of the nervous system as a way to advance a theological agenda, specifically, as a way to reassert the authority of the Book of Common Prayer and justify the postures it prescribed.[103]

Three hundred years later, scientists remained committed to splitting up the world into good and bad religion. In *God and Golem: A Comment on Certain Points Where Cybernetics Impinges on Religion* (1964), Norbert Wiener argued that the new science that he had named sixteen years earlier in *Cybernetics* should serve as a reform movement to rationalize religion. He

---

cate dance at the center of the human machine. As Fernando Vidal has demonstrated, the brain assumed, at this time, a "definite ontological primacy, well beyond its function as the mind's cause, material foundation, or condition of possibility" (Vidal, *Sciences of the Soul*, 34–35, 37ff., 76–77).

102. George Rousseau, "'Brainomania': Brain, Mind and Soul in the Long Eighteenth Century," *Journal for Eighteenth-Century Studies* 30, no. 2 (2007): 161–91. During this period, as the brain began to exceed itself, it aided and abetted all manner of imperial ambition and theories of racial difference. On the early stirrings of race science and polygenetic theories of human origins as filtered through neural metaphors, see John Atkins, a British naval surgeon and slave trader whose medical textbooks and travelogues circulated heavily in the early eighteenth century; John Atkins, *The Navy Surgeon; or, Practical System of Surgery: With a Dissertation on Cold and Hot Mineral Springs; and Physical Observations on the Coast of Guiney* (London: J. Hodge, 1742). For a contemporary treatment of the brain as marker of civilized difference, see M. Meckel, "Recherches anatomiques," *Histoire de l'Académie Royale des Sciences et Belles Lettres* (1753): 79–103. See also Londa Schiebinger, "The Anatomy of Difference: Race and Sex in Eighteenth-Century Science," *Eighteenth-Century Studies* 23, no. 4 (1990): 387–405.

103. Louis Caron, "Thomas Willis, the Restoration and the First Works of Neurology," *Medical History* 59, no. 4 (2015): 525–53; Rina Knoeff, "'The Reins of the Soul': The Centrality of the Intercostal Nerves to the Neurology of Thomas Willis and to Samuel Parker's Theology," *Journal of the History of Medicine and Allied Sciences* 59, no. 3 (2004): 413–40.

WILLIS

De Cerebro, Tab. III

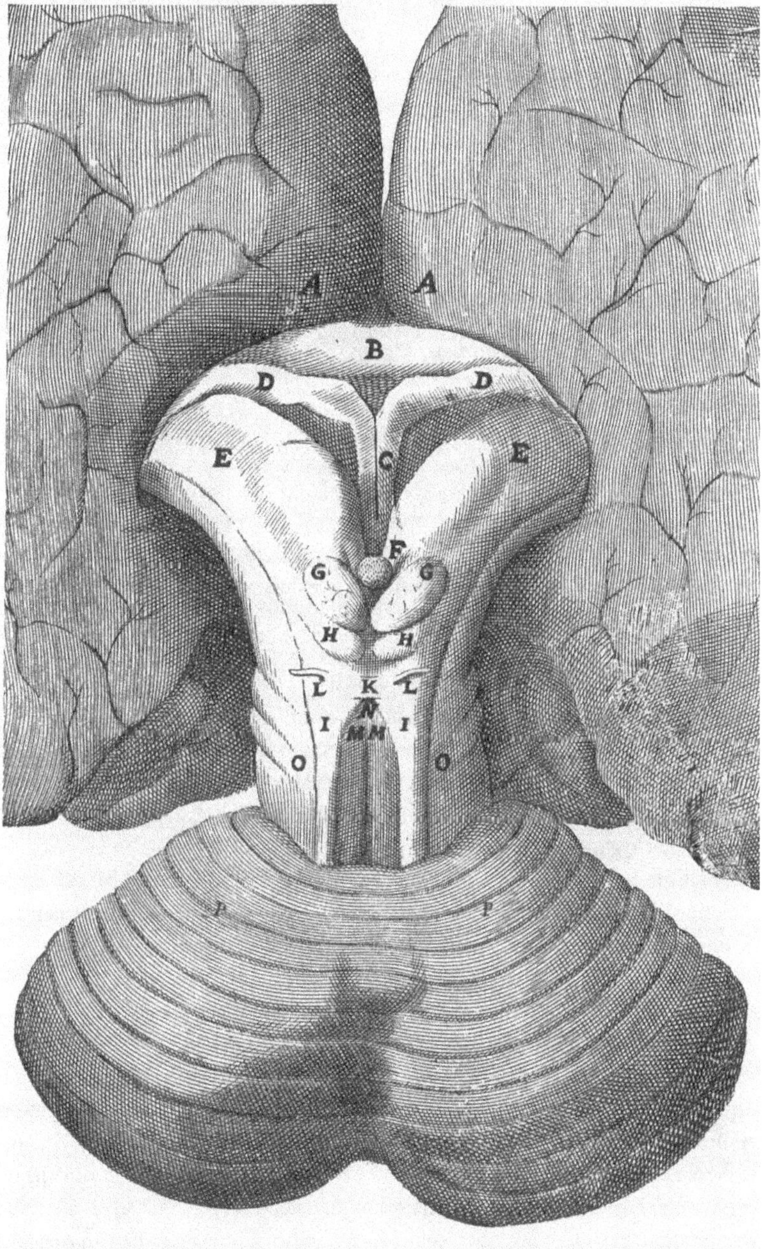

FIGURE 7. Plate of Thomas Willis's anatomy of the brain, ca. 1664, from Swedenborg's *Cerebrum* (1882).

called on the cybernetically inclined to create a space of ethical contemplation where traditional concerns over omnipotence and omniscience would be newly subject to mathematical approaches. According to Wiener, cybernetics was also the ethical alternative to the "gadget worshippers"—scientists and technocrats who saw in artificial intelligence the possibility to be freed from human limitations, conscience, and moral responsibility. Cybernetics, by contrast, was a "policy as far removed from that of the gadget worshipper as it is from the man who sees only blasphemy and the degradation of man in the use of any mechanical adjuvants."[104]

Cybernetics, according to Wiener, was straight down the middle. And it was postsecular in that it had overcome the limitations of both science and religion and their long-standing disciplinary dustup. Another way of putting this would be to say that cybernetic theses of secularization such as Wiener's could be remarkable for their imperial ambition (both figurative and literal).[105] The pioneering neurophysiologist Warren McCulloch, for example, shared Wiener's sense of having moved beyond the concerns of tra-

104. Norbert Wiener, *God and Golem: A Comment on Certain Points Where Cybernetics Impinges on Religion* (Cambridge, MA: MIT Press, 1964), 6–7, 53–55, 73. Wiener's rehearsal of secularization, despite its talk of technological wonders, was a familiar one in terms of its gendered and racial coding. Pitched as a necessary assault on the sentimentality of religion, Wiener's heroic masculinity engaged in "real probing of real questions." Whereas Wiener likened his method to the surgeon's scalpel—a violence in the service of truth—religion for him, by contrast, was "squeamish," "prejudiced," emotional and unable to look beyond its own self-interests. Religion remained a prisoner of its own domestic proclivities. "Religion, whatever else it contains," wrote Wiener, "has often something in itself of the closed front parlor of a New England farmhouse, with drawn blinds, wax flowers under a bell jar on the mantlepiece, gilded bulrushes surrounding grandfather's portrait on an easel, and a harmonium in black walnut, never played except at weddings or funerals." Wiener's call for "an independent study of systems involving both human and mechanical elements" was also aimed at those who misconstrued cybernetic ideas in a pathetic attempt at mimicry—like the Africans who mindlessly adopted the clothing of their colonizers without proper discrimination (*God and Golem*, 3, 7–9, 4, 73, 89–90).

105. Wiener, *God and Golem*, 93. In his notes from March 1951, psychiatrist and fellow cybernetic traveler W. Ross Ashby alludes to the white mythology undergirding his own cybernetic quest to design a brain, and the delirium it entails—"If I can, by this method, develop a machine that imitates advanced brain activities without my being able to say how these activities have arisen, I shall be like an African explorer who, having heard of Lake Chad, and having sought it over many months, stood at last with it at his feet and yet, having long since lost his reckoning, could not say for the life of him where in Africa Lake Chad was to be found." The W. Ross Ashby Digital Archive, http://www.rossashby.info/journal/page/3151.html.

ditional science and religion both. McCulloch described his journey of passing through the last stage of metaphysics in which his "spine" would allow him to resist the temptations of godly speculation and the transcendental ego. "For the first time in the history of science," announced McCulloch, "we know how we know and hence are able to state it clearly." McCulloch's triumphal declaration of secular subjectivity was dependent on figures like Wiener and von Neumann—"men [who] have altered our metaphysics by altering our physics."[106] For McCulloch, whose undergraduate mentor was the Quaker mystic Rufus Jones, secularization was a matter of becoming all-knowing in a world devoid of any spirit save for the mathematical. Mysticism in McCulloch's secular age was an algorithmic proposition. Mathematics, in turn, as he admitted, could have "cosmic significance."[107] Trusting, as it were, the numbers and their spawn to dissolve whatever contradictions arose, McCulloch understood his method superior to forms of bad science *and* bad religion. "Our adventure," announced McCulloch, "is actually a great heresy. We are about to conceive of the knower as a computing machine."[108]

The cybernetic fold, the cybernetic apparatus, the cybernetic moment, the cybernetic thread—these scholarly terms, in their strategic vagueness, aim to capture how a set of beliefs and techniques revolving around projects of communication and control became a horizon of epistemic and political possibility.[109] For the diffusion of cybernetics served to redefine categories and reimagine entire spheres of human being—and human being itself—as

106. Warren S. McCulloch, "Through the Den of the Metaphysician," *British Journal for the Philosophy of Science* 5, no. 17 (1954), 31, 18. McCulloch's was a hyperrational masculinity ginned up on metaphors of military precision. Within the cybernetic fold, in general, the computating subject was, more often than not, coded as male. See, for example, the male codes of the Turing Test explored by Elizabeth A. Wilson in *Neural Geographies: Feminism and the Microstructure of Cognition* (New York: Routledge, 1998), 112–13.

107. Quoted in Tara Abraham, *Rebel Genius: Warren S. McCulloch's Transdisciplinary Life in Science* (Cambridge, MA: MIT Press, 2017), 170.

108. McCulloch, "Through the Den," 28, 19. For a reading of McCulloch's reanimation of German Idealism, see Leif Weatherby's "Digital Metaphysics: The Cybernetic Idealism of Warren McCulloch," *Hedgehog Review* 20, no. 1 (Spring 2018): 19–35.

109. See Bernard Dionysius Geoghegan, "From Information Theory to French Theory: Jakobson, Levi-Strauss, and the Cybernetic Apparatus," *Critical Inquiry* 38 (Autumn 2011): 96–126; George P. Richardson, *Feedback Thought in Social Science and Systems Theory* (Philadelphia: University of Pennsylvania Press, 1991), 169ff.

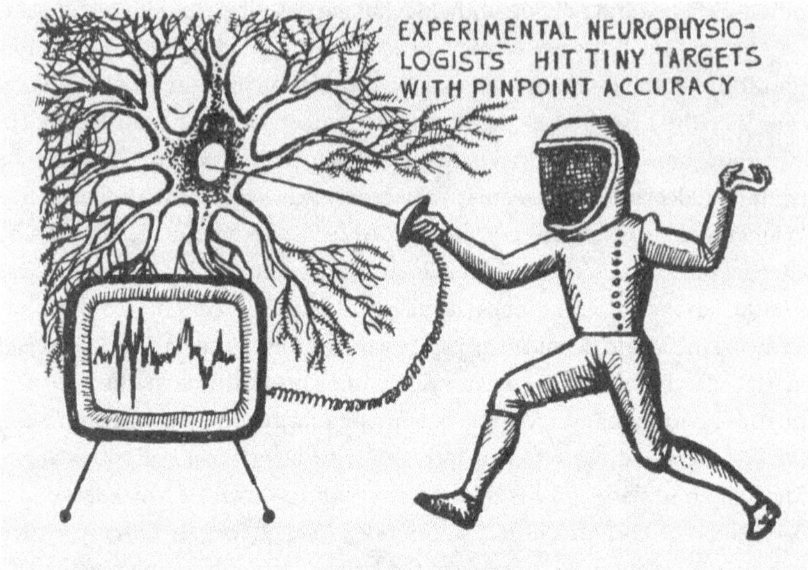

FIGURE 8. Cartoon from Y. Saparina, *Cybernetics within Us* (North Hollywood: Wilshire Book Co., 1967).

reducible to (and manipulable by means of) a neural mechanics of information processing.[110]

By the early 1960s, a consensus was emerging about how best to understand what Wiener had called that "special sort of machine known as a human being." Descriptions of systems in terms of information theory were, "broadly speaking, interchangeable with descriptions that are couched in logical net terms."[111] This informational staging of neural processing spread

110. Jurgen Ruesch and Gregory Bateson, *Communication: The Social Matrix of Psychiatry* (New York: W. W. Norton, 1951); Allen Newell and Herbert A. Simon, "The Logic Theory Machine: A Complex Information Processing System," *IRE Transactions on Information Theory* 2, no. 3 (1956): 61–79. In contrast to Kline's intellectual history of "cybernetics in crisis," mine is not an argument for waning prestige and/or influence. I argue that the cybernetic moment has not yet ended. Its atmospheric diffusion is, rather, a problem of analytic perspective (Kline, *Cybernetics Moment*, 229). So whereas Kline sees the diffusion of cybernetics, in part, as evidence for its decline in coherence and political relevance, I would suggest a more capacious understanding of political influence as discursive, that is, as a matter of discipline rather than persuasion.

111. Wiener, *Human Use*, 79; F. H. George, *The Brain as a Computer* (Oxford: Pergamon Press, 1962), 5. See also Anotol Rapoport, "Application of Information Networks to a Theory of Vision," *Bulletin of Mathematical Biophysics* 17 (1955): 15–33.

rapidly and was codified in Ulric Neisser's *Cognitive Psychology* (1967), a popular textbook that suggested that "the task of a psychologist trying to understand human cognition is analogous to that of a man trying to discover how a computer has been programmed." Such work, he added, would allow the psychologist to "account" for "the way information is processed by men."[112]

In the heady days of the cybernetic boom, instrumentation such as electroencephalography (EEG) provided scientists such as Neisser proof of the neuromatic concept. For the brain wave signaled the activity of neural nets, which, in turn, signaled the transmission of signal from one neuron through a specific channel to another.[113] Electrical signal, moreover, was the medium of information flow, binding humans to themselves, to other humans, and to other systems (biological, mechanical, cultural) capable of processing information. In the postwar human sciences, for example, self and society were increasingly considered to be "systems [that] *store, process, and communicate information* about themselves and their environments."[114] McCulloch's quip that his was the "timeliest [of] applications of the quantitative theory of information"[115] inspired the use of information theory to solve problems

112. Ulric Neisser, *Cognitive Psychology* (New York: Appleton-Century-Crofts, 1967), 6–8. See also Jerome S. Bruner, Jacqueline J. Goodnow, and George A. Austin, *A Study of Thinking* (New York: Wiley, 1956) and George A. Miller, Eugene Galanter, and Karl H. Pribram, *Plans and the Structure of Behavior* (New York: Henry Holt, 1960), and Michael A. Arbib, *Brains, Machines, and Mathematics* (New York: McGraw-Hill, 1964).

113. See, e.g., D. O. Walter and M. F. Gardiner, "Some Guidelines from System Science for Studying Neural Information Processing," *International Review of Neurobiology* 13 (1970): 343–74.

114. Hunter Heyck, *Age of System: Understanding the Development of Modern Social Science* (Baltimore: Johns Hopkins University Press, 2015), 11. According to historian Lily Kay, the enthusiastic adoption of the informational paradigm at midcentury was central to a "technoepistemic transformation across the disciplinary landscape and the culture at large." Over the past seventy years, as the biology of the central nervous system has become, first and foremost, a matter of information processing, the techniques of neurophysiology have become integral to any study of mind, in general. Lily E. Kay, "From Logical Neurons to Poetic Embodiments of Mind: Warren S. McCulloch's Project in Neuroscience," *Science in Context* 14, no. 4 (December 2002): 593; Margaret A. Boden, *Mind as Machine: A History of Cognitive Science*, vol. 2 (Oxford: Clarendon Press, 2008), 1114. See also Alexander G. Dimitiov et al., "Information Theory in Neuroscience," *Journal of Computational Neuroscience* 30, no. 1 (2011): 1–5.

115. Warren S. McCulloch, "Mysterium Iniquitatis of Sinful Man Aspiring into the Place of God" [1955], in *Embodiments of Mind* (Cambridge, MA: MIT Press, 1970), 158.

of communication and control wherever they might manifest—like noise in a channel, madness in the brain, or pollution in an environmental system.[116] The coming secular age, in other words, would be optimized and automatic and utterly liberating. For some, however, optimization and automaticity conjured a more ominous algorithmic existence. In 1960, for example, the philosopher C. A. Muses, who collaborated with cybernetic luminaries such as Wiener and McCulloch, noted the will to power that had infected neuromatic investigations at midcentury. "What has become historically evident as man's dominating aim," warned Muses, is "the replication of himself by himself by technological means."[117]

In the epilogue, I turn to this audacious literalization of the human-machine metaphor. As the model for building computers, the neuromatic brain has become both superstructure and base for a society bent on actualizing, fine-tuning, and profiting from the automaticity of the human.[118] For as an aspirational mathematics fuels the construction of machines designed for a world that is imagined to accord with this very same mathematics,

116. Indeed, as the pollution crisis took hold in the popular imagination in the late 1960s and 1970s, a technologically induced problem yielded a technological solution. Instead of environment there was "environmation," a system of inputs and outputs to be addressed for purposes of prediction. Frederick F. Gorschboth (IBM), "Environmation," in *Cybernetics, Artificial Intelligence, and Ecology: Proceedings of the Fourth Annual Symposium of the American Society for Cybernetics*, ed. Herbert W. Robinson and Douglas E. Night (New York: Spartan Books, 1972), 291–302.

117. C. A. Muses, "The Logic of Biosimulation," in Muses, ed., *Aspects of the Theory of Artificial Intelligence: Proceedings of the First International Symposium on Biosimulation* (New York: Plenum, 1962), 116. And if that was not disturbing enough, Muses then doubles down in his elaboration of the viral "form of this dominating aim." It has become "a supermachine," he continues, "the psychological and the human content of that aim is control, mastery, the ability to impose his whims at will upon as much of the rest of the material universe as possible." Muses, who had written his dissertation on the German mystic Jakob Boehme, became interested in cybernetics as a way to understand various esoteric practices such as tantra and Buddhist meditation. He would later go onto collaborate with Joseph Campbell and founded, under the pseudonym Musaios, an international society, The Lion Path, dedicated to the revival of a global shamanism. Musaios, *The Lion Path: You Can Take It with You: A Manual of the Short Path to Regeneration for Our Times* (Golden Sceptre Publishing, 1988).

118. Andy Clark, *Natural-Born Cyborgs: Minds, Technologies, and the Future of Human Intelligence* (New York: Oxford University Press, 2004). It has long been, of course, a reasonable wager to make, or rather, a "not-too-unreasonable first-order assumption [to make] that there is a correlation of some sort between the number of [nerve] impulses and the number of bits per second, [and that] input-output performance curves of neurons in units

those machines prompt their users to imagine the world in terms of *their* algorithmic design.[119] "Modern computers," prophesied Neisser the year before he published his foundational textbook on the psychology of the neuromatic brain, "can be programmed to act unpredictably and adaptively in complex situations. That is, they are intelligent." Similarly, humans "can be manipulated, 'brain washed,' and apparently controlled without limit" given their newly discovered mechanical "properties."[120]

6. POETICS

Neuromatic addresses the discourse of neuromation as a living thing—a generative, self-organizing system that manifests a natural resistance to entropy. It is, in other words, a black box—a happening that can be described in its effects but not necessarily explained in terms of its exact mechanics. For the discourse of neuromation is a distributed phenomenon in which ideas about, practices around, and demands of the brain increasingly serve to reinforce each other's authority. The effects of discourse do not register, first and foremost, as stemming from the same source or central cause. Discourse does not persuade directly but does its work through dispersion. Discourse is not determinative per se but disciplinary. It alters, unannounced, the relationalities within the sensorium and patterns of perception. Consequently, discourse is difficult to talk about. It goes without saying because it came without saying.[121]

Michel Foucault's notion of discourse and his genealogical method, from which my analysis draws much inspiration, was born from within the cybernetic fold. Foucault's style of tracing resemblances and recognizing patterns was a Nietzschean proposition run through the conceptual architecture

of impulses per second may be assumed to have a similar shape to curves calculated in bits per second." James Grier Miller, "Information Input Overload and Psychopathology," *American Journal of Psychiatry* 116 (1960): 697.

119. On programming "robot beliefs," see Nils J. Nilsson, *Understanding Beliefs* (Cambridge, MA: MIT Press, 2014).

120. Ulric Neisser, "Computers as Tools and Metaphors," in *The Social Importance of Cybernetics*, ed. C. R. Dechert (New York: Simon and Schuster, 1966): 74–75.

121. Michel Foucault, "The Order of Discourse," in *Language and Politics*, ed. Michael J. Shapiro (New York: New York University Press, 1984), 108–38. See also Pierre Bourdieu, *Outline of a Theory of Practice* (Cambridge: Cambridge University Press, 1977), 167.

of neural networks and information processing.[122] Indeed, the concept of cybernetic control was central to Foucault, a concept of power as information, the circulations of which were disembodied yet totalizing.[123] In Foucault's vision of power, as media theorist Lydia Liu and others have noted, discourse is a capacious analytic container, power circulating and surging in all directions through networks, individuals living as discrete fixtures of a gigantic communication machine.[124] For Foucault, discourse was the solving term that connected individuals to their environments. It is a category that seeks to capture the relations of the parts it names and, more importantly, how circular causality between parts added up to more than their sum. Different from previous solving terms in the social sciences such as the culture concept, the mechanics of discourse, à la Foucault, were wholly informational—a system of relations, of signal formations with circles and a self-organizing bent.[125]

Consequently, *Neuromatic* not only contemplates the prospect of total incorporation but considers it to be a working condition. Rather than engage in a polemic about our evaporating humanity in the age of spiritual machines or point, fingers crossed, to a world in which our humanity (or analysis of it) is consummated in digital advance, I offer a view from within the belly of this particular Leviathan. For it is now well-nigh impossible to be, let alone think, outside of a neuromatic paradigm.[126] Which is to say that

122. In "Message or Noise," a short talk Foucault delivered in 1966, he plays with the notion of signal and noise as a way to broach the representational force of information theory. He is taken with its movement beyond meaning, beyond the "linear representations" that connect body and soul, organism and tissues. Foucault here is thinking in terms of feedback—"new kinships, either diagonal or lateral"—that mark both the phenomenon to be investigated and the authority of the investigative pose. Michel Foucault, "Message ou Bruit," *Concours Médical* 88 (October 22, 1966): 6285–86; reprinted in Michel Foucault, *Dits et Écrits*, vol. 1 (Paris: Gallimard, 1994), 557–60.

123. Hayles, *How We Became Posthuman*, 192–99; Céline Lafontaine, "The Cybernetic Matrix of 'French Theory,'" *Theory, Culture and Society* 24 (2007): 36.

124. Lydia H. Liu, *The Freudian Robot: Digital Media and the Future of the Unconscious* (Chicago: University of Chicago Press, 2010), 23.

125. The looping authority of discursive formations, conceived of as the patterned compatibility between "objects, modes of statement, concepts, [and] thematic choices" would be central to Foucault's eventual articulation of biopolitics; Michel Foucault, *The Archaeology of Knowledge and the Discourse of Language*, trans. A. M. Sheridan Smith (New York: Pantheon Books, 1982), 38.

126. On the enclosures endemic to "Telematic Man," see Jean Baudrillard, *xerox and infinity* (London: Touchepas, 1988).

critique is ever compromised. For I must admit, when I look askance at the traffic in black market crania in the nineteenth century[127] or excessive efforts in the present to turn the world and ourselves into information processing devices—from elite neuroscience laboratories to popular websites such as www.BranHQ.com, I do not do so innocently. For I, too, am participating in the spectacle of the brain and images of the brain and images of images of the brain. I, too, have an excessive attraction to the claims that the brain continues to make on me. My guilt is only heightened when I consider the disturbing fact that what you, dear reader, hold in your hand (or touch-screen) at this very moment is yet another scheme to address, isolate, and lay claim to the space within the skull.

Such a priori complicity, I contend, requires the historian to dwell in the relationship between critical intent, the archive, and the environment, to struggle to consider the maneuvers enacted and the rules that ordered them even before intent, itself, is intended. This, of course, is easier said than done. Consequently, *Neuromatic* wrestles throughout with a difficult, perhaps even impossible, question: How has the conceptual order of the neuromatic brain come to define not just the world around us, not just our critical encounters with it, but the very atmosphere in which those critical encounters could potentially occur?

Neuromatic self-consciously employs the thematic choices of neural networks and information theory. As a formal experiment this book's inner logic imitates, from an odd angle, the contemporary neurological consensus about the brain as network—tremulating threads, recurring operations, advantageous redundancies, loops folding in upon themselves and emerging anew. My narrative is neither linear nor driven by the logic of cause and effect. Laden with footnotes that circle and expand the system of the text, my book seeks, instead, to connect different times, different spaces, different people, and different themes in order to explore the grey spaces (and grey matters) of unexpected diffusion.[128] The index serves as the logical calculus of this particular network.

I am aware that I may be accused of technological determinism (an academic diagnosis of paranoia) in my attempt to consider discourse on its own

127. Cathy Gere, "A Brief History of Brain Archiving," *Journal of the History of the Neurosciences* 12, no. 4 (2003): 396–410.

128. "Genealogy is grey," wrote Foucault, "meticulous and patiently documentary." See "Nietzsche, Genealogy, History," in *Language, Counter-memory, Practice: Selected Essays and Interviews*, ed. D. F. Bouchard (Ithaca, NY: Cornell University Press, 1977), 139.

terms, that is, *as* an organic totality. My formalism may even be charged with replicating the cybernetic proclivities under scrutiny. But in moving from the evidence of effect to a barely recognizable purposive system, in intuiting a system's feedback and relays, I do so with the objective of modeling this system in order to intervene, deliberately and definitively. How else to keep up with it, analytically or even politically? For the discourse I am attempting to grasp, to quote Stafford Beer, is "difficult to recognize because of its apparent dispersion through another system, or because its purpose is not the apparent dispersion through another system, or because its purpose is not the *apparent* purpose of its own system."[129] Yet to recognize discourse as systematic, however fleetingly, is the task at hand. For the closure that it portends may be a precondition for moving across and perhaps even beyond it. Which is to say, I do not situate my critical resistance in the pure potentiality of the brain, nor in its unpredictable propulsions, nor in the limits of contemporary neuroscience.[130]

So rather than tell a linear story of change over time, *Neuromatic* insists on the genealogical gathering of disparate moments. Indeed, given those processes by which epistemics and politics come to coalesce on a vast scale, dutiful historical descriptions of change over time often serve to take the focus away from the more disturbing elements of any story. My particular history is scenic rather than synthetic—archivally expansive and full of errant juxtapositions, dramatic temporal shifts, and recurring characters (like Grey Walter who appears in different chapters as witness to the first demonstration of the alpha wave, as a metaphysician of cognition, as close friend of Eileen Garrett, as interlocutor with William S. Burroughs, respectively). Rather than a trove of data meant to be skimmed and downloaded, *Neuromatic* is a provocation against the forgetfulness that accompanies the fantasy of absolute storage and instant recall. The "synaptic gaps" between chapters serve as periodic reminders of this provocation.

My goal, then, is not to provide *the* definitive account, or even to claim some kind of archival totality. I seek instead, tragically I fear, to write across a neuromatic imaginary, turning its words and its expectations against themselves. For genealogy is a narrative frame that seeks to appreciate the conditions of *its* telling and to build into *its* documentation a process of critical

129. Stafford Beer, *Cybernetics and Management* (New York: Wiley, 1967), 39.

130. Catherine Malabou, *What Should We Do With Our Brain?*, trans. Sebastian Rand (New York: Fordham University Press, 2008).

reflection. Another way to put this is to say that genealogical critique "bears witness." It assumes that immanent analysis of the world is but an effect of that world, an effect that in those rare cases may be fed back in such a way as to inflect rather than affirm what already is.[131]

131. According to Theodor Adorno, a significant figure in the Frankfurt School, "a successful work [of] immanent criticism is not one which resolves objective contradictions in a spurious harmony, but one which expresses the harmony negatively by embodying the contradictions, pure and uncompromised, in its innermost structure. Confronted with this kind of work, the verdict 'mere ideology' loses its meaning. At the same time, however, immanent criticism holds in evidence that the mind has always been under a spell. On its own it is unable to resolve the contradictions under which it labours. Even the most radical reflection of the mind on its own failure is limited by the fact that it remains a reflection, without altering the existence of which its failure bears witness." From "Cultural Criticism and Society," *Prisms* (Cambridge, MA: MIT Press, 1983), 32.

SYNAPTIC GAP: MEASURING RELIGION

> The Cognitive Science of Religion (CSR) is a burgeoning and highly interdisciplinary enterprise. . . . What unites these researchers is a shared focus on the role of human cognition in religious thought and behaviour, which they study by importing axiomatic assumptions from the cognitive revolution and their respective disciplines. The rich array of culturally postulated supernatural agents and supernatural realms, and the associated diversity of culturally prescribed and proscribed behaviours, are assumed to be constrained and canalized by genetically endowed cognitive capacities and structures shared by all typically developing humans. These structures are assumed to govern the types of information that is attended to, the contexts in which information is attended to, and the manner in which information is stored, processed and acted upon.
>
> DIMITRIS XYGALATAS AND RYAN MCKAY, "Announcing the *Journal for the Cognitive Science of Religion*" (2013)

OVER THE PAST FEW YEARS I have drawn much critical inspiration from time spent at MindLab at the University of Aarhus, Denmark. MindLab (2009–14) was a "cross-cutting neuroscience and cognition research framework" funded by the Danish Ministry of Science, Technology, and Innovation.[1]

1. In contrast to the strict privileging of neural activity in the experimental design of neuroscientists, a focus on neural activity was supplemented within MindLab by other

At MindLab researchers studied links between human interaction and brain functioning and examined "how religion and cognition intersect to constrain and motivate agency, interaction, identity, moral values and symbolic meanings."[2]

During my visits to MindLab I took notes and pictures. I made friends with the scientists and postdocs and gained their trust. I learned about their experimental agenda. I interviewed them and retreated with them to Sandbjerg Manor in southern Jutland. In the space where Prussian military leaders strategized during the Second Schleswig War of 1864, I listened to leading figures discuss big pictures and future trends in the cognitive study of religion and tried to see the world from their point of view. The members of MindLab, like members of all tribes, are motivated by conceptions of who they are not, of anxious fears of those who might threaten their peaceful and well-funded way of life. I learned this lesson on my first night at MindLab when I was standing around with folks after dinner, drink in hand, dank plumes of hash wafting in the wind. As we talked in a circle I uttered the name of Walter Benjamin in passing. Immediately the group burst into laughter. I had hit a nerve. I smiled, uneasily, as if I, too, knew the truth of the matter. I did not bring Benjamin up again in conversation because I

explanatory frames. The Religion, Cognition, and Culture project at MindLab, for example, involved "a number of models and methods including fieldwork, textual studies, functional neuroimaging and experimental psychology, and it expands on ongoing collaborations on brain imaging studies of religious practices, e.g. of praying and meditation." Cognitive science at MindLab may best be understood as a more humanistic version of neuroscience—drawing on the research stream of highly technical studies (and not always uncritically) to talk about much larger questions of mind and body (Uffe Schjødt, "The Neural Correlates of Religious Experience," *Religion* 41, no. 1 [March 2011]: 91–95). As cofounder Armin Geertz declares, cognition is the coin of MindLab's realm—a realm that includes not only what "goes on in the individual mind" but anything and everything that might come into reach of its interdisciplinary splay: "In adapting our approach to contemporary research in neurobiology, archaeology, anthropology, comparative religion and philosophy of science, we [at MindLab] hold that cognition is embrained, embodied, encultured, extended and distributed" (Armin W. Geertz, "Brain, Body, and Culture: A Biocultural Theory of Religion," *Method and Theory in the Study of Religion* 22 [2010]: 304).

2. https://mindlab.au.dk/menu5-en. In 2018, a project similar in texture and tone to MindLab was initiated by the Danish Ministry of Business's Disruption Taskforce. In addition to facilitating a digital transformation across government agencies, the Taskforce (initially led by the former policy chief at Uber) will initiate scientific collaborations with the tech industry while also ensuring best practices. "MindLab 2.0: Denmark Establishes Its Next-Generation Innovation Lab," *apolitical* (December 11, 2018), https://apolitical.co/solution_article/mindlab-2-0-denmark-establishes-its-next-generation-innovation-lab/.

soon learned that people like Walter Benjamin make no sense whatsoever and that kind of thing is fine if you are a poet or trust-fund hipster or something like that, but if you are going to say anything meaningful about the world well, then, you need to say it clearly and concisely. For the folks at MindLab Benjamin was a symbol, a taboo figure who reactively generated many boundaries and bonds of solidarity. At MindLab I heard much talk about the dangers of relativists, postmodernists, feminists, cultural anthropologists, and critical historians such as myself who were guilt-ridden and weak-kneed, too afraid to believe in truth or else traumatized by the confidence of the strong scientific pose. These barbarians at the gates, I was told, were misguided in the way they saw science as but one among many possible rationalities. These weak-willed academics did not follow standard epistemic protocols of objectivity, hypothesis, and experimentation in order to produce real, tangible, and useful results about reality.[3]

Deficiencies in the gears of reason. Sentimental renderings without sufficient evidence.

Rather than challenge these claims directly or even bring up what I felt to be the strong categorical presence of secular modernity, I offered my interlocutors stony riffs on why I was interested in them in the first place. Religion, I said, has long been a site of fascination and an object of relentless measurement and theorization—from early modern investigations into the brain as a "chapel of deity" giving way to phrenological theories of localization and diagnoses of religious insanity[4] giving way to the discovery of the motor cortex and its relation to hypnosis[5] and telepathy giving way to ruminations on the electrical emanations of neural activity to the focus on hallucinations, epilepsy,[6] and other extreme states among midcentury neu-

3. Walter Benjamin, *On Hashish*, ed. Howard Eiland (Cambridge, MA: Harvard University Press, 2006).

4. A. L. Wigan, *A New View of Insanity: The Duality of the Mind Proved by the Structure, Functions, and Diseases of the Brain, and by the Phenomena of Mental Derangement and Shown to be Essential to Moral Responsibility* (London: Longman, Brown, Green, and Longmans, 1844).

5. William McDougal, "The State of the Brain during Hypnosis," *Brain* 31 (1908): 242–58. See also the relatively noninvasive methods of measuring the mind in William Brown and Godfrey H. Thompson, *The Essentials of Mental Measurement* (Cambridge: Cambridge University Press, 1921).

6. Cathy Gere, "'Nature's Experiment': Epilepsy, Localization of Brain Function and the Emergence of the Cerebral Subject," in *Neurocultures: Glimpses into an Expanding Universe*, ed. Fernando Vidal and Francisco Ortega (Frankfurt: Peter Lang, 2011), 235–47.

rophysiologists giving way to the isolation of religion as a neural phenomenon[7] giving way to more recent invocations of neuroscience as fodder for theological renewal and so-called new atheist claims alike.[8]

Without name-dropping Benjamin again or mentioning the dialectic of Enlightenment, I argued that MindLab was a culmination of a much longer trajectory. For within this history there is good religion and there is bad religion, the former increasingly mirroring the values of the present secular order.[9] What was being produced at MindLab, I suggested, was religion. Here, in ecstatic acts of attention, was the reification of *the* secular category, par excellence—religion, defined *primarily* as a mode of cognitive processing and specified neural activity. Cognitive studies of religion conducted at MindLab and elsewhere serve to articulate a boundary between the religious and the secular that can be either quite stark[10] or rather fluid.[11] To parse individual cognitive studies of religion based on the degree to which they see religion as exceptional or on their alleged sympathy (or lack thereof) with religion would be to miss much about their compatibility under the sign of secularism. For the significance of these scientific studies lies in the ways in which they manufacture religion and its natural distinction, study after study, regardless of results, turning religion into a measurable difference:[12]

7. B. K. Anand et al., "Some Aspects of Electroencephalographic Studies in Yogis," *Electroencephalography and Clinical Neurophysiology* 13, no. 3 (June 1961): 452–56; Akira Kasamatsu and Tomio Hirai, "An Electroencephalographic Study of the Zen Meditation (Zazen)," in *Altered States of Consciousness: A Book of Readings*, ed. Charles T. Tart (New York: John Wiley and Sons, 1969), 489–501; Roland Fisher, "Cartography of the Ecstatic and Meditative States," *Science* 174, no. 412 (November 26, 1971): 897–904.

8. Daniel C. Dennett, "Thank Goodness!," *The Edge: The Third Culture* (November 2, 2006), https://www.edge.org/conversation/thank-goodness.

9. Uffe Schjoedt, "The Religious Brain: A General Introduction to the Experimental Neuroscience of Religion," *Method and Theory in the Study of Religion* 21 (2009): 312.

10. J. T. Chibnall et al., "Experiments on Distant Intercessory Prayer: God, Science, and the Lesson of Massah," *Archives of Internal Medicine* 161, no. 21 (2001): 2529–36.

11. Justin L. Barrett, "How Ordinary Cognition Informs Petitionary Prayer," *Journal of Cognition and Culture* 1 (2001): 259–69; Marcel Mauss, *On Prayer*, ed. W. S. F. Pickering (New York: Durkheim Press/Berghahn Books, 2003).

12. Uffe Schjoedt, Hans Stødkilde-Jørgensen, Armin W. Geertz, and Andreas Roepstorff, "Highly Religious Participants Recruit Areas of Social Cognition in Personal Prayer," *Social Cognitive and Affective Neuroscience* 4, no. 2 (2009): 199–207; Uffe Schjoedt, Hans Stødkilde-Jørgensen, Armin W. Geertz, Torben E. Lund, and Andreas Roepstorff, "The Power of Charisma—Perceived Charisma Inhibits the Frontal Executive Network of Believers in Intercessory Prayer," *Social Cognitive and Affective Neuroscience* 6, no. 1 (2011): 119–27.

the brain and images of the brain and images of images of the brain generating the criteria for what constitutes religion as a *natural* difference.[13] Religion, here, becomes a feature of the world, built-in as it were, that persists beyond the vicissitudes of time and culture. Regardless of particular conclusions about, say, the efficacy of prayer or the function of fire-walking rituals or the practice of spirituality or the emergence of new social movements or large complex societies, cognitive studies conjure the difference between the religious and the secular as natural, that is, as something that can be ascertained through the scientific method.[14] Cognitive studies of religion are religion considered without considering its history, religion considered *instrumentally*, in terms of what is valuable, therapeutic,[15] or pathological about its presence.

13. Paul Bloom, "Religion Is Natural," *Developmental Science* 10, no. 1 (2007): 147–51.

14. Which is to say that in contrast to their narrow methodological purview, the cognitive sciences of religion are quite expansive in terms of subject matter. See, e.g., Dimitris Xygalatas, *The Burning Saints: Cognition and Culture in the Fire-Walking Rituals of the Anastenaria* (London: Routledge, 2012); Kevin S. Seybold, "Cognitive Sciences: A Perspective on Spirituality and Religious Experience," in *Oxford Textbook of Spirituality in Healthcare* (Oxford: Oxford University Press, 2012); Ann Taves, "'Magical Thinking' and the Emergence of New Social Movements," *Journal of Cognitive Historiography* 1, no. 2 (2015): 146–70; Ara Norenzayan, *Big Gods: How Religion Transformed Cooperation and Conflict* (Princeton, NJ: Princeton University Press, 2013); Jennifer Larson, *Understanding Greek Religion* (New York: Routledge, 2016); Olympia Panagiotidou with Roger Beck, *The Roman Mithras Cult: A Cognitive Approach* (New York: Bloomsbury, 2016); David Wengrow, *The Origins of Monsters: Image and Cognition in the First Age of Mechanical Reproduction* (Princeton, NJ: Princeton University Press, 2013); István Czachesz and Risto Uro, eds., *Mind, Morality and Magic: Cognitive Science Approaches in Biblical Studies* (London: Routledge, 2014); Helen De Cruz and Johan De Smedt, *A Natural History of Natural Theology: The Cognitive Science of Theology and Philosophy of Religion* (Cambridge, MA: MIT Press, 2015); Lisa Raphals, "Body and Mind in Early China and Greece," *Journal of Cognitive Historiography* 2, no. 2 (2017): 132–82; Robert N. McCauley, George Graham, and A. C. Reid, "Theory of Mind, Religiosity, and Autistic Spectrum Disorder: A Review of Empirical Evidence Bearing on Three Hypotheses," *Journal of Cognition and Culture* 19, no. 5 (2019): 411–31; Wesley J. Wildman, Joseph Bulbulia, Richard Sosis, and Uffe Schjoedt, "Religion and the COVID-19 Pandemic," *Religion, Brain and Behavior* 10, no. 2 (2020): 115–17; Adam E. Tratner et al., "Fear the Unseen: Supernatural Belief and Agency Detection in Virtual Reality," *Religion, Brain and Behavior* 10, no. 2 (2020): 118–31.

15. During a MindLab retreat in 2015 I sat beside a chemist who worked for a pharmaceutical company in Berlin. The chemist informed me that he often attended scientific conferences on religion and cognition for insights into the neural mechanisms of religion. His company, he continued, was interested the possibilities of catalyzing neural processes

Within the vast majority of cognitive studies of religion, the analytic difference between the religious and the secular is premised on their continuum. There is a founding and necessary assumption, namely that "the very existence of religion requires such cognitive mechanisms that also function outside of religion." These shared mechanisms include "commitment gadgets," "coalition psychology," and "constructing databases about the reputational effects of one's own and others' actual behavior and inferred dispositions."[16] Religion, as a natural human activity whose sources and consequences are bound up in society, can be advantageous or not, depending on the scene or evolutionary circumstance. But what to make of circumstances that are historical, cultural, or otherwise contingent? For in addition to information passing through the front parietal lobe, might there also be discursive forces flowing through the cognitive and neuroscientific fields that are integral to the making of a human subject who is truly, naturally, and ordinarily religious?

Within the neuro- and cognitive sciences, so proudly complicit in this secular age when the natural difference between the religious and the secular goes unquestioned precisely because it has become unquestionable, there is neither room nor necessity for asking questions about the conditions that have made the measurement of religion possible, let alone so alluring a proposition.[17]

---

involved in religious experiences for purposes of treating depression and anxiety. See also Elisabet Domínguez-Clavé et al., "Ayahuasca: Pharmacology, Neuroscience and Therapeutic Potential," *Brain Research Bulletin* 126, no. 1 (2016): 89–101; Dave E. Nichols and Benjamin R. Chemel, "The Neuropharmacology of Religious Experience: Hallucinogens and the Experience of the Divine," in *Where God and Science Meet: How Brain and Evolutionary Studies Alter Our Understanding of Religion*, vol. 3: *The Psychology of Religious Experience*, ed. Patrick McNamara (Westport, CT: Praeger, 2006), 1–33.

16. Ilkka Pyysiäinen, "Cognitive Science of Religion: State of the Art," *Journal for the Cognitive Science of Religion* 1, no. 1 (2012): 11.

17. While acknowledging the influence of "nurture" on nature, for example, scientists often view "culture" as an influence on preexisting cognitive systems as opposed to constitutive of the questions being asked to secure their existence. See, for example, the foundational work of E. Thomas Lawson and Robert N. McCauley in *Rethinking Religion: Connecting Cognition and Culture* (Cambridge: Cambridge University Press, 1990).

01

THINKING ABOUT COGNITIVE SCIENTISTS THINKING ABOUT RELIGION

> The mind of man is the noblest work of God which reason discovers to us, and therefore, on account of its dignity, deserves our study. It must indeed be acknowledged, that although it is of all objects the nearest to us, and seems the most within our reach, it is very difficult to attend to its operations, so as to form a distinct notion of them; and on that account there is no branch of knowledge in which the ingenious and speculative have fallen into so great errors, and even absurdities.
>
> THOMAS REID, *Essays on the Intellectual Powers of Man* (1785)

1. FALSE POSITIVES

In 1948 at Tufts University, psychologists turned to the new science of electroencephalography (EEG) and discovered what they called the "kappa wave." This newfound rhythm had been detected by electrodes placed just back of the eyes on the side of the head. Kappa waves were thought to make visible the process of thinking itself—making decisions, reading, discriminating, conducting simple mathematical tasks—a baseline rhythm of cognition. The Tufts study, conducted within an institution whose lineage runs through the Universalist Church and P. T. Barnum both, claimed nothing less than to have isolated the antiphonal swing of thinking and seeing, fixed

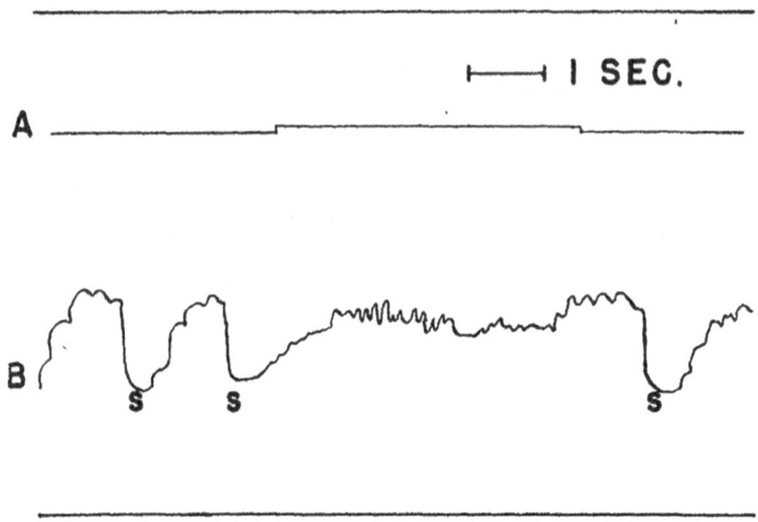

FIG. 1. Kappa bursts during reading.

FIGURE 9. Kappa wave, from John L. Kennedy, Robert M. Gottsdanker, John C. Armington, and Florence E. Gray, "A New Electroencephalogram Associated with Thinking," *Science* 108, no. 2811 (Nov. 12, 1948): 527–29.

concentration combined with visual tracking, that allowed you to take it all in for the express purpose of thinking about it all. Here were the inner jagged harmonies within the human frame that did not necessarily align with human will and control but nonetheless made them possible.

> Kappa intrudes occasionally when the subject is trying not to think. Introspective reports suggest that the intrusions of kappa correspond to "thoughts" during the period of attempted voluntary inhibition of thinking.[1]

It should be noted, however, that later studies discovered that kappa waves had little to do with thinking and everything to do with noisy corruptions of signal, otherwise known as artifacts. A few artifacts and their corresponding patterns had already been corroborated among scientists—a loose rubber tube, shuffling feet, a misapplied electrode, troubles in the amplifier, and

1. John L. Kennedy et al., "A New Electroencephalogram Associated with Thinking," *Science* 108, no. 2811 (November 12, 1948): 527–29.

issues with electrical output.[2] Kappa waves, as it would be discovered fourteen years later, corresponded to the rapid fluttering of the eyelids, sometimes so minute that it was difficult to see. Such all-but-invisible minutiae produced a rhythmic energy pattern picked up by the leads connecting the scalp to the amplifier.[3] Science—advancing its evolutionary scheme and overcoming, at once, "biological noise" as well as the intrusive hum of our machines.[4]

The Tufts study was part of the early enthusiasms surrounding discovery, measurement, and decipherment of the "brainscript." By midcentury the brain wave—that most cybernetic of lines—promised direct access to the material language of the brain.[5] As part of the initial cybernetic splay, EEG was the newly nominated science's most visible and recognizable technology. It signified a future in which normal cognition could be mapped and codified, when scientists would be able to predict and manage violent behavior through early EEG detection. According to the neurophysiologist

2. F. A. Gibbs and E. L. Gibbs, *Atlas of Electroencephalography* (Cambridge, MA: Lew A. Cummings Co., 1941), 192ff.

3. Stanley A. Lorens Jr. and Chester W. Darrow, "Eye Movements, EEG, GSR and EKG during Mental Multiplication," *Electroencephalography and Clinical Neurophysiology* 14 (1962): 739–46. See also Frank J. McGuigan, *Cognitive Psychophysiology: Principles of Covert Behavior* (New York: Prentice Hall, 1978), 279.

4. K. S. Lion and D. F. Winter, "A Method for the Discrimination between Signal and Random Noise of Electrobiological Potentials," *Electroencephalography and Clinical Neurophysiology* 5 (1953): 109–11. According to a typical EEG training manual, "artifacts come in many different forms and have diverse causes," including chewing and tongue movements, the popping of electrodes, the hum of ventilation systems, perspiration, etc. But "the major underlying problem is the enormous amplification required to record brain waves" (A. James Rowan and Eugene Tolunsky, *Primer of EEG (with Mini-Atlas)* [Philadelphia: Butterworth-Heinemann, 2003], 20–21, 22f.).

5. Cornelius Borck, "Recording the Brain at Work: The Visible, the Readable, and the Invisible in Electroencephalography," *Journal of the History of the Neurosciences* 17 (2008): 367–79. See also his *Brainwaves: A Cultural History of Electroencephalography*, trans. Ann M. Hentschel (New York: Routledge, 2018). As Norbert Wiener observed in his memoir, brain waves "speak a language of their own, but this language is not something that one can observe precisely with the naked eye, by merely looking at the ink records of the electroencephalograph. There is much information contained in these ink records, but it is like the information concerning the Egyptian language which we had in the days before the Rosetta Stone, which gave us the clue" (*I Am a Mathematician: The Later Life of a Prodigy* [Garden City, NY: Doubleday, 1956], 289).

William Grey Walter, EEG would soon be able to measure proclivities for self-indulgence and self-control, "together with the effects of encouraging or discouraging them."[6]

Walter was an integral figure in the deployment of electroencephalography (or EEG). Walter cofounded the journal *Electroencephalography and Clinical Neurophysiology* in 1949 and was coeditor from 1953 to 1957. As a graduate student in 1934, Walter had been witness to the first public demonstration of the existence of a brain wave.[7] As Walter described the scene:

> Pads soaked with salt solution to make them conducting were placed on the subject's head. . . . The pads were electrically connected to a powerful amplifier, and this to an instrument in which the amplified electrical changes were made to move a pen across a strip of paper which was drawn along at right angles to the direction of motion of the pen.[8]

Here was a moment—the moment of application—in which a correlation between cognition and the waveform tightened. A twitch with logic, reason, and sustained vitality. The quest for the "physiological basis of thought" had found a new medium—a "deus ex machine [*sic*]," gushed Walter, "it is electricity." A new horizon had been properly identified. The language of human nature, in all of its ornate precision, had revealed itself. "The physiological background of our perception and thinking may be neither peace nor chaos," Walter continued, "but a deep all-embracing rhythm of which we are unconscious, perhaps because we are so used to it." Walter's enthusiastic witness was tempered, however. For some things, he cautioned, "may be beyond our power to know. As so often, our apparatus will probably fail us just when we want to generalize."[9]

For the most part Walter's warning went unheeded. As epileptologist and EEG pioneer William G. Lenox wrote in the first volume of Walter's newly founded *Electroencephalography and Clinical Neurophysiology*, the brain was on the cusp of revealing the secrets we most wanted it to disclose.

6. W. Grey Walter, *The Living Brain* (New York: W. W. Norton, 1953), 257. On the futurity built into the recognition and framing of brain waves, see Stefan Helmreich, "Potential Energy and the Body Electric: Cardiac Waves, Brain Waves, and the Making of Quantities into Qualities," *Current Anthropology* 54 (October 2013): S139–S148.

7. Borck, "Recording," 369–70.

8. On Walter's magnification of heretofore infinitesimal differentials, see his "Thought and Brain: A Cambridge Experiment," *Spectator* (October 4, 1934): 478–79.

9. Walter, "Thought and Brain," 478–79.

Despite the relative youth of electroencephalography, "the thrill of watching the brain write its own confession of guilt on moving paper is not yet lost. Descriptions of changes in the height and frequency of surface waves in relation to clinical symptoms is of practical value, but we must not stop there. . . . We sense an approach to the meaning and origin of the electrical activity of the nervous system." Even by the end of his own career, Walter was admitting to the "eerie experience [of] discern[ing] through an electric machine the genesis of a person's intentions."[10]

The legitimacy of the brain wave lay in the epistemic promise of technology—the EEG as a truth-telling device, brain waves as unmediated signs of the real, accessible frequencies to a heretofore unknown. With the brain abuzz with continuous electric pulsation, there was the accompanying frenzy of pattern recognition when it came to brain waves. Here was an empirical approach to think about thinking, a way to figure out what we did not know that enabled us to know. Here was a way to distinguish between and make inferences about significant and insignificant data, normal and abnormal patterns. Here was a way to imitate how the brain made those very same demarcations and built a life for itself out of those inferences. Here was a cognitive-inspired approach to cognition.

Electroencephalograms were surveyed like a newly discovered country, mapped and keyed and turned into an atlas for the burgeoning neural cartographer. Brain waves were soon aligned with gender, age, activity, and pathology. To each, his or her own encephalogram. Patterns, patterns everywhere: normal and abnormal, for all ages, in infants, boys and girls—restless, aggressive, pyromaniacal, and prone to tantrums—in women and men, with eyes open and closed, brain waves of Europeans, Anglo Americans, and African Americans; brain waves during nighttime sleep and afternoon naps; in epileptics and the parents of epileptics, and family members of epileptic parents; brain waves in patients hopped up on mescaline or hallucinating on LSD; in patients with meningitis, Parkinson's and Alzheimer's, schizophrenia and dementia; in professors and so-called mongoloids; in the delusional and comatose, in the brain damaged and lobotomized; brain waves in those who stuttered, had migraines, or were afflicted with multiple sclerosis, as well as those who suffered brain tumors, manic depression, and

10. William G. Lennox, "Influence of Drugs on the Human Electroencephalogram," *Electroencephalography and Clinical Neurophysiology* 1 (1949): 45; W. Grey Walter, *Observations on Man, His Frame, His Duty and His Expectations* (Cambridge: Cambridge University Press, 1969), 36–37.

thyroid disease.[11] With such standardization, EEG became a useful tool in clinical diagnosis as disorders that had previously been thought of as predominantly psychic (like epilepsy) assumed new physiological and electrical dimensions.[12]

In the authoritative *Atlas of Electroencephalography* (1944), there was a precise sculpting of the normal and abnormal by making space for exceptions to a strict reading of categorical difference—"questionably normal EEGs in normal adults," "abnormal EEGs in normal children," "morons with normal EEGs." These unexpected findings served to refine the difference between normal and abnormal and to prove the spectrum of natural difference between them. The volume's editors, Frederic A. Gibbs and Erna L. Gibbs, believed that "an increased knowledge of the mechanisms operating in the human brain strengthened the hope of controlling not only the 'abnormal' mental states which manifest themselves as psychoses, but also those 'normal' mental states which manifest themselves in crime, oppression, and war." They also believed that, with advances in EEG technology, they would soon be able to distinguish, at a mathematical level, all manner of physiological characteristics, including racial and sexual differences.[13]

11. The Gibbs's *Atlas of Electroencephalography* (1941) contained stunningly beautiful pages of EEG scripts. The power of these representations, like romantic works of art, was to convince the viewer of a reality that existed but could not be seen without extraordinarily enhanced sensitivities. See also Richard L. Masland, George Auston, and Francis C. Grant, "The Electroencephalogram Following Occipital Lobectomy," *Electroencephalography and Clinical Neurophysiology* 1 (1949): 273; George G. Merrill and Elwood E. Cook, "The Electroencephalogram in the Negro: A Comparison of Electrical Activity of the Brain in White and Negro Patients," *Electroencephalography and Clinical Neurophysiology* 9 (1957): 531–32; A. C. Mundy-Castle, "The Electroencephalogram and Mental Activity," *Electroencephalography and Clinical Neurophysiology* 9 (1957): 643–55.

12. For a meticulous discussion of these materializing tendencies, see Borck, *Brainwaves*, 184ff.

13. Gibbs and Gibbs, *Atlas*, v–vi; in 1968, for example, EEG was used to confirm the physiological fundament of "hypnotic susceptibility"—a sentimental quality of impressibility long associated with the religious nature of femininity; Perry London et al., "EEG Alpha Rhythms and Susceptibility to Hypnosis," *Nature* 219 (July 6 1968): 71–72. See also the concurrent attempt to locate "gonadal-brain mechanisms" and to place sexual difference on a grid of neural intelligibility; William Vogel et al., "EEG, Physique, and Androgens," *Perceptual and Motor Skills* 26 (1968): 419–29. More recently, EEG generates such intelligibility in order to aid and abet the demographic calculations of advertisers. See, e.g., Giulia Cartocci et al., "Gender and Age Related Effects While Watching TV Advertisements: An EEG Study," *Computational Intelligence and Neuroscience* (2016): 1–10.

While individual results of an EEG could be differently interpreted, there was little, if any, critique of its fundamental mechanics, promise, and, to paraphrase Isabelle Stengers, the social relationships that it established.[14] The relations that EEG generated at midcentury—to researchers, to the foundations that funded their work, to the patients who were promised answers about and leverage on their cognitive quirks and long-term illnesses, to the scientists in related fields who drew inspiration from the advances in EEG, to the general public invested in the promise of brain waves—soon became, for all intents and purposes, irreversible.[15] Amid this triumph, neurophysiologists utilized EEG to probe normal and abnormal brains and their respective relationship to their environments. In doing so, they were also setting a research agenda for large swaths of the human sciences in the postwar years.[16]

2. THE COGNITIVE SCIENCE OF RELIGION

> It is precisely because humans create their gods that these gods embody anthropologically specific characteristics drawn from a particular place and time.
>
> SABA MAHMOOD, *Religious Difference in a Secular Age* (2015)

Electroencephalograms remain central to doctors and scientists in various fields, including the cognitive science of religion.[17] In what amounts

14. Isabelle Stengers, *The Invention of Modern Science*, trans. Daniel W. Smith (Minneapolis: University of Minnesota Press, 2000), 103.

15. Roland Barthes, "The Brain of Einstein," in *Mythologies*, trans. Annette Lavers (New York: Farrar, Strauss, and Giroux, 1972), 68–70, and "Einstein's Brain Waves," *Life Magazine* (February 26, 1951): 40. See also the science fiction romp *Brain Wave* by Poul Anderson (New York: Ballantine Books, 1954).

16. In "Adaptiveness and Equilibrium," *Journal of Mental Science* 86 (May 1940): 478–83, Ross W. Ashby posited that the relations within the brain and between brain and environment were matters of negative feedback. Ashby would soon embrace information theory as requisite for designing a brain.

17. See, e.g., Michiel van Elk, "An EEG Study on the Effects of Induced Spiritual Experiences on Somatosensory Processing and Sensory Suppression," *Journal for the Cognitive Science of Religion* 2, no. 2 (2015): 121–57.

to similar practices under the sign of secularism, for example, brain waves have been used to both explain and explain away the mechanics of religion. As a privileged tool within cognitive inquiries into religion, EEG has aided and abetted a host of questions that seek leverage on the nature of religious cognition.[18]

Cognitive science—with its phalanx of journals[19] and seemingly endless

18. For an overview of recent, unpublished and somewhat flawed but now useful-in-hindsight EEG studies on religion, see Michiel van Elk, "What's Hidden in My Filedrawer and What's in Yours? Disclosing Non-published Findings in the Cognitive Science of Religion," *Religion, Brain and Behavior* (2020): 1–12. See also Claudio Imperatori et al., "Neurophysiological Correlates of Religious Coping to Stress: A Preliminary EEG Power Spectra Investigation," *Neuroscience Letters* (2020): 134956; Xuzhou Li et al., "A Diffusion Tensor Imaging Study of Brain Microstructural Changes Related to Religion and Spirituality in Families at High Risk for Depression," *Brain and Behavior* 9, no. 2 (2019): e01209, also DOI: 10.1002/brb3.1209; Ahmed Izziden and Srivas Chennu, "A Neuroscience Study on the Implicit Subconscious Perceptions of Fairness and Islamic Law in Muslims Using the EEG N400 Event Related Potential," *Journal of Cognition and Neuroethics* 5, no. 2 (2018): 21–50; Craig E. Tenke et al., "Association of Posterior EEG Alpha with Prioritization of Religion or Spirituality: A Replication and Extension at 20-year Follow-up," *Biological Psychology* 124 (2017): 79–86; Xinmei Deng et al., "Differences in Frontal EEG Asymmetry during Emotion Regulation between High and Low Mindfulness Adolescents," *Biological Psychology* 158 (2021): 107990; Tim Lomas et al., "A Systematic Review of the Neurophysiology of Mindfulness on EEG Oscillations," *Neuroscience and Biobehavioral Reviews* 57 (2015): 401–10; Mahsa Vaghefi et al., "Spirituality and Brain Waves," *Journal of Medical Engineering and Technology* 39, no. 2 (2015): 153–58; Craig E. Tenke et al. "Neuronal Generators of Posterior EEG Alpha Reflect Individual Differences in Prioritizing Personal Spirituality," *Biological Psychology* 94, no. 2 (2013): 426–32; Mario Beauregard and Vincent Paquette, "EEG Activity in Carmelite Nuns during a Mystical Experience," *Neuroscience Letters* 44 (2008): 1–4; J. P. Banquet, "Spectral Analysis of the EEG in Meditation," *Electroencephalography and Clinical Neurophysiology* 35, no. 2 (1973): 143–51. See also works in the parapsychological vein such as Robert A. Charman, "Has Direct Brain to Brain Communication Been Demonstrated by Electroencephalographic Monitoring of Paired or Group Subjects?," *Journal of the Society for Psychical Research* 70.1, no. 882 (January 2006): 1–24; Bruce E. McDonough et al., "EEG Frequency Domain Analysis during a Clairvoyance Task in a Single Subject Design: An Exploratory Study," in *Research in Parapsychology 1988*, ed. L. A. Henkel and R. E. Berger (Metuchen, NJ: Scarecrow Press, 1989), 38–40.

19. Established journals in the cognitive science of religion include *Journal of Cognition and Culture* (Brill, 2001), *Religion, Brain and Behavior* (Taylor and Francis, 2011), *Journal for the Cognitive Science of Religion* (Equinox), and *Journal of Cognitive Historiography* (Equinox, 2014). *Method and Theory in the Study of Religion* (Brill) is a general journal devoted to methodological and theoretical issues that includes many articles (and entire issues) devoted to questions of cognition.

funding opportunities—claims to have instituted a paradigm shift in the study of religion.[20] Pascal Boyer, one of the most visible figures in the cognitive science of religion (hereafter CSR), bluntly states that past theorizing in the field of religious studies has consisted "in the half-hearted adoption of particular academic fads." Most scholars of religion, he writes, pay "lip-service to the current fad, while carrying on with the[ir] erudition projects" of mere cataloguing. Boyer, who has been considered something of an architect of CSR since the publication of *Religion Explained* in 2001 (a book that biologist E. O. Wilson suggests was written "in the spirit of the French Enlightenment"), pulls no punches when it comes to describing the descriptive mode of contemporary religious studies: "As a consequence of this lackadaisical approach to explaining religious thought and behavior, the field has become theoretically amorphous, and unresponsive to actual scientific proposals."[21] The solution, according to Boyer and his colleagues, is to take up the calling of the brain to explain itself to itself, that is, to offer an explanation of human cognition that conforms to the actual mechanics of cognitive processing.[22]

In this chapter I historicize the conceptual framework of CSR and tend, as any empirically minded scientist would, to ecological confounds, cultural artifacts, and the conditions that make possible (and increasingly

20. For a summary of the revolution, see Steven Pinker, "The Evolutionary Psychology of Religion," in *Where God and Science Meet: How Brain and Evolutionary Studies Alter Our Understanding of Religion*, ed. Patrick McNamara (Westport, CT: Praeger, 2006), 1–9.

21. Pascal Boyer, "Explaining Religious Concepts: Levi-Strauss the Brilliant and Problematic Ancestor," in *Mental Culture: Classical Social Theory and the Cognitive Science of Religion*, ed. Dimitris Xygalatas and William W. McCorkle Jr. (Durham, UK: Acumen Publishing, 2013), 168.

22. Similar to Boyer, Edward Slingerland, a cognitive scientist and self-described "empirically responsible intellectual," dismisses those whom he considers not stuck, as he puts it, in "the endless cycle of contingent discourses and representations of representations." Anxious to announce his "pragmatic" credentials, his common sense naturalism, and "embodied approach to culture," Slingerland bristles at those who would claim that humans are estranged or their knowledge incommensurable with or even limited by the natural world. For "the process of evolution," he argues, "ensures that there is a tight fit between our values and desires and the structure of the world in which we have developed." There is a security to be found in such common sense pleas, a safe space of immunity from the culture that contains us, which then serves as an excuse to set aside the messy work of immanent critique. But I digress . . . Edward Slingerland, "Who's Afraid of Reductionism? The Study of Religion in the Age of Cognitive Science," *Journal of the American Academy of Religion* 76 (2008): 404, 378, 382.

legible) cognitive investigations into religion. CSR is an international and self-consciously interdisciplinary subfield with competing claims about its object of study.[23] But differences in opinion about the origins of religion or arguments about whether, for example, religion is an evolutionary adaptation or an evolutionary effect, betray an underlying epistemic coherence. A style of reasoning is present and practiced among those who might disagree about the primacy of belief or ritual, the exact relationship between cognition and culture, or the preferred means of measuring that relationship. One point of substantive agreement across the various corners of CSR is that whatever religion is, it involves the superimposition of intentional agency on natural entities, events, or even groups.[24] I refer here, of course, to that curious engine of religious belief that goes by the name of the "hyperactive agency detection device" (HADD).

This chapter begins with a survey of the recent science of HADD and follows a series of trap doors that serve to contextualize, but also disturb, the story cognitive scientists often tell themselves in order to be themselves. This chapter tacks between the present state of CSR and the "season of revivals" that occurred in Northampton, Massachusetts, in the mid-1730s. These revivals, spearheaded and theorized by Jonathan Edwards, were a founding moment of Protestant religious history in America. During the revivals the concept of hypersensitivity to divine agents came to the fore—as a bludgeon for critics of enthusiastical excess and, for defenders, a new rationale. Rivals such as Charles Chauncy dismissed the season as a loss of emotional control to be avoided. But in Edwards's attempt to experience *and* classify *and* adjudicate the real presence of spirit, he countered Chauncy's parry by turning Northampton into a vast scientific laboratory for considering, objectively, how God's agentive presence could be detected.

Between the present and Edwards's discovery of a "new spiritual sense,"

23. CSR operates within the afterglow of the "social brain hypothesis" put forward by neuroscientist Michael Gazzaniga in *The Social Brain: Discovering the Networks of the Mind* (New York: Basic Books, 1985). Accordingly, social life not only has neural correlates but the cognitive demands made by living within complex groups serve as an evolutionary engine. Consequently, the brains that meet these computational demands have an evolutionary advantage. See also Chris D. Frith, "The Social Brain?," *Philosophical Transactions of the Royal Society* 362 (2007): 671–78.

24. Christine Ma-Kellams, "When Perceiving the Supernatural Changes the Natural: Religion and Agency Detection," *Journal of Cognition and Culture* 15 (2015): 337–43. See also Paul Bloom and Csaba Veres, "The Perceived Intentionality of Groups," *Cognition* 71 (1999): B1–B9.

two mid-twentieth-century moments resonate rather intensely with the epistemics and politics of agency detection. The first is Fritz Heider and Marianne Simmel's "An Experimental Study of Apparent Behavior," published in 1944. At Smith College in Northampton, Massachusetts, Heider and Simmel were pioneers in the critique of behaviorism that would eventually be known as cognitive psychology. Heider and Simmel claimed to have demonstrated how humans ascribed human characteristics, motives, and narrative to situations that were anything but. The second moment is the emergence of electroencephalography as it was theorized by the cybernetic pioneer William Grey Walter in the 1950s. I want to call attention to the degree to which CSR has drawn inspiration from Heider and Simmel's work and has internalized much of Walter's language of cognitive scanning and prediction. In doing so, CSR has promoted a particular vision of the human that lends itself uniquely (and disturbingly) to present concerns and desires. Consequently, CSR is not only part and parcel of a contemporary secular imaginary but, in its argument that the "whole function of the brain is summed up in error correction," has extended this imaginary in a cybernetic key.[25]

Inspired by recent work in "critical neuroscience" and "embodied cognitive science," I am interested in the historical conditions and discursive compatibilities of CSR.[26] Where does the desire to locate religion in the brain

25. The words of W. Ross Ashby, another cybernetic trailblazer from midcentury, cited by Andy Clark in his seminal "Whatever Next? Predictive Brains, Situated Agents, and the Future of Cognitive Science," *Behavioral and Brain Sciences* 36, no. 3 (June 2013): 181. Inspired by Claude Shannon's theory of redundancy, John von Neumann spent much time thinking about the "question regarding errors, foreseeing errors, and recognizing and correcting errors" in both computers and the brain. "An artificial machine," von Neumann suggested, "may well be provided with organs which recognize and correct errors automatically. In fact, almost every well-planned machine contains some organs whose function is to do just this" (John von Neumann, "A General and Logical Theory of Automata" and discussion in *Cerebral Mechanisms in Behavior: The Hixon Symposium*, ed. Lloyd A. Jeffress [New York: Wiley, 1951], 35).

26. Melissa M. Littlefield and Jenell M. Johnson, eds., *The Neuroscientific Turn: Transdisciplinarity in the Age of the Brain* (Ann Arbor: University of Michigan Press, 2012); Anthony Chemero, *Radical Embodied Cognitive Science* (Cambridge, MA: MIT Press, 2009). Consequently, my critique of CSR moves beyond the mild suggestion that CSR extend its inquiry into the diversity of religious and secular experiences. For example, historian and CSR fellow traveler Ann Taves thoughtfully suggests that CSR would benefit from a more comparative methodology in which experiences deemed religious are set beside those considered "pathological and/or imaginary." As Taves writes: "If we want to know

come from? What story is CSR telling itself? What makes its approach so persuasive and cultus so inviting? In what remains I situate CSR and its will to explain religion within a history that extends far beyond CSR's dutiful citation of canonical discoveries.[27]

3. THE HYPERACTIVE AGENCY DETECTION DEVICE

Over the past twenty years the concept of a hyperactive agency detection device (HADD) has become a central component in the cognitive study of religion. Cognitive scientists use this felicitous phrase to discuss the bundle of cognitive processes that prime humans to scan for and believe in supernatural agents. HADD is the flipside of ADHD (Attention Deficit Hyperactive Disorder).[28] Whereas in ADHD, attention is attenuated by extraneous physiological activity, HADD is a corruption of attention brought about by

why—evolutionarily speaking—humans postulate and engage with superhuman agents, we need to look at a much wider range of phenomena, including psychopathology, experiences induced by drugs or computer simulations or fiction and art, and experiences associated with creative inspiration." This is all well and good but my critical inquiry has a sharper edge. I am interested in exploring how and why cognitive scientists postulate the object of religion—how they take its measure and engage in questions about supernatural agents as considered by others (never themselves, of course!). Ann Taves, "'Religious Experience' and the Brain," in *Evolution of Religion: Studies, Theories, and Critiques*, ed. Joseph Bulbulia et al. (Santa Margarita, CA: Collins Foundation, 2008), 218.

27. D. Jason Sloane and William M. McCorckle, *The Cognitive Science of Religion: A Methodological Introduction to Key Empirical Studies* (London: Bloomsbury, 2019).

28. Jonathan Williams and Eric Taylor, "The Evolution of Hyperactivity, Impulsivity and Cognitive Diversity," *Journal of the Royal Society Interface* 3, no. 8 (2006): 399–413. Indeed, HADD has emerged alongside increasingly authoritative diagnoses of inattention—Minimal Brain Damage morphing into Minimal Brain Dysfunction in the 1970s morphing into Attention Deficit Disorder in the 1980s and, most recently, into Attention Deficit/Hyperactivity Disorder (ADHD). Steven Rose, *The Future of the Brain: The Promise and Perils of Tomorrow's Neuroscience* (New York: Oxford University Press, 2005), 254. On the conflation of attention and the will, see the groove laid down by William James in his conclusion that "effort of attention is thus the essential phenomenon of will" (*Principles of Psychology*, vol. 2 [New York: Henry Holt, 1890], 562). On the history of ADHD as a "psychological fact" rooted in earlier pathologies of the will, see Andrew Lakoff, "Adaptive Will: The Evolution of Attention Deficit Disorder," *Journal of the History of the Behavioral Sciences* 36, no. 2 (2000): 149–69.

too much attention paid to particular things.[29] HADD is, and has always been, the source of our overdeterminations—from the worship of animal spirits to contemporary conspiracy theorists. HADD serves the universal human tendency to project animacy onto the world:

> This mechanism is triggered by very minimal cues. We see faces in the clouds and detect predators in rustling bushes because such ambiguous perceptions easily trigger the postulations of agency. . . . [A] normally functioning HADD is hyperactive by its very nature—hyperactivity is not something special. From an evolutionary point of view, this is plausible, insofar as the costs of false positives that an overreacting detector produces are lower than the benefits it brings.[30]

To be clear, the hyperactive agency detection device is not part of a crass claim about where religion is located, à la phrenology, but a more subtle attempt to understand the consequential integration of scanning and computational capacities of the brain. Indeed, the hyperactive agency detection device is not thought to involve one mental faculty or neural system but is rather "a result of the coordinated activity of many automatic mental systems."[31] Different parts of the brain come together to form a sustained statistical stare. The accumulation of statistics allows the stare to surveil. The

29. Tomasso Bertolotti and Lorenzo Magnani, "The Role of Agency Detection in the Invention of Supernatural Beings," in *Model-Based Reasoning in Science and Technology: Studies in Computational Intelligence*, ed. Lorenzo Magnani et al. (Berlin: Springer, 2010), 239–62. Although, technically, agency detection can be triggered by sound or smell, vision takes precedence in most discussions of HADD.

30. Ilkka Pyysiäinen, *Supernatural Agents: Why We Believe in Souls, Gods, and Buddhas* (New York: Oxford University Press, 2009), 13. On the insistence on evolving mechanisms of cost-benefit analysis, see Scott Atran, "The Cognitive and Evolutionary Roots of Religion," in *Where God and Science Meet*, ed. Patrick McNamara (Westport, CT: Praeger, 2006): 181–208. On the beneficial relationship between agency detection and memory, see Joshua E. VanArdsall et al., "Adaptive Memory: Animacy Processing Produces Mnemonic Advantages," *Experimental Psychology* 60, no. 3 (2013): 172–78.

31. Specifically, HADD is the coordination of "theory-of-mind systems and agency-detection and contagion-avoidance and social exchange." Pascal Boyer, "Religious Thought and Behavior as By-products of Brain Function," *Trends in Cognitive Science* 7, no. 3 (March 2003): 23. For a more recent refinement of HADD "as necessary but insufficient in explaining religious culture," see Andrew Ross Atkison, "HIDD'n HADD in Intelligent Design," *Journal of Cognition and Culture* 20, no. 3–4 (2020): 304–16.

eye is kept relatively mobile amid a swirl of information and data stimulations, tracking patterns across the surface of things.

Since the 1940s there has been an abundance of work on "the precise stimulus conditions that give rise to these percepts" of agency and the "perceptual 'grammar' of causality."[32] Religion, however, as a misguided form of agency detection—animism—has a rich anthropology behind it—think Ludwig Feuerbach, Karl Marx, E. B. Tylor, and Sigmund Freud. It was not until 1980, however, that the anthropologist Stewart Guthrie made the explicit case for a cognitive theory of religion as agency detection.[33] Guthrie argued that religion could be explained as an efficacious anthropomorphism—the process by which humans projected their own humanity on the world, for better or for worse. Guthrie's question was essentially this: if animism is a conceptual mistake (for who among us believes in ghosts?), how then to explain the situation in which many people still detect agents when in fact they have not?

Guthrie's *answer?*

Such detections, asserts Guthrie, "arise inevitably, as by-products—namely, as false positives—of our scanning an uncertain world for what matters most. What matters most is agency."[34] The question of why agency matters is, of course, assumed. According to Guthrie and other cognitive scientists, the category of agency and their interest in it is a natural by-product of evolutionary history and not, for example, wrapped up in the directives of management theorists or business consultants interested in workplace and

32. Brian J. Scholl and Patrice D. Tremoulet, "Perceptual Causality and Animacy," *Trends in Cognitive Sciences* 4, no. 8 (2000): 299–309. This work has increasingly focused on how the perception of intentional agency is precipitated by inferences we make about pattern, form, and movement rather than substance. Jakob Hohwy, *The Predictive Mind* (Oxford: Oxford University Press, 2013), 19.

33. Stewart E. Guthrie, "A Cognitive Theory of Religion," *Current Anthropology* 21, no. 2 (1980): 181–203. See also Guthrie's *Faces in the Clouds: A New Theory of Religion* (New York: Oxford University Press, 1993).

34. Stewart Elliott Guthrie, "Anthropology and Anthropomorphism in Religion," in *Religion, Anthropology and Cognitive Science*, eds. Harvey Whitehouse and James Laidlaw (Durham, NC: Carolina Academic Press, 2007), 37. In his definition of religion, Scott Atran declared, "Supernatural agency is the most culturally recurrent, cognitively relevant, and evolutionarily compelling concept in religion. The concept of the supernatural is culturally derived from an innate cognitive schema" (*In Gods We Trust: The Evolutionary Landscape of Religion* [New York: Oxford University Press, 2004], 57).

market efficiency. In terms of larger trends, however, agency detection has become a key concept for management researchers who seek to conceptualize and frame economic problems in terms of corporate design and the management of reward systems.[35]

One marvels at the lack of reflexivity when the analytic of agency detection is naturalized amid a culture dense with measures of "attention regulation," "attentional expertise," and "attention disorders."[36] And then there is the academic industry of "Error Management Theory" in which evolutionary insight serves to frame good and bad decisions, overconfidence and acts of deception—among businessmen, pilots, doctors, economists, and, of course, the pious whose decision to believe in God is taken "seriously" and subject to analysis.[37]

4. DISTINGUISHING MARKS ON A SCREEN

The hyperactive agency detection device, or something like it, has been historicized and popularized[38] and tested in the laboratory using, more often than not, two-dimensional simulations—television screens essentially, used

35. See, e.g., Nicolai Foss and Diego Stea, "Putting a Realistic Theory of Mind into Agency Theory," *European Management Review* 11, no. 1 (Spring 2014): 101–16, and Stefan Linder et al., "Epistemics at Work: The Theory of Mind in Principal-Agent Relations," in *The Oxford Handbook of Strategy Implementation*, ed. Michael A. Hitt et al. (New York: Oxford University Press, 2017), 101–26.

36. Pyysiäinen, *Supernatural Agents*, 98.

37. For an explicit application of Boyer, Barrett, and HADD in terms of optimizing efficiency within the workplace, see Dominic D. P. Johnson, "The Error of God: Error Management Theory, Religion, and the Evolution of Cooperation," in *Games, Groups, and the Global Good*, ed. S. A. Levin (Berlin: Springer, 2009), 169–80. On the exchange between CSR studies and error management theorists, see Ryan McKay, "Religion and Agency," *Journal for the Cognitive Science of Religion* 2, no. 2 (2014), 93–95. See also Andy Clark's declaration that he is interested in "the way that spending metabolic money to build complex brains pays dividends in the search for adaptive success" ("Whatever Next?," 181–82).

38. On the popular currency of HADD, see Daniel C. Dennett, *Breaking the Spell: Religion as a Natural Phenomenon* (New York: Penguin, 2006), 109, 116, 151, and Richard Dawkins, *The God Delusion* (New York: Houghton Mifflin Harcourt, 2008), 214. See also Gabriel Andrade, "Medical Conspiracy Theories: Cognitive Science and Implications for Ethics," *Medicine, Health Care and Philosophy* (2020): 1–14.

to stimulate responses in experimental subjects.[39] Indeed, the "proper evolutionary domain" of agency "encompasses animate objects relevant to hominid survival"—such as predators, protectors, and prey—"but which actually extends (as an inadvertently but spontaneously activated evolutionary by-product) to moving dots on computer screens, voices in wind, and faces on clouds." Just as pornographic pictures or drawings can arouse measurable sexual pleasure, humans are easily fooled to experience signs for reality when it comes to staring at a flat surface.[40]

The hyperactive agency detection device is a machine of sorts, inside your head right now. *No matter who you are*. It is a screening mechanism. All "typically developing humans" possess one. It is the machine that is at the base of religion. It is the machine that quite literally makes your prayers possible. It scans the horizon for movement and pattern and alerts us to forces of otherness, variously construed. As the machine that runs the program of prayer, the hyperactive agency detection device is responsible for sensing God as a fully realized agent. It is on all of the time, a form of troubleshooting the lines of transmission between you and what is on the other side of the screen.[41]

I use the word "screen" deliberately to mark the particularity of mechanical metaphor that is being used to designate a universal human characteristic. For how else would agency be detected if it were not wholly televisualized? Here is cognitive scientist Ilkka Pyysiäinen citing the relevant sources—Boyer and Barrett, among others, in order to list the "most common direct cues for agency":

> 1. Animate motion that has as its input such things as nonlinear changes in direction, sudden acceleration without collision, and change of physical shape that accompanies motion. . . .
> 2. An object reacting at a distance.
> 3. Trajectories that only make sense if the moving entity is trying to reach or avoid something, which leads to goal-ascription.

39. Lakoff, "Adaptive Will," 161. See also H. E. Rosvold et al., "A Continuous Performance Test of Brain Damage," *Journal of Consulting Psychology* 20 (1956): 343–50.

40. Scott Atran, "Religion's Innate Origins and Evolutionary Background," in *The Innate Mind*, vol. 2: *Culture and Cognition*, ed. Peter Carruthers, Stephen Laurence, and Stephen Stich (New York: Oxford University Press, 2006), 305.

41. On the instructions for troubleshooting one's scanning mechanism, see John F. Rider, ed., *Television: How It Works* (New York: John F. Rider, 1948), 107

4. An entity appearing to be moving by conscious intention to an apparent end result (intention-ascription).[42]

Such cues, argues Pyysiäinen, can "trigger the feeling or intuition of agency spontaneously and automatically, in the sense that this intuition can neither be rationally controlled nor initiated or terminated at will."[43]

The cues involved here speak to a subject who witnesses from afar, seeing in terms of physical laws, expectation, and counterintuition. For scanning yields incredibly detailed knowledge but is ever at one remove. Scanning is a privileged form of vision that assumes a state of immunity or, at the very least, idealizes it, from the beginning. And it is precisely because the scanner does not get involved, physically, in that which it scans, that there is much room for error correction above the fray. Scanning, then, is said to be scientific, statistical, and akin to how difference is accounted for in nature.[44] As a matter of real-time processing, improvement and growth are built in. Self-correction is premised on the scanner keeping *its* distance.[45] Scanning, here, is said to be liberated vision, a process of seeing without categories or preexisting taxonomies, extracting patterns in an algorithmic glance.[46]

CSR's specific deployment of the scanning metaphor as natural occludes from consideration what goes into the making of that metaphor and that deployment.[47] For by 1980, at the dawn of the cable age and the twenty-

42. Pyysiäinen, *Supernatural Agents*, 13.

43. Pyysiäinen, *Supernatural Agents*, 13.

44. On earlier efforts to naturalize scanning within the cybernetic archive, see Alan Turing, "On Computable Numbers, with an Application to the Entscheidungsproblem," *Proceedings of the London Mathematical Society* 42, no. 1 (1937), 230–65. As Mary A. B. Brazier summarized the neural mechanism of scansion, "chains of neurons in the central nervous system . . . are the structure for a circulating process which functions as a scanning mechanism. This idea of scansion . . . is analogous to the time-sweep in television; it has a definite period: the time taken for an impulse to circulate through the entire loop" ("Neural Nets and the Integration of Behavior," in *Perspectives in Neuropsychiatry*, ed. Derek Richter [London: H. K. Lewis, 1950], 35–36).

45. On the way in which scientific agency is often enhanced through recourse to masculine virtue, see Donna J. Haraway, *Modest_Witness@SecondMillennium. FemaleMan_Meets_OncoMouse: Feminism and Science* (New York: Routledge, 1997), 32.

46. Orit Halpern, *Beautiful Data: A History of Vision and Reason since 1945* (Durham, NC: Duke University Press, 2014), 204.

47. For a brief history of the metaphors used to discuss mental phenomena, see Dedre Gentner and Jonathan Grudin, "The Evolution of Mental Metaphors in Psychology: A 90-Year Retrospective," *American Psychologist* 40, no. 2 (February 1985): 181–92.

four-hour news cycle (and the same year in which the American Psychiatric Association renamed the disorder of hyperkinesis "Attention Deficit Disorder"), scanning was becoming a thing.[48] It was the logic of television. And it was how we detected the agency of others and exercised our own. Scanning also became part and parcel of a conception of the human, particularly a cognitive conception of the human bent on pattern recognition.[49] As a physiological process, scanning serves to evacuate everything that weighs down the present save for its mathematical truth. Better living through scanning is sensing this truth of present conditions and continually producing probabilistic scenarios that may stem from those conditions. *Better, stronger, faster.*

5. BREAKING THE SPELL

According to the cognitive psychologist Justin Barrett (who coined the term), the hyperactive agency detection device was an evolutionary advantage in a hostile environment. Our ancestors were those individuals whose vigilance bordered on paranoia. They "scanned" the horizon for potential threats and predators and, in the process, attributed agency and purpose to trees, rocks, the wind, and whatever else struck their fancy as alive. "We constantly scan our environment for the presence of other people and nonhuman agents," argues Barrett. "If you bet that something is an agent and it isn't, not much is lost. But if you bet that something is not an agent and it turns out to be one, you could be lunch."[50]

48. Robert L. Spitzer et al., eds., *Diagnostic and Statistical Manual of Mental Disorders*, 3rd ed. (Washington, DC: American Psychiatric Association, 1980), 41–45.

49. For a slightly earlier iteration of this humanism, see Leonard Uhr, ed., *Pattern Recognition: Theory, Experiment, Computer Simulations, and Dynamic Models of Form Perception and Discovery* (New York: John Wiley and Sons, 1966).

50. Justin L. Barrett and Frank C. Keil, "Conceptualizing a Nonnatural Entity: Anthropomorphism in God Concepts," *Cognitive Psychology* 31 (1996): 219–47; Justin L. Barrett, "Exploring the Natural Foundations of Religion," *Trends in Cognitive Sciences* 4 (2000): 29–34; Justin L. Barrett and Amanda Hankes Johnson, "The Role of Control in Attributing Intentional Agency to Inanimate Objects," *Journal of Cognition and Culture* 3, no. 3 (2003): 208–17; Justin L. Barrett, *Why Would Anyone Believe in God?* (Lanham, MD: Alta Mira Press, 2004), 31. Barrett is drawing, in part, from physiologist Walter Cannon's articulation of the fight or flight response over a century ago. See Cannon's *Bodily Changes in Pain, Hunger, Fear and Rage: An Account of Recent Researches into the Function of Emotional Excitement* (New York: D. Appleton, 1922), 187–88, 211.

Anthropologist Pascal Boyer has offered a friendly amendment to Barrett's argument, writing that it makes sense to "over-detect" agents only if you can quickly discard false positives. "Otherwise," notes Boyer, "you would spend all your time recoiling in fear, which is certainly not adaptive." Like Guthrie and Barrett, Boyer insists that religion is a natural attribute of the human that demands explanation. According to Boyer, humans are expert at scanning the horizon in order to produce as many inferences as possible. They are always computing, learning from their mistakes, and maximizing inferences for the sake of future efficiency, survival, and sociality.[51]

Boyer writes that religion is bound up in this complex process of computational scanning and pattern recognition. In the hands of CSR, perception becomes disembodied. This lends itself to a conception of religion as bloodless belief or, more precisely, as a matter of looking at the sacred from a distance but doing little else by way of interaction. Yet Boyer insists that his is an embodied approach, going so far as to critique Guthrie for being too abstract in his theorizing of HADD: "What happens in religion," Boyer contends, "is not so much that people see 'faces in the clouds' (in the way described by Guthrie) as 'traces in the grass.' That is, people do not so much visualize, concretely, what supernatural agents must be like as detect traces of their presence in many circumstances of their existence."[52] These traces, although not necessarily objectified in the mind, add up to a kind of object status because each of the traces is grounded in the material world. As inferences accumulate, supernatural agents are eventually imagined to have full access to any and all information that pertains to you: God as a strange phenomenal entity that is created and projected and internalized by way of scanning for surface signals.

Such access, then, is integral to Boyer's understanding of religion, evident, for example, in his privileging of prayer as an instance of agency detection.[53] "People," writes Boyer, "tend to construe gods and spirits as agents

51. Pascal Boyer, *Religion Explained: The Evolutionary Origins of Religious Thought* (New York: Basic Books, 2001), 147. HADD, for example, is comprised, in part, of an algorithm of fear. As Boyer writes, "fear is not just what we experience about it; it is also a program, in some ways comparable to a computer program" (22). Boyer, too, is drawing from Walter Cannon's rendering of the fight-or-flight response. See Cannon, *Bodily Changes*, 187–88, 211.

52. Boyer, *Religion Explained*, 145.

53. Prayer is a direct by-product of our tendency to detect agents in our environment and to then project human characteristics. The routinization of anthropomorphism. It is a

with strategic knowledge, and therefore explain their moral intuitions as the fact that supernatural agents are monitoring their actions." According to Boyer, "Concepts of full-access agents do not just require less effort but also generate richer inferences than other supernatural concepts."[54] These are agents worth praying to precisely because doing so seems to make sense of the world. These are agents worth praying to because the results are beneficial in the evolutionary sense, that is, in hindsight.[55]

Whereas Guthrie wrote of prayer as central to the religious imagination, Boyer situates prayer in a larger history of cognition, a larger history of seeking better odds and individuation. So whereas prayer may be confessional at base and a strategy of animation, it is also an exemplary statistical practice once an animated world has been established. So prayer has had a number of evolutionary advantages—it served as an intervention into a world defined by probabilities; it activated mental systems, like the penchant for cost-benefit analysis; and it fed into our natural disposition to frame reality in terms of cognition. Prayer has served its purpose, according to Boyer, and one that we can look back on with various degrees of patronizing respect.

As part of a complex predictive apparatus, HADD is key to an ongoing process of secularization in which the passions of religion subside, its disciplines forgotten, and its traditions recede into memory.[56] The secularization thesis embedded in Boyer's explanation of prayer, a story of cognitive progress, shares much in common with nineteenth-century anthropologists

---

process of coming to act on that detection and confirming to ourselves that we are thinking with another entity and that the agent in question "understand[s] not only our language but also the way we use it" (*Religion Explained*, 142, 156).

54. Boyer, *Religion Explained*, 283, 165.

55. Kevin R. Foster and Hanna Kokko, "The Evolution of Superstitious and Superstition-like Behavior," *Proceedings of the Royal Society* 276 (2009): 31–37; Paul Seabright, "On the Origins of Enchantment: Not Such a Puzzle," *Religion, Brain and Behavior* 10, no. 3 (2020): 345–57.

56. Or perhaps just certain religions. For when the future is no longer full of surprises it will entail a kind of secularized bliss. As noted approvingly by Clark ("Whatever Next?," 192), David Mumford writes that the end-goal of cognitive evolution is an "ultimate stable state" in which each layer of processing neurons would "predict . . . what each lower layer is sensing." Mumford adds that "In some sense, this is the state that the cortex is trying to achieve: perfect prediction of the world, like the oriental Nirvana, as Tai-Sing Lee suggested to me, when nothing surprises you and new stimuli cause the merest ripple in your consciousness." David Mumford, "On the Computational Architecture of the Neocortex.

such as E. B. Tylor.[57] Like his nineteenth-century precursors, Boyer's cognitive view lends itself to aspirations of normalizing the human, of taming its excesses, of offering a cure for "irrational and dysfunctional cognition."[58] And, as with Tylor, a subtle strain of paranoia laces Boyer's recognition of such dysfunction, a pose he also shares with new atheists writing in an age of simmering Islamophobia.[59] Not only do religious concepts and activities impinge on our agentive potential but, to quote Boyer, they "hijack our cognitive resources."[60] For religion in Boyer's scheme may have had its local advantages, but, at the end of the day, it is a form of laziness or obsessive compulsive pathology or both. Scientific atheism, on the other hand, is a hard job. For unlike religion, argues Boyer, "disbelief is generally the result of deliberate, effortful work against our natural cognitive dispositions—hardly the easiest ideology to propagate," he adds.[61] Cognitive scientists, Boyer triumphantly claims, can scan and see more clearly than anybody else. Accord-

---

II. The Role of Cortico-Cortical Loops," *Biological Cybernetics* 6, no. 3 (1992): 247. Clark, it should be noted, is a CSR fellow traveler whose work is approvingly cited by those investigating agency detection. For what amounts to a magisterial statement on the cognitive mechanics of predictive processing, see Clark's *Surfing Uncertainty: Prediction, Action, and the Embodied Mind* (New York: Oxford University Press, 2016).

57. See, e.g., my take on Tylor's story of cognitive progress in "In the Men's Room: E. B. Tylor and the Will to Systematize," *Social Text* 32, no. 3 (2014): 87–107.

58. Pascal Boyer and Brian Bergstrom, "Threat-detection in Child Development: An Evolutionary Perspective," *Neuroscience and Biobehavioral Reviews* 35, no. 4 (2001): 1034–41.

59. Glenn Greenwald, "Sam Harris, The New Atheists, and Anti-Muslim Animus," *Guardian*, April 3, 2013, https://www.theguardian.com/commentisfree/2013/apr/03/sam-harris-muslim-animus.

60. Pascal Boyer, "Religion: Bound to Believe?," *Nature* 455 (2008): 1038–39. To his defense, Boyer notes that "this hijacking occurs simply because religion provides some form of what psychologists would call super stimuli." Religious concepts are not singular in their deception. They "hijack our cognitive resources, as do music, visual art, cuisine, politics, economic institutions and fashion." But then again, religious concepts have immediate and deleterious effects. For as Boyer notes, "hijacking also occurs because religions facilitate the expression of certain behaviours. This is the case for commitment to a group, which is made all the more credible when it is phrased as the acceptance of bizarre or non-obvious beliefs." Richard Dawkins, in a more bombastic mode, refers to "the nerve endings of transcendent wonder that religion [has] monopolized in past centuries" (*The God Delusion* [Boston: Houghton Mifflin, 2008], 33).

61. Boyer, "Religion: Bound to Believe?," 1039.

ingly, cognitive inquiry has the power to disenchant, to transform religion from a mystery into a problem to be solved.[62]

This notion of religion as a naturally occurring pathology has assumed a central place in the securitization of the world.[63] Indeed, the white-hot righteous madness of religion has become the object of containment for those who pen the metaphysics of the war on global terrorism. Scott Atran, for example, is a leading cognitive scientist and founder of ARTIS International, a nonprofit group "that uses social science research to help resolve seemingly intractable political and cultural conflicts." Atran and his team of researchers have recently addressed the problem of "devoted actors" who "adhere to sacred, transcendental values that generate actions disassociated from rationally expected risks and rewards." Such actors are dangerous and unpredictable precisely because they have failed to achieve their own version of agentive reason. Having failed to negotiate the communal pressures on them, the identity of such actors is a product of insufficient detection of their own agency and overinvestment in the agency of divinities that define the coherence of their local communities.[64] Rather than being an effect of calculating costs and consequences, the actions of terrorists have been the result of "costly commitment to idiosyncratic and apparently absurd beliefs and associated values, cued by sartorial and corporeal markers (e.g., veils, beards, and especially more indelible marks, such as the *zabiba* on the forehead of pious Muslims generated by repeated friction with the prayer mat)."[65]

62. In doing so, Boyer has moved well beyond his initial theorizations into event-related fMRI studies that have demonstrated that the detection of "animate contingency" involves "automatic, bottom-up neural processing . . . largely confined to parietal networks dedicated to complex visuo-spatial detection" (S. J. Blakemore, Pascal Boyer, et al., "The Detection of Contingency and Animacy from Simple Animations in the Human Brain," *Cerebral Cortex* 13, no. 8 (2003): 843).

63. On the explicit link between HADD and mental illness, see Robert N. McCauley and George Graham, *Hearing Voices and Other Matters of the Mind: What Mental Abnormalities Can Teach Us about Religions* (New York: Oxford University Press, 2020).

64. CSR is often fueled by a bemused and perhaps even benevolent concern for those subjects "prone to self-induced spiritual experiences [because they] under perceive the extent of their own agency in the world" (McKay, "Religion and Agency," 94).

65. Scott Atran, "The Devoted Actor: Unconditional Commitment and Intractable Conflict across Cultures," *Current Anthropology* 57 (June 2016): S192–S203. Luke Savage, "New Atheism, Old Empire," *Jacobin* (December 2, 2014), https://jacobinmag.com/2014/12/new-atheism-old-empire; and Mohammad Hassan Khalil, "'We Are at War with Islam': The

As Boyer declared shortly after the publication of *Religion Explained*, "Atran's work is a brilliant exposition of the evolutionary by-product interpretation [of religion] as well as a mine of references for empirical research into the psychology of religion."[66] Drawing inspiration equally from the current clash of civilizations and the evolutionary refinement of our hyperactive agency detection devices, Boyer insists on the pressing need "to establish why and how religious thought is so pervasive in human societies." This "understanding," he adds, "is especially relevant in the current climate of religious fundamentalism." Boyer insists that this close-mindedness can be resisted through clear thinking about the brain and its by-products. Here is a place where politics has been transcended. Having arrived at the deep fundament of the human, Boyer is calculating the odds, just like the brains he studies are doing. All in the service of instantiating a secular imaginary of a particular kind.[67] All for the possibility of "hazard[ing] a guess at what the realistic prospects are for atheism."[68]

From Boyer's post-Protestant perch, atheism becomes the most heroic, the most agentive, the most human position one can assume in a universe

Case of Sam Harris," in *Jihad, Radicalism, and the New Atheism* (Cambridge: Cambridge University Press, 2017), 97–129. Harris, it should be noted, received his PhD in neuroscience from UCLA and has contributed, in addition to his incendiary writings on Islam, articles in the cognitive study of religion. See, e.g., P. K. Douglas, Sam Harris, et al., "Performance Comparison of Machine Learning Algorithms and Number of Independent Components Used in fMRI Decoding of Belief vs. Disbelief," *NeuroImage* 56, no. 2, (May 2011): 544–53. As historian Kathryn Lofton argues, the new atheistic pose is part of a longer tradition of "critical misogyny." Referring to Harris and others, Lofton writes that "this minstrel virility plays out in demonstrations of protective strength, plowing away at the big two nemeses (Christianity and Islam) in the interest of protecting the little guy. It is also exhibited in grand tours of scientific proof, or plodding expulsions of religious duplicity"; https://tif.ssrc.org/2009/11/16/so-you-want-to-be-a-new-atheist/.

66. Front matter, Atran, *In Gods We Trust*.

67. Michel Foucault, *The Order of Things: An Archaeology of the Human Sciences* (New York: Vintage Books, 1994), 57. On the utility of CSR in an age of terrorism, see the chapter by Harvey Whitehouse (a sometime collaborator with Boyer) on cognition and rites of terror ("Terror," in *The Oxford Handbook of Religion and Emotion*, ed. John Corrigan (New York: Oxford University Press, 2008), 259–75). On how the insights of CSR are used to gain explicit leverage on terrorism, see Kumar Ramakrishna, *Islamist Terrorism and Militancy in Indonesia: The Power of the Manichean Mindset* (New York: Springer, 2014), 64–68.

68. Boyer, "Religion: Bound to Believe?," 1038–39.

that has only begun to be properly understood.[69] Boyer's evolutionary frame assumes that humans, in general, and Boyer, in particular, have gotten better at detecting agency and more accurate in their capacity to discard false positives. Boyer, for example, claims to be getting beyond the "surface of religious concepts" and "moving from the table to the kitchen and observing how the concepts are concocted in human minds." Religion is neither special nor should it be protected. Subsequently, Boyer calls for a recovery of "scientific ambition" in order to move beyond mere, but necessary, "description of those aspects of human nature that lead people to adopt certain ideas or beliefs rather than others."[70] In his fervent belief in the neuromatic, Boyer seems to be playing a zero-sum game. For the more accurate and less hyperactive your detection device becomes, the more agency that will be attributed to you and the less agency that will be attributed to entities that, *for God's sake*, are not agents after all is said and done.[71]

As our evolutionary inheritance, the hyperactive agency detection device offers necessary leverage on the future. For the device conjures something essential to reason, itself. It is objectivity in a state of vital growth, becoming more efficient through the ages, more accurate, more sedate, more streamlined for the long haul. Rest assured, the hyperactive agency detection device is strategically diminishing its own hyperactivity, *becoming its true, even-tempered detection device self*. And the more it checks its own excess,

69. See also Pierre Lienard and Pascal Boyer, "Whence Collective Rituals: A Cultural Selection Model of Ritualized Behaviors," *American Anthropologist* 108, no. 4 (December 2006): 814–27.

70. Boyer, *Religion Explained*, 57, 172, 48, 31.

71. This is also a fair characterization of a scientist like Barrett, a committed evangelical who, unlike Boyer, has no grudge with what he identifies as religion. For Barrett, "believing in other minds and believing in God are comparably natural beliefs. One is not markedly more strange or bizarre than the other." See his chapter "The Naturalness of Believing in Minds: An Analog for Understanding Belief in God" from *Why Would Anyone Believe in God?*, 95–105. There is a degree to which Barrett's unorthodox defense of science as a rigorous belief in other minds preserves and protects religion from Boyer's politics. But more importantly, Barrett shares Boyer's disenchanting gaze in so much as he has come out the other side of Enlightenment with the ability to have privileged access (via science) to the physiological fundament of human being. In naturalizing religion for different political agendas, Barrett and Boyer share the same secular imaginary and are, together, a formation of what I have been calling neuromation. For both, religion comes easy for humans because believing in God is bound up in natural predispositions, in a fundament that is neurological. For Barrett religion should be a privileged site of exploration, for Boyer a privileged site of explanation and future evacuation.

the more we become the only agents deserving of the designation. Yet even as Boyer and others make the case for the fact that the hyperactive agency detection device contains the seeds of its own refinement, they do so in such a way that begs a question that has little to do with the time of evolution but instead inquires into the artifactual, and perhaps even agentive, pressures of our secular age.[72]

6. NORTHAMPTON

In the midst of World War II, our natural penchant for anthropomorphism was discovered in Northampton, Massachusetts. An experiment was conducted, facts were gathered, and questions were asked:

> What kind of person is the big triangle? . . . Why did the two triangles fight? . . . Why did the circle go into the house? . . . What did the big circle do when it was in the house with the big triangle? . . . Why did the big triangle break the house?[73]

In 1944, at Smith College in Northampton, Professor Fritz Heider and Marianne Simmel (granddaughter of Georg Simmel) published their foundational study, "An Experimental Study of Apparent Behavior." In it they explored the discrepancy between the laws of physics and the misperception and mischaracterization of these laws by human observers.[74] Heider and

72. A recent study has suggested that such artifactual pressures exist within the algorithms themselves but have gone undetected because of our inflated belief in the precision of technology. On bugs in the most common software packages for fMRI analysis, see Anders Eklund et al., "Cluster Failure: Why fMRI Inferences for Spatial Extent Have Inflated False-Positive Rates," *Proceedings of the National Academy of Sciences* 113, no. 28 (2016): 7900–7905.

73. Fritz Heider and Marianne Simmel, "An Experimental Study of Apparent Behavior," *American Journal of Psychology* 57, no. 2 (1944): 248–51.

74. Other foundational works include Albert Michotte, *The Perception of Causality*, trans. T. R. Miles and E. Miles (1946; London: Methuen, 1963); Howard E Gruber et al., "Effects of Experience on Perception of Causality," *Journal of Experimental Psychology* 53, no. 2 (February 1957): 89–93. Extensions of this line of inquiry include Gunnar Johansson, "Visual Perception of Biological Motion and a Model for Its Analysis," *Perception and Psychophysics* 14, no. 2 (1973): 201–11.

Simmel demonstrated that there were common biases in which our perception was consistently at odds with the way things were in essence. References to Heider and Simmel are pervasive in the scientific literature on religious cognition and agency detection, and Boyer himself claims to be updating their study when he identifies the neural networks involved in hyperactive agency detection.[75]

In "An Experimental Study of Apparent Behavior," Heider and Simmel demonstrated, among other things, how different perceptual cues catalyzed different degrees of agency attribution. The attribution of movement in an anthropomorphic key, for example, was linked with the attribution of motivation. They conducted their anatomy of agency detection—on a flat screen—by exposing test subjects to a "moving picture-film . . . in which geometrical figures (a large triangle, a small triangle and a circle) were shown moving in various directions at various speeds."[76]

The Heider and Simmel film had no sound, only animated geometry

75. For the citational authority of the Heider and Simmel experiment, see also Guthrie, *Faces in the Clouds*, 95; Barrett, "The Naturalness of Religious Concepts," 406; Pyysiäinen, *Supernatural Agents*, 13; Ilkka Pyysiäinen, "Cognitive Science of Religion: State-of-the-Art," *Journal for the Cognitive Science of Religion* 1, no. 1 (2012): 5–28; Nicholas Epley, Adam Waytz, and John T. Cacioppo, "On Seeing Human: A Three-Factor Theory of Anthropomorphism," *Psychological Review* 114, no. 4 (2007): 864–86; Anondah R. Saide and Rebekah A. Richert, "Socio-Cognitive and Cultural Influences on Children's Concepts of God," *Journal of Cognition and Culture* 20, no. 1–2 (2020): 22–40. On the use of the Heider and Simmel film to diagnose cognitive abnormality, see Ami Klin, "Attributing Social Meaning to Ambiguous Visual Stimuli in Higher-functioning Autism and Asperger Syndrome: The Social Attribution Task," *Journal of Psychology and Psychiatry and Allied Disciplines* 41, no. 7 (2000): 831–46; Dermot M. Bowler and Evelynne Thommen, "Attribution of Mechanical and Social Causality to Animated Displays by Children with Autism," *Autism* 4, no. 2 (2000): 147–71; Bloom and Veres, "Perceived Intentionality of Groups."

Boyer draws on the self-evident authority of Heider and Simmel in articles published during the writing and production of *Religion Explained*. In these coauthored studies, Boyer used fMRI and manipulated visual stimuli to investigate the neural networks involved in agency detection. S. J. Blakemore, Pascal Boyer, et al., "How the Brain Perceives Causality: An Event-Related fMRI Study," *Brain Imaging* 12, no. 17 (December 4, 2001): 3741–46; J. L. Schafroth et al., "No Evidence That Monkeys Attribute Mental States to Animated Shapes in the Heider-Simmel Videos," *Scientific Reports* 11 (2021): 3050. fMRI has also been used to localize neural functioning while viewing animations based on the Heider and Simmel film in Naoyuki Osaka et al., "Effect of Intentional Bias on Agency Attribution of Animated Motion: An Event-Related fMRI Study," *PLoS One* 7, no. 11 (2012): 1–6.

76. Heider and Simmel, "Experimental Study," 243–59.

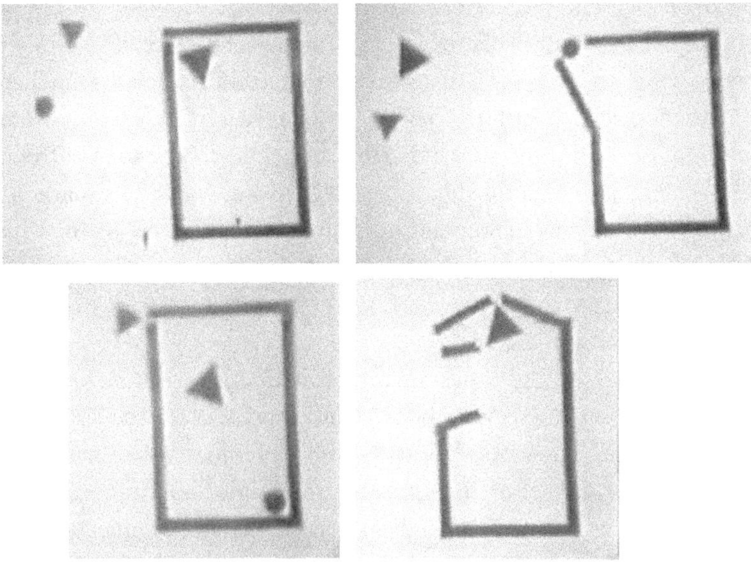

FIGURE 10. Four still images from the 1944 film used in Fritz Heider and Marianne Simmel, "An Experimental Study of Apparent Behavior," *American Journal of Psychology* 57, no. 2 (1944): 243–59.

and the clackety hum of a 16 mm projector in a room.[77] Heider's laboratory surveyed the reactions of 114 undergraduates attending Smith College. The students were divided into three groups and shown the film twice. The first group of thirty-four was given a specific command: "write down what happened in the picture." The second group of thirty-six was given the anthropomorphic prompt "to interpret the movements of the figures as actions of persons." The third group of forty-four was shown the film in reverse and asked fewer questions.

An overwhelming majority (97%) "interpreted the picture in terms of actions of animated beings, chiefly of persons." A narrative pattern emerged among these groups—the big triangle was often typecast male—"aggressive,

77. In both content and form or, rather, in its will to arrive at the space of pure form—on the screen but also in the mind—the Heider and Simmel film anticipates cinematic turns of the 1950s in which the sensibility of abstract expressionism (and its desire to secure, materially, the solitary vision of the artist in the process of making art) was taken up by filmmakers such as Robert Breer. In *Form Phases I–IV* (1952–56) Breer used hand-cut figures and stop animation to distill the essence of the human that was beyond and/or unsullied by content. See, e.g., *Form Phases #4*, http://vk.com/video17060051_168161860.

warlike, belligerent, pugnacious, quarrelsome, troublesome, mean, angry, bad-tempered . . . bully, villain, taking advantage of his size . . . dominating, power-loving, possessive." Two-thirds of the second group attributed femininity to the circle, describing "her" as "frightened, afraid, fearful, cowardly, shy, timid, meek, not too sure of herself . . . helpless, dependent." The gendered ecology in which this film was screened was leaving its mark in the meaning attributed to it. Students here, in becoming accustomed to living vicariously through a screen, were also being induced to narrate sexual difference as a scripted drama: *a way of seeing becoming increasingly compatible with a way of being in the world.*

Looking back on the experiment that made his career, Heider cannot help but signal the universality of the truth that he had discovered. "It has been impressive," he writes, "the way almost everybody who has watched it has perceived the picture in terms of human action and human feelings." This, of course, is not surprising given that Heider and Simmel had rigged the system. In their composition and editing they had produced a film that drew from narrative expectations, domestic hierarchies, and gendered conventions of the time. They had infused drama into the flatness of lines and arcs. It was, indeed, a story about humans. For as Heider later admitted, he had loaded his film with all kinds of anthropomorphic cues—"As I planned the action of the film," Heider recalled, "I thought of the small triangle and the circle as a pair of lovers or friends, and I thought of the big triangle as a bully who intruded on them. The rectangle served as a room with a door, which could be opened or closed. The movements of the three characters were such that the two smaller ones in the end defeated and eluded the bully." Geometry was meant, here, to be alive. And, indeed, geometry was manifest—in "apparently" conscious shapes and movements, gendered prompts, and scenes of domestic conflict. Despite his tampering, Heider concluded that "movements or behavior, if you like, of even unchanging forms can produce an impelling impression of a network of interpersonal events and relations involving love, hate, power connections, fights, and happy reunions."[78] Universal human emotions generated by particular marks on a screen.

78. As Heider recalled, "We had cardboard figures of our three characters and the room, and we placed them on a sheet of glass, on which the camera was focused from below. I figured out in advance exactly how far I had to move a figure to give the impression of slow or rapid movement, and we set out to work. I would place a figure; Marianne would then make the exposure; I would move the figure to the next position; and so on. It took us

And the elephant in the room—the cognitive dissonance experienced by those primed, according to Heider's script, to humanize cardboard shapes moving around a pull-down screen in a small room. At least one Smith undergraduate noted these artificial constraints (and the possibility that the results of this experiment were present from the very beginning): "The first thing we see in this little episode," she remarked, "is triangle number-one closing the door of his square. Let's insist that the action of the play is on a two-dimensional surface (not that it makes much difference) and we will undoubtedly start calling the square in which triangle number-one seems to make his dwelling, a house, which infers three dimensions. But we are not sticking to the theme of our story." The words of the anonymous Smith undergraduate hint at a pressing demand to narrativize what has been unnaturally internalized. For in the extended riff and hesitation of this particular undergraduate one detects a reflexive recognition on her part of the power of mediation, that is, the capacity for an image on the screen to make its way in without leaving a mark.[79]

In studying the interpretive habits of these young women, Heider and Simmel had discovered a flaw in *their* cognitive processing that was also universal. This flaw, in other words, was a feminized trait from the beginning, produced in the screening room where crass narratives of sexual difference were built into the film and, by extension, the diagnosis. From this perspective, Heider and Simmel's "An Experimental Study of Apparent Behavior" was an implicit call to overcome what amounted to the gendered pathology of hyperactive agency detection and to temper this feminine proclivity in all of us—and all in the service of a universal type, beyond sexual difference.[80] No wonder, then, that Heider and Simmel lamented how "experiments on the perception of the behavior of others" have been stalled by an overemphasis on physical characteristics and facial recognition. There was a deeper layer, they inferred, not readily accessible but nonetheless where our shared

---

about six hours, working in this exposure-by-exposure fashion, to make a film that gives a perception of lively movement." Fritz Heider, *The Life of a Psychologist: An Autobiography* (Lawrence: University of Kansas Press, 1983), 148–49.

79. Heider and Simmel, "Experimental Study," 247.

80. Perhaps it is not all that surprising that the Heider and Simmel film was infused with site-specific yet unannounced norms of masculinity, femininity, and sexual hierarchy. What is surprising, however, is how researchers cite Heider and Simmel, without critique, and in particular, the sexual drama of their film, as supporting evidence for religion being a mode of agency detection. Guthrie, *Faces in the Clouds*, 95–96.

humanity occurred. Consequently, in order to access this level, "we have presented situations and activities without the face." The individuality of individuals, in other words, had a formal essence. And it was precisely this formal essence that conditioned its recognition.[81]

7. JONATHAN EDWARDS, HYPERACTIVE AGENCY DETECTOR

> It has been slanderously reported and printed concerning me, that I have often said that the millennium was already begun and that it began at Northampton . . . but the report is very diverse from what I say now, that I looked upon the late wonderful revivals of religion as forerunners of those glorious times so often prophesied of in the Scripture, and that this was the first dawning of that light.
>
> JONATHAN EDWARDS, letter denying a charge made by Charles Chauncy (1744)

The "enthusiastical town of Northampton" was also ground zero for the evangelical surge centered on Jonathan Edwards and the "season of revivals" that occurred there in the 1730s.[82] Over the course of six months, beginning in late 1734, Edwards confirmed the testimony of over three hundred individuals who had been "savingly wrought upon." By the summer of 1735, Edwards concluded that "the town seemed to be full of the presence of God. . . . There were remarkable tokens of God's presence in almost every house. . . . The

81. Heider and Simmel, "Experimental Study," 243–44.

82. Jonathan Edwards, *Some Thoughts Concerning the Present Revival of Religion in New-England, and the Way in Which It Ought to Be Acknowledged and Promoted, Humbly Offered to the Publick, in a Treatise on That Subject: In Five Parts; Part I. Shewing That the Work That Has of Late Been Going on in This Land, Is a Glorious Work of God. Part II. Shewing the Obligations That All Are Under, to Acknowlege* [sic], *Rejoice in and Promote This Work, and the Great Danger of the Contrary. Part III. Shewing in Many Instances, Wherein the Subjects, or Zealous Promoters, of This Work Have Been Injuriously Blamed. Part IV. Shewing What Things Are To be Corrected or Avoided, in Promoting This Work, or in Our Behaviour under It. Part V. Shewing Positively What Ought To be Done to Promote This Work* (Boston: S. Kneeland and T. Green, 1742), 38.

goings of God were then seen in his sanctuary."[83] Edwards would spend a lifetime defending the legitimacy of the "goings of God" and the human role in them. During the revivals and their aftermath, Edwards insisted that God's agency had been truly and properly detected in Northampton.

In responding to critics like Charles Chauncy, who dismissed the revivals as a feminized spectacle—a "Grand Delusion" in which "*Passions* [were] rais'd to such an *extraordinary* heights without a proportionate Degree of Light in the Understandings"—Edwards struck a defensive yet measured tone.[84] Looking back in 1751 at the season of revivals, Edwards acknowledged excesses but, more importantly, errors overcome. For Edwards, "full calmness and impartiality of mind" were matters of long-term observation and error management.[85] Over time, in sermons and publications Edwards would identify false positives and deceptive data—clear signs of overzealous imaginations and conversion stories that did not add up.[86] But Edwards cautioned his critics not to dismiss the whole for the parts but rather to appreciate the general truth of revivals in order to correct the particular errors that occurred in their midst. Continuing his own effort to move beyond the strictures of Calvinist theology and mechanical philosophy both, Edwards wrote that it was precisely such error correction that also occurred within the season of conversion:

> With respect to the late season of revival of religion amongst us for three or four years past, it has been observable, that in the former part of it, in the years 1740 and 1741, the work seemed to be much more pure. . . . Persons seemed to be sensible of their former errors, and had learned more of their own hearts, and experience had taught them more of the tendency and consequences of things. They were now better guarded, and their affections were not only stronger, but attended with greater solem-

83. Jonathan Edwards, *The Great Awakening*, in *The Works of Jonathan Edwards*, vol. 4, ed. C. C. Goen (New Haven, CT: Yale University Press, 1972), 151.

84. Charles Chauncy, *Seasonable Thoughts on the State of Religion in New-England* (Boston: Rogers and Fowle, 1743), 320.

85. Edwards, *Great Awakening*, 565–66.

86. See, e.g., *A Faithful Narrative about the Surprising Work of God in the Conversion of Many Hundred Souls in Northampton, Massachusetts* (1737) and *The Distinguishing Marks of a Work of the Spirit of God, Applied to That Uncommon Operation That Has Lately Appeared on the Minds of the People of New England: with a Particular Consideration of the Extraordinary Circumstances with Which This Work Is Attended* (1741).

> nity, and greater humility and self-distrust, and greater engagedness after holy living and perseverance: and there were fewer errors in conduct.[87]

Both Edwards and his critics, albeit from different angles and for different purposes, sought to explain this so-called Surprising Work of God. Whereas Edwards sought to explain the revivals, critics sought to explain them away. Both sides, however, framed the revivals in terms of agency detection. Chauncy sought to discredit Edwards and the so-called New Lights as hubristic. They were delusional, according to Chauncy, in their failure "to know the true Spring of other Errors." God, in other words, was not the source of the revival. Passions were overwhelming a "due Ballance between Light and Knowledge in their minds."[88] Edwards, however, saw in Northampton a laboratory space where religion, defined in terms of the phenomenon of agency detection, could, indeed, be understood, refined, and deployed. In providing a list of "distinguishing marks" and justification for them, Edwards was nothing but a systematic observer of the social and psychological worlds of revival.

Edwards suggested that more was going on in the revivals than could be accounted for with a sole focus on God's will. Expanding his theological focus with anthropological insight, Edwards provided a morphology of conversion in which he laid out the typical stages that a person underwent, and to a certain degree initiated, during the conversion experience. Edwards also acknowledged his own rhetorical sway, admitting not only to have played a role in stoking the fires of revival but also defending that role as providential. In carving out a role for human agency within a situation defined absolutely by God's will, Edwards offered a theology of human affect that immediately served to justify the form and content of his appeal:

> I should think myself in the way of my duty to raise the affections of my hearers as high as possibly I can, provided that they are affected with nothing but truth, and with affections that are not disagreeable to the nature of what they are affected with.[89]

87. Edwards, *Great Awakening*, 555. On the development of Edwards's pairing of religious hypocrisy and erroneous judgment, first broached in "The Heart of Man Is Exceedingly Deceitful" (1733), see Sarah Rivett, *The Science of the Soul in Colonial New England* (Chapel Hill: University of North Carolina Press, 2011), 271, 277–78, 299ff. Rivett offers an expansive take on the secular blur within Puritan modes of theological inquiry.

88. Chauncy, *Seasonable Thoughts*, 320, 324.

89. Edwards, *Great Awakening*, 160–91, 402, 386–88.

For Edwards, conversion was premised on the activation of the affections and not their mere exploitation (as his critics contended). Consequently, in his bracing sermons Edwards claimed to be generating something quite real in his listeners, helping to catalyze their inborn spiritual senses with "nothing but truth." For Edwards's sermons were vehicles of divine *and* human will.[90] They were meant to evoke new sensations because they spoke simultaneously in two overlapping registers—of human life *and* God's providence. Painting a moving picture with language, Edwards evoked time's passage and icons from the local scene: dead bodies, grazing sheep, and roses in the briar patch. Yet his goal was to conjure the archetype amid them—not just God in the abstract but the living principle. God on the move. God having touched the ground. God assuming his status. God becoming a source of human freedom, still beyond representation yet possessing a natural personality.[91]

Such personality could best be recognized once one's "new spiritual sense" (NSS) was activated. The NSS, whose foundation was laid in nature, was built in and ready to go. This new spiritual sense, and the "new dispositions that attend it," declared Edwards,

> are no new faculties, but new principles of nature: I use the word principles, for want of a word of a more determinate signification. By a principle of nature in this place, I mean that foundation which is laid in nature, either old or new, for any particular manner or kind of exercise of the faculties of the soul; or a natural habit, or foundation for action, giving a person ability and disposition to exert the faculties in exercises of such a certain kind; so, that to exert the faculties in that kind of exercises, may be said to be his nature.[92]

As a culmination of Edwards's theology of revival, the NSS was a mix of Lockean sensationalism cut with Calvinist orthodoxy, Arminianism cut

90. Jonathan Edwards, *Images or Shadows of Divine Things*, ed. Perry Miller (1948; reprint, Westport, CT: Greenwood Press, 1977), 90–92.

91. Rivett, *Science of the Soul*, 295. There is something wonderfully perverse, then, in recent cognitive approaches to the sermons of Edwards and the biological impact of his words and imagery on his listeners—see Michal Choinski, "A Cognitive Approach to the Hermeneutics of Jonathan Edwards's Sermons," *Jonathan Edwards Studies* 4, no. 2 (2014): 215–27. See also Willem van Vlastuin, "A Retrieval of Jonathan Edwards's Concept of Free Will: The Relevance for Neuroscience," *Jonathan Edwards Studies* 4, no. 2 (2014): 198–214.

92. Jonathan Edwards, *A Treatise Concerning Religious Affections in Three Parts* (Philadelphia: James Crissy, 1821), 138.

with Newtonian insights into the physics of crowds. Edwards offered swirling descriptions of a human freedom in light of God's absolute sovereignty. Moreover, the NSS was part of the nervous system and an intervention into nature on God's part.[93] The NSS was justified by both "reason and Scripture." It was a term that solved the paradoxical relationship of human will and divine providence, "temporal things" and "spiritual things."[94]

Just as the NSS could become overly stimulated and, therefore, error-prone, the proper regulation of the NSS was essential for not only detecting God but for maintaining social order, in general.[95] As the historians Ann Taves and Finbarr Curtis have both demonstrated, Edwards's theology was animated by sociological speculation.[96] In taking the NSS as the object of his scrutiny, within a bubble of his own creation, Edwards honed in on the

93. Edwards, here, was granting the brain, or more precisely, the effects of the brain, evidentiary status in the argument for soul and its access to divinity. In long wavering on whether or not the brain was the seat of the soul itself, the NSS may be seen as a maturation of Edwards's ambivalent musings on the nervous system. See, for example, Edwards's youthful notes (collected and first published in the 1790s) that circulated in the early Republic as "The Natural History of the Mental World, or the Internal World: Being a Particular Enquiry into the Nature of the Human Mind." In "SEEING," noted Edwards, that "the brain exists only mentally, I therefore acknowledge that I speak improperly when I say the soul is in the brain. . . . But we have got so far beyond those things for which language was chiefly contrived that, unless we use extreme caution, we cannot speak (except we speak exceedingly unintelligibly) without literally contradicting ourselves. Corollary. No wonder, therefore, that the high and abstract mysteries of the deity, the prime and most abstract of all beings, imply so many seeming contradictions" (Appendix to "The Mind" in *The Works of President Edwards with a Memoir of His Life*, vol. 1 (New York: S. Converse, 1829), 679). Although Edwards never confirmed outright that the brain was a font of divinity in this world, he did consider it to be an active organ in cultivating piety and warding off evil impressions. "It is by impressions made on the brain," wrote Edwards, "that any ideas are excited in the mind, by the motion of the animal spirits, or any changes made in the body" (*Treatise Concerning Religious Affections*, 236).

94. Edwards, *Great Awakening*, 402; Perry Miller, *Jonathan Edwards* (Toronto: William Sloane Associates, 1949); on the scientific leanings of Edwards and surrounding context, see Avihu Zakai, *Jonathan Edwards's Philosophy of Nature: The Re-enchantment of the World in the Age of Scientific Reasoning* (London: T. and T. Clark, 2010).

95. Edwards, *Great Awakening*, 402.

96. Ann Taves, *Fits, Trances, and Visions: Experiencing Religion and Explaining Experience from Wesley to James* (Princeton, NJ: Princeton University Press, 1999); Finbarr Curtis, "Locating the Revival: Jonathan Edwards's Northampton as a Site of Social Theory," in *Embodying the Spirit: New Perspectives on North American Revivalism*, ed. Michael J. McClymond (Baltimore: Johns Hopkins University Press, 2004): 47–66.

psyche with an empirically minded approach toward the study of the mind making (and living) the case for divine presence. In assuming a position later taken up by psychology and the cognitive sciences, Edwards examined the fabulous stories about how and why and when supernatural agency becomes an attribute in the world. For in gathering the testimony of the converted, surveying the reports of observers of revivals, and "observing" the signs over time, Edwards reformulated the question of whether God was present or not. Rather than focus on God or the individual convert alone, Edwards considered two things in relation to one another: God's indeterminate effects on society *and* the human ability to detect them.[97]

Edwards's notion of the NSS addressed how God's actions stirred religious passions and how those passions, in turn, were the conduit of God on earth. *Social order divinely generated.*[98] In a state of grace, a people and divinity looped in upon themselves so that the loop took precedence over individual parts, a process defined by Edwards in terms of divine love.[99] It took the revivals to bring this loop to the fore for observation. Circles within circles. A popular swirl that was self-sustaining. Here was a self-organizing system that could not be defined in causal terms.[100] It could not be measured,

97. Edwards, *Great Awakening*, 254–55.

98. In his counter to the anonymous divinity of mechanical philosophy, Edwards argued that God did not operate, nor could he be described, according to mere mathematics. There was, then, an arbitrariness to divine providence—at least from a merely human perspective. The arbitrariness of God, then, was not random but was constituted by the limits of the human. Or, to put this another way, by the difference that the human made in this self-organizing system. On the intellectual history of self-organization and the role it played in the development of a secular imaginary, see Jonathan Sheehan and Dror Wahrman, *Invisible Hands: Self-Organization and the Eighteenth Century* (Chicago: University of Chicago Press, 2015). As they note, the mid-eighteenth century witnessed an explosion of names and metaphors of self-organizing forces as a particular kind of modernity began to set in.

99. And Edwards often spoke of such love in terms of perpetual motion. "The whole material universe is preserved by gravity, or attraction, or the mutual tendency of all bodies to each other. One part of the universe is hereby made beneficial to another. The beauty, harmony and order, regular progress, life and motion, and in short, all the well-being of the whole frame, depends on it. This is a type of love or charity in the spiritual world" (Edwards, *Images or Shadows of Divine Things*, 79). Edwards moved beyond metaphors of mere revolution to capture a sense of cosmic undulation—comprised of human life and divine providence ("Notes on the Bible," *The Works of President Edwards with a Memoir of His Life in Ten Volumes*, vol. 9 [New York: G. and C. and H. Carvill, 1830], 401). See also Edwards, *Images or Shadows of Divine Things*, 85–86.

100. Edwards, *Great Awakening*, 256.

exactly, but it could be studied in terms of its effects, of input and output ratios. Edwards was engaged in identifying traces rather than first principles of this system. No hard proof but a simmering kind of certainty built on the accumulation of inferences. For "spiritual things" could be detected with conviction but not necessarily measured with precision:

> If we would get a right notion of what is spiritual, we must think of thought or inclination or delight. How large is that thing in the mind which they call thought? Is love square or round? Is the surface of hatred rough or smooth? Is joy an inch, or a foot in diameter? These are spiritual things. And why should we then form such a ridiculous idea of spirits, as to think of them so long, so thick, or so wide; or to think there a necessity of their being square or round or some other certain figure?[101]

Edwards neither posited nor demanded conclusive evidence of God's presence. Yet in insisting that the agency of God could not be measured, Edwards assumed that the human will to detect persisted and could be fruitfully harnessed and, moreover, observed. The natural mechanics of the NSS was precisely what Edwards was interested in documenting and charting: "a new kind of perception or spiritual sensation, which is in its whole nature different from any former kinds of sensation of the mind, as tasting is diverse from any of the other senses."[102] For the privileged mark of God's presence was not solely his own mark. It was also bound up in the physiological conditions of responsive human interiors.

In the end, Edwards shifted attention away from God as object toward the observable effects of the human perception of God's agency. Moreover, it was Edwards's own capacity for detecting these subtle effects that was part and parcel of his piety. For example, in his sermon, "The Most High: A Prayer-Hearing God," delivered in Northampton in 1735, Edwards suggested that the practice of agency detection was synonymous with true religion. Prayer, according to Edwards, was a blessed attunement of our NSS. God, in "present[ing] himself as the object of prayer" and "in [God's] giving such *free access* to him by prayer," Edwards preached that our attributions of agency were a logical consequence of God's presence. "God in his Word manifests himself ready at all times to allow us this privilege." True religion, consequently, revolved around prayerful communication with God. "Let us live

101. Edwards, Appendix to "The Mind," 678.

102. Edwards, *Treatise Concerning Religious Affections*, 137–38.

prayerful lives," intoned Edwards, "continuing instant in prayer, watching thereunto with all perseverance. Praying always, without ceasing, earnestly, and not fainting."[103]

Edwards put forward fundamental questions familiar to this secular age—what is the living principle of religion, how to identify it, and how to persuade others of the authenticity of that identification? For example, in laying out nine "negative signs" of God's presence (epiphenomena that were, themselves, insufficient to prove the presence or absence of divinity), Edwards shifted the debate over the legitimacy of the revivals. He turned away from the question of the presence of God to the complex process by which his agency was detected by humans but not necessarily known.[104] "God in some instances seems to have gone quite beside the ordinary laws of nature," insisted Edwards in the md-1730s.[105] Yet Edwards could not resist the will to measure these instances. In using a solving term like NSS, he gave license to a new terrain of assessment and new ways of framing the relationship between human and divine. For when Edwards claimed that "the things of religion take place in men's hearts," he was speaking of processes below the surface—natural processes of attention, reflection, and recall.[106] In doing

103. "The Most High," Edwards intoned, "is a God that hears prayer. Though he is infinitely above all and stands in no need of creatures, yet he is graciously pleased to take a merciful notice of poor worms of the dust. He manifests and presents himself as the object of prayer, appears as sitting on a mercy-seat, that men may come to him by prayer." "The Most High: A Prayer-Hearing God," *The Works of President Edwards in Four Volumes: A Reprint of the Worcester Edition, with Valuable Additions and a Copious General Index*, vol. 4 (New York. Leavitt, Trow, 1844), 561 72. Coming full circle, see the triumphal tone of Edwards scholar David W. Kling in "Jonathan Edwards, Petitionary Prayer, and the Cognitive Science of Religion," *Theology and Science* 18, no. 1 (2020): 113–36.

104. Edwards, *Great Awakening*, 228–48. This was a deft rhetorical move on Edwards's part. Instead of focusing solely on God's capacity to affect the heart, Edwards also put up for consideration the mind's capacity to attribute meaning to empty signs. And it was here, in this potential for exaggeration and error, that he forced his critics to reconsider their insufficient reading of the signs.

105. Jonathan Edwards, "Images of Divine Things," ca. 1735, in *The Works of Jonathan Edwards*, vol. 11: *Typological Writings*, ed. Wallace E. Anderson and Mason I. Lowance (New Haven, CT: Yale University Press, 1993), 61.

106. For Edwards, the soul was intimately connected to mental perception at an *almost* anatomical level. "So the soul may be said to be *in the brain*," wrote Edwards, "because ideas that come by the body immediately ensue only on alterations that are made there, and the soul most immediately produces effects nowhere else" (Edwards, Appendix to "The Mind," 679).

so he was generating a new kind of theological reasoning, one that may be recognized, for all intents and purposes, as scientific.

8. DETECTING THE LIFE OF THE BRAIN

It may be only a slight exaggeration to claim that hyperactive agency detection, given its lineages that run through Northampton, is a conceit both psychological in origin and theological in its demand. At midcentury, the neurophysiologist William Grey Walter acknowledged as much. Born in Kansas City and educated at Cambridge University, Walter was an integral figure through the 1960s in the diffusion of EEG. "The machines that flash and click in our laboratories now," wrote Walter (in reference to studies that he was conducting at the Burden Neurological Institute), "are the first forms of the living brain's extended life, the rudiments of racial understanding, as Gutenberg's first printing presses were the forerunners of the Reformation."[107] Framing the living brain in terms of evolutionary theory, Walter mused that despite the stubborn complexity of our environment, the brain was becoming ever more precise. Or rather, by way of the neurophysiologist, the brain was becoming ever more objective, ever more capable of building a mirror by which it could finally see itself plain and whole.

In *The Living Brain*, published in 1953, Walter begins by writing that the brain is "something more than the pinko-grey jelly of the anatomist." And the *more* is precisely what Walter spends the rest of the book making the case for, namely, that the brain is alive, purposeful perhaps, and agentive in its physiological makeup. Walter was quite self-conscious about attributing agency to the brain itself, ever hedging and preferring the phrase "intimations of personality" as the distinguishing mark of its vitality. The brain, according to Walter, was involved in probabilistic thinking that was independent of conscious decision-making. Moreover, such thinking was dependent on the recognition of pattern. "The animal," warned Walter, "that cannot learn by experience how" to "avoid or overcome repeated dangers cannot survive. [Their brains] cannot deal with uncertainty [or] reckon the odds." According to Walter, our "ancestral organ . . . evolve[d] a mechanism capable of . . . new . . . processes" of statistical analysis that could

107. W. Grey Walter, *The Living Brain* (New York: W. W. Norton, 1953), 278.

FIGURE 11. William Grey Walter and a 16-channel EEG, 1964. Science Museum Group.

be "recorded as electrical eddies swirling in subtle patterns through the brain."[108]

In the context of war machines, madness, and epilepsy research, Walter developed a brain topography machine that pictured the brain as an electrical network of responses that decoded "the EEG records automatically, plainly displaying the frequencies and amplitudes every ten seconds, [in a] kind of wave analysis." Here, at last, was a method by which the abstract truths of the brain could be laid bare. It was possible, wrote Walter, "to chart the distribution of these responses not only over the cortex, but also as they involve the depths of the hemispheres. The intricacy of such charts is bound to be very great, it seems probable in fact that much cerebral activity consists of *interrelated* patterns."[109]

108. Walter, *The Living Brain*, 15, 38, 16; Grey Walter, "Patterns in Your Head," *Discovery: A Monthly Popular Journal of Knowledge* (February 1952), 58.

109. Grey Walter and H. W. Shipton, "A New Toposcopic Display System," *EEG Clinical Neurophysiology* 3 (1951): 281–92. On the performance of the heroic masculine gaze that

The pattern that most intrigued Walter was the alpha rhythm. This was a surface pattern that was thought to go deep. Alpha rhythms are the principal background feature in the EEG of a normal adult. They have a rhythmic frequency at 8 to 13 Hz, and no two people's alpha rhythms are exactly the same. Alpha rhythms are best measured in a relaxed, waking state with eyes closed, for when normal subjects open their eyes the alpha rhythms approach a flat line. "The more regular" alpha rhythms, according to Walter, indicate an internal "process of hunting for information. . . . Their rhythmicity is a sign of the perpetual quest, their arrest is a mark of its ending." "Alpha rhythms are a process of scanning—[a] search for a pattern—which relaxes when a pattern is found."[110]

In defining alpha rhythms in terms of scanning, Walter offers insight into a pathology that had yet to be called hyperactive attention. According to Walter, if your alpha rhythms do not dissipate when you have your eyes open and focused on a task, then your hunt for information becomes rather intense. Those individuals with hyperactive alpha rhythms tend to be lost in erratic focus. Their attention tends toward paranoia, reasoned Walter, because their brains scan incessantly without finding anything that triggers the stoppage of the scan. They are fidgety, nervous, weak-willed. Prone to surprise. As Walter writes in diagnostic mode, "an alpha rhythm which persists when the eyes are open and the test subject is apparently fully occupied—such a state is usually suggestive of some isolation from reality."[111]

Walter was taken with the possibility of inducing such isolation by way of

---

predominated at Burden Neurological Institute during Walter's tenure, see David Saunders, "Wired-up in White Organdie: Framing Women's Scientific Labour at the Burden Neurological Institute," *Science Museum Group Journal* 10, no. 10 (2018); http://dx.doi.org/10.15180/181003.

110. Walter, *The Living Brain*, 153, 109.

111. Walter, *The Living Brain*, 222. Since the 1970s EEG has been used to diagnose the abnormality of children with ADHD. More recently, EEG has become a treatment device for ADHD, allowing patients to hone and habitualize their skills of concentration as they stare at a screen of their own brain waves. In one study, for example, the more a child is able to maintain a desired frequency (coordinated with an ideal state of attention), the more rewards they will receive at the end of the session. James E. Hastings and Russell A. Barkley, "A Review of Psychophysiological Research with Hyperkinetic Children," *Journal of Abnormal Child Psychology* 6 (1978): 413–47; V. J. Monastra et al., "The Effects of Stimulant Therapy, EEG Biofeedback, and Parenting Style on the Primary Symptoms of Attention Deficit/Hyperactivity Disorder," *Applied Psychophysiology and Biofeedback* 27 (2009):

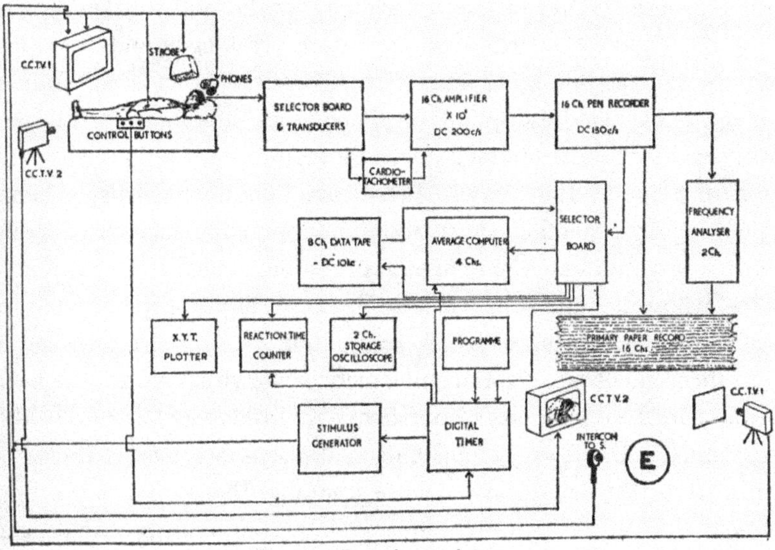

Fig. 1. Experimental setup.

FIGURE 12. Grey Walter's laboratory. W. Grey Walter, "The Contingent Negative Variation and Its Significance for PSI Research," in *PSI Favorable States of Consciousness: Proceedings of an International Conference on Methodology in PSI Research*, ed. Roberto Cavanna (New York Parapsychology Foundation, 1970), 171. "The subject, in a secluded part of the room, can be submitted to a battery of stimuli by means of television, flash and sound generators, touch capsules, etc. He is surveyed by a closed-circuit television system, and has access to a little box by which he can control the operation of the machinery. Then there are the conventional amplifiers and recorders (we use 16 channels) providing a primary paper record of the usual type. We also use various types of storage devices and a timing system, an elaborate clock which synchronizes various stimuli with the operation of the computers. In our present rig, all the analytical components have been replaced by a single LINC 8 computer" (170).

electrical manipulation. Using techniques of sensory overload, Walter demonstrated time and again how test subjects exposed to stroboscopic light reported "see[ing] things which [were] not present in the stimulus."

Test subjects could, in other words, be convinced that agents had been detected when in fact they had not. Moreover, the descriptions had a physiological correlate. Flashes of light "evoked in the brain a characteristic elec-

231–49. On neurofeedback as "efficacious and specific," see Martijn Arns et al., "Efficacy of Neurofeedback Treatment in ADHD: The Effects on Inattention, Impulsivity and Hyperactivity: A Meta-analysis," *Clinical EEG and Neuroscience* 40, no. 3 (2009): 180–89.

trical response." Walter surmised that "the stimulus of the flicker received in the visual projection area of the cortex was breaking bounds; its ripples were overflowing into other areas."[112] A disruption of form that produced content. Subjective descriptions of the content, however fanciful, were clearly an effect of disrupting lines of communication within the brain. In measuring "the overflow of visual responses into other sensory systems," Walter theorized that the hallucinations induced in his laboratory revealed a formal program underlying human cognition. And the program of the brain—whose operations were revealed by way of stroboscopic bursts—was itself engaged in projects of pattern recognition.

In theorizing the network connectivity of the visual cortex, the hallucinations induced in Walter's laboratory were instances of intense reading for scientist and test subject alike. In the laboratory the brain learned to overcome its penchant for hyperactive scanning. This was where the brain learned to identify false positives and thereby become more probabilistic. As Walter gushed: "the brain must, quietly, unobtrusively, incessantly" practice "statistical analysis" and "reckon the odds in favor of one event or one set of events implying another."[113] Within a virtual space where agency detection was designed to happen (as was the acceleration of evolutionary time in which a single brain refines its own skills of detecting agents and false positives), all involved could become overwhelmed by signal. *All within the confines of the laboratory set-up.*

In his 1964 article in *Nature*, Walter relied on the stroboscope to make the case that the intensity of the flashing signals mattered little, if at all, when it came to priming expectations.[114] Pattern was everything for this "system of extracting signals from noise." Walter wrote that despite the fact that the brain was a source of incredible amounts of background noise, he had identified "consistent patterns of response, particularly in the extensive nonspecific regions of the frontal lobes."[115] What these patterns suggested was that the brain signaled its readiness to detect new patterns when properly

112. Walter, *The Living Brain*, 247, 91.

113. Walter, *The Living Brain*, 103, 169.

114. W. Grey Walter et al., "Contingent Negative Variation: An Electric Sign of Sensorimotor Association and Expectancy in the Human Brain," *Nature* 203 (July 25, 1964): 380–84.

115. W. Grey Walter, "The Contingent Negative Variation and Its Significance for PSI Research," in *PSI Favorable States of Consciousness: Proceedings of an International Conference on Methodology in PSI Research*, ed. Roberto Cavanna (New York Parapsychology Foundation, 1970), 172.

and repeatedly primed. Moreover, the brain could learn from its expectations when they were confirmed (or not) by the experimenter who was in charge of initiating clicks that foretold of a series of flashes (or flashes that foretold of a series of clicks). Here was how the brain set down its epistemic grooves, context free, arriving at decisions before they were consciously made.

What Walter called the Contingent Negative Variation (CNV) was a real-time measurement of the brain making statistical inferences, learning, and arriving at testable hypotheses. This capacity for expectation was, by definition, a matter of pattern recognition that occurred below conscious awareness. Clicks followed by flashes, flashes followed by clicks. Neither the content of the stimuli (clicks and flashes were interchangeable) nor their intensity mattered as much as the fact that the stimuli were patterned. The brain, here, rather than the individual per se was involved in probabilistic thinking that was independent of conscious decision-making. Here was the brain habituating itself, learning to be certain through repetitive stimuli, affected by attitude and social cues. "The CNV," argued Walter, within a bubble of his own creation, "is very sensitive to subtle changes in human environment."[116]

Not surprisingly, this is where it begins to get a little weird. For in discovering that the uptick in alpha frequency happened in conjunction with the perception of patterned stimuli, Walter made a further inference. The change in alpha patterns, he wrote, was also "the movement of some hitherto unsuspected mechanism of the brain." "What is this mechanism?," asked Walter.[117] His answer opens up, for us, a conceptual space for thinking about hyperactive agency detection and, perhaps, too, a discursive condition that makes the detection of that hyperactivity possible.

It was, Walter reasoned, "as though the brain were working as a very accurate probability computer." For the "behavior of the spontaneous and artificial rhythms," insisted Walter, is "strictly in accordance with the effects to be expected from a scanning mechanism." Signs of an entity that sees from above, as it were. Or, as Walter wrote, "the most familiar example of such mechanism is in television, where a space-pattern is most economically

116. This work was supported, in part, by Eileen Garrett's Parapsychology Foundation of New York, and was pursued in light of a future of "electronic ESP"—when "by looking at a record, it is possible to determine when a person is going to do something. The experimenter can know before the subject actually takes action that he is going to do so." Walter, "Contingent Negative Variation and Its Significance," 181, 186.

117. Walter, *The Living Brain*, 107.

converted for transmission into a time sequence of impulses by the scanning mechanism of the camera."[118] *Vision shorn of lushness but rich in range, rationality, and reach.*[119]

So it was that in the same year *TV Guide* began publication and the coronation of Queen Elizabeth II was broadcast on the BBC, Walter suggested that the alpha rhythm could be understood as evidence of a cerebral scanning mechanism, ever moving back and forth, capturing signals and projecting them but also making sense of the disruptions.[120]

Scanning, here, hints at the complex conditions within the brain as it interacts with its environment—the means to gather data, to extract patterns, to make predictions based on these patterns, to gather more data, "to communicate every item of information received in any one part of it to all its other parts." Moreover, the "millions" of neural connections that make scanning possible exceed themselves in the process. In light of his TV metaphysics, the metaphor Walter lands on to describe the authority of this network is not at all surprising: "*The director, in effect, does not go to the projection room; he has too many other things to direct, what with the other senses and memories and all his own reflections; rather he has some arrangement by which he is kept informed about anything that might interest him during his waking hours*" (italics his, underlining mine).

118. Walter, "The Contingent Negative Variation and Its Significance," 174; Walter, *The Living Brain*, 111, 108.

119. Walter's description of this part of the brain not involved in statistical inference clearly reflects gendered hierarchies operative within this particular version of the scientific gaze: "The brain is a great chatterbox, and this is one of the main difficulties of our research, as the gossip is ineluctable. The ten thousand million cells are constantly interacting about all kinds of trivialities, some of which are housekeeping functions" (Walter, "Contingent Negative Variation and Its Significance," 172).

120. Walter, *The Living Brain*, 82. See also Rhodri Hayward, "The Tortoise and the Love-Machine: Grey Walter and the Politics of Electroencephalography," *Science in Context* 14 (2001): 625. It is curious to note how television, even before Walter's ode to the technology of scanning, could itself be understood in terms of neural metaphors. By 1948, television scanning had become integral to the process of breaking up and reassembling "a great many elements or elemental areas and information. . . . In viewing a scene the image is carried to the brain by the eye over a huge network of transmission lines which tells the brain the intensity and the color of the light at every point in the field of vision"; Rider, ed., *Television: How It Works*, 3. See also R. D. Kell, A. V. Bedford, and M. A. Trainer, "Scanning Sequence and Repetition Rate of Television Images," in *Television: Collected Addresses and Papers on the Future of the New Art and Its Recent Technical Developments* 1 (New York: RCA Institutes Technical Press, 1936): 355–74.

"How is the director kept *informed*?" asks Walter, a coy recognition of his own heated prose and the kind of agency he has detected, by his own scanning of the signs.[121]

Who or what is this director?[122]

Everything, it seems, pivots on how one reads "in effect." Elsewhere Walter describes the brain as an "instrument more infinitely wonderful than television." In conjuring the notion of infinity, here, Walter slips in a sly note of theological specificity. For in Walter's prose one is struck by the simultaneous move of alluding to the underlying mechanisms of the brain as agentive *and* a steadfast denial that such agency could ever be sufficiently accounted for. So, on one hand, the brain is an "instrument infinitely more wonderful than television." Behind this instrument lie the intentions of all manner of actors and overseers, including the director in charge of the various processes of sensory capture. On the other hand, Walter is committed to the idea of the brain as a black box—studied not by opening it up or revealing its contents in toto but by simply and judiciously observing what goes in and what comes out. So while "future research may well carry us," Walter insisted, "into vistas of ever increasing enchantment," such enchantment will inevitably be "describable only in the convention of mathematical language" and through recourse to inputs and outputs.[123]

An interesting subtext of *The Living Brain* is Walter's warning to the field of electroencephalography to remain steadfastly secular. He urged fellow neurophysiologists to be vigilant in tempering the human proclivity to over-read, to ascribe pattern where there, in fact, is none. "We are daily reminded," wrote Walter, "how readily living and even divine properties are projected into inanimate things by hopeful but bewildered men and women; and the scientist cannot escape the suspicion that his projections may be psychologically the substitutes and manifestations of his own hope and bewilderment." For Walter to recognize this primitive proclivity, however, was a matter of objective assessment, ever the sign of one's civilized maturity. Such recognition, by way of brain-imaging machines, was an essential move in achieving the eradication of this pesky cognitive survival of hyperactive attention.

121. Walter, *The Living Brain*, 82, 104, 107.

122. A power both immanent and external to the system in question, or what a pulpy version of cybernetics referred to as "an elaborate monitoring system, something like a closed-circuit television [that] keeps the brain informed of events in all parts of the body." Y. Saparina, *Cybernetics within Us* (North Hollywood: Wilshire Book Co., 1967), 85.

123. Walter, "Patterns in Your Head," 57; Walter, *The Living Brain*, 60.

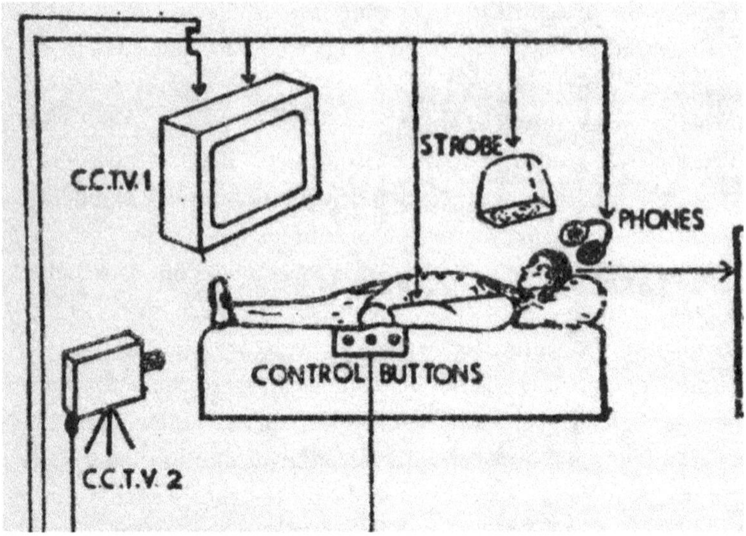

FIGURE 13. Grey Walter's Laboratory (close-up). W. Grey Walter, "The Contingent Negative Variation and Its Significance for PSI Research," in *PSI Favorable States of Consciousness: Proceedings of an International Conference on Methodology in PSI Research*, ed. Roberto Cavanna (New York Parapsychology Foundation, 1970), 171.

So there is a fantasy simmering between the lines of Walter's work—of being able to train our brains and to temper (and inevitably overcome) our penchant for hyperactive agency detection—for such hyperactivity is a matter of the "inefficiency or inadequacy of the selective mechanisms" in our brain—too many stimuli and not enough mature discernment. But as we learn about our brain learning about the world, the epistemic future looks bright indeed. We are on the right path, concludes Walter, for it is already possible to imagine that "were these built-in selectors perfectly efficient, we could learn all we need to by just letting things happen to us."[124]

There is a TV ease to Walter's argument—not simply about how the brain resembles that cutting-edge technology but also about how the power of its persuasion depends on its representational prowess and promise of direct

124. Walter, "The Future of Clinical Neurophysiology," *Electroencephalography and Neurophysiology* 27, no. 7 (1969): 645; *The Living Brain*, 61, 115, 164.

access via scanning.[125] Like an intelligent scanning mechanism, the brain is invested in recognizing patterns on the horizon, capturing those patterns, learning from those patterns, and becoming better and more efficient in scanning the horizon. For as the logic of television plays itself out, the space between us and the director remains incredibly vast. Yet there is a payoff to Walter's attention—described by him as a "reverent attitude"—"modest, thoughtful, religious." For there is control to be had in such passive repose. For if we gaze at the television long enough our brains will eventually be able to turn it off. The fantasy of passivity, then, is instrumental. Throughout *The Living Brain*, Walter suggests that submission to neural explanation will beget a new era of the human in which freedom is based not on execution of a decision but the management of *potential* decisions. Note, for example, how Walter draws on the TV metaphor not only to understand the logic of the brain but as an example of the kind of human agency that was on the horizon and made possible by the neuroscientific gaze. By looking at an encephalogram, declared Walter,

> it is possible to determine when a person is going to do something. The experimenter can know before the subject actually takes action that he is going to do so. . . . One can program a computer to recognize a rise of this shape and size and accordingly switch on a television picture. The potential rises to turn the clicks off, so the subject himself is actually doing nothing. He is lying absolutely passive, now and then wishing or willing that a particular event would occur (that the television picture would appear).[126]

Walter, like Boyer after him, was operating on the far end of the evolutionary spectrum. Both noted the natural human capacity to hyperactively detect the purpose of pattern in the world. Both acknowledged its evolutionary advantages. And both used technical knowledge of that advantage to readjust the intensity of this capacity, thereby perfecting its operation and making possible the detection of the essential rhythms that correlate with both the initial state of hyperactive scanning and a present state of sober analytic overcoming. The mystery of one's own agency becomes a problem

125. On the EEG imagined as a television broadcast, see Barbara B. Brown, *New Mind, New Body: Bio-feedback: New Directions for the Mind* (New York: Harper and Row, 1974), 308.

126. Walter, *The Living Brain*, 12, 181.

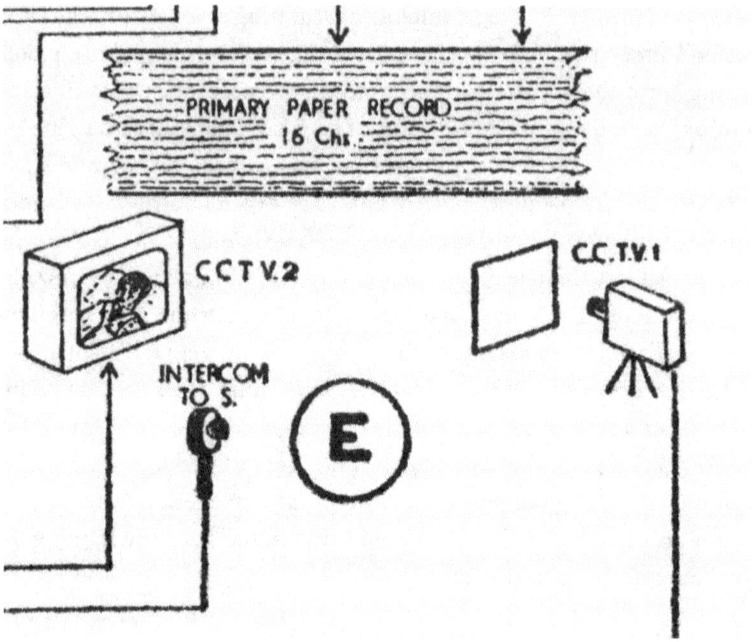

FIGURE 14. Grey Walter's Laboratory (closer-up). W. Grey Walter, "The Contingent Negative Variation and Its Significance for PSI Research," in *PSI Favorable States of Consciousness: Proceedings of an International Conference on Methodology in PSI Research*, ed. Roberto Cavanna (New York Parapsychology Foundation, 1970), 171.

to be worked on and worked out.[127] "I think in order to do great things with a machine," confessed Walter at the end of his career, "you must somehow become part of it; we should know what it *feels like* to be a computer."

For Walter, then, the story of the brain is two stories—a story of *its* emancipation and, eventually, a story of *our* submission. Both stories evoke an evolutionary frame. For it is the evolutionary frame that makes sense of Walter's own desire to acquire leverage on his always already strangely mechanized self—for this is the way he will achieve his humanity—corresponding with the author of his being, holding a mirror up to his brain so he can see it seeing itself. This moment of conflating brains, selves, and machines is post-

127. Boyer, too, stands in awe of an evolutionary process in which "input that produces comparatively richer inferences or produces them with less computational effort (or both) will, other things being equal, be better acquired, stored and communicated." Pascal Boyer, "Why Do Gods and Spirits Matter at All?," in *Current Approaches to the Cognitive Science of Religion*, ed. Ilkka Pyysiäinen and Veikko Anttonen (New York: Bloomsbury, 2002), 89.

human, postsecular, postreligious—the moment when—and here I amplify Walter's own voice: *the moment when "the master brain can discover its own place and settle down at last to its proper work."*[128]

9. AGENTS LIKE US

> Twenty-first century machines—based on the design of human thinking—will do as their human progenitors have done—going to real and virtual houses of worship, meditating, praying, and transcending—to connect with their spiritual dimension.
>
> RAY KURZWEIL, *The Age of Spiritual Machines* (1999)

Like all the other machines that detect patterns of agency in and around us, the hyperactive agency detection device is engaged in perpetual prediction.[129] It is calculating how to act in an information-rich environment, that is, an environment made up of "statistical structures."[130] Scanning, here, becomes the base mechanics of probability theorizing—the fact that our agency detection devices are bent toward hyperactivity, of erring on the side of overestimating the presence of agents, is for our long-term evolutionary good.[131] Tamping down such hyperactivity comes at a price, however—for reason, rightly arrived at, portends the jettisoning of our illusions of comfort

128. Walter, *The Living Brain*, 278.

129. In 1969, Walter had concluded that "cerebral computation" was stochastic based on the computation of machines (*Observations on Man*, 32).

130. In research funded by the Office of Naval Research, computer scientist Andrew S. Gordon has extended the utility of Heider and Simmel's ninety-second animated film into conversations about the predictive brain and optimizing the success of human-computer interactions. Gordon uses the film in order to think more productively about the problems of "automating the processes of narrative interpretation." In so doing, Gordon naturalizes the "intelligence" exhibited in anthropomorphic proclivities and presents the film as a resource for humanizing the algorithms of artificial intelligence. Andrew S. Gordon, "Interpretation of the Heider-Simmel Film Using Incremental Etcetera Abduction," *Advances in Cognitive Systems* 7 (2018): 27.

131. Pascal Boyer and Brian Bergstrom, "Threat-detection in Child Development: An Evolutionary Perspective," *Neuroscience and Biobehavioral Reviews* 35 (2010): 1034–41.

and divine connection, of being watched over, of seeing the world straight and mean and just as it is.

Accuracy and access from a distance.

Since the publication of *The Living Brain*, the logic of scanning has only intensified with the proliferation of screens.[132] There is little question that information processing has now become the natural logic of the brain and scanning the condition that makes human thought and human life possible.[133] This is the kind of humanism that creeps up, that is embedded in the products that we consume, the media through which we socialize, and the devices that we find ourselves in front of at any given time. *We scan. We count data points and calculate. We predict. We detect false positives. We learn. We repeat. We move forward. We function executively. We are human according to an elaborately detailed cognitive blueprint.* And let us not forget that scanning is also an ideal of scientific sight. Scanning is to perceive as an information-processing machine, without bias or subjective inflection. *As in a telephone wire or a computer.* Scanning is to aspire to a vision whittled down to form, striving to live in the future tense when (and where) one need not be bothered by questions of substance. Scanning is to repeat without difference. Scanning is to draw from the differential between present and past, every moment, in order to produce the future.

Relentlessly.

But I insist that the hyperactive agency detection device, for whatever else it might be, is also a historical object. For the claim that the brain is involved in processes of scanning, statistical inference, and prediction may well be true, but that truth (and our arrival at it) is not unrelated to the desire for that claim *to be true* (not to mention the cultures and contexts that fuel that desire). This brain may scan the horizon for a deviation in pattern and it may then anthropomorphize that deviation into a fully realized agent, but what other machines and what other conventions also make this possible?

132. The language of scanning, data gathering, and religious attention is now commonplace. See, e.g., Antoine Lutz et al., "Attention Regulation and Monitoring in Meditation," *Trends in Cognitive Sciences* 12, no. 4 (2008): 163–69, and J. A. Brefczynski et al., "Neural Correlates of Attentional Expertise in Long-Term Meditation Practitioners," *National Academy of Sciences* 104, no. 27 (July 3, 2007): 11483–88.

133. Such naturalization is, perhaps, most evident in those who seek a Grand Unified Theory of neural functioning. See also Michael L. Anderson and Tony Chemero's sharp critique of universalism in the guise of "epistemic internalism": "The Problem with Brain GUTs: Conflation of Different Senses of 'Prediction' Threatens Metaphysical Disaster," *Behavioral and Brain Sciences* 36, no. 3 (2013): 204–5.

For rather than grapple with the complexities of life as it is lived or, for that matter, the contingencies of science as it is practiced, CSR investigators adopt a rather clean (and rather unified) picture of cognition (of their test subjects as well as their own).[134] Despite acknowledging the role that culture plays in the evolution of cognition, there is very little attention paid to mediation, discourse, and so on, and much authority invested into the prime mover of the story told by cognitive scientists—"selective pressures over evolutionary history."[135]

So we accept the fact that we are scanning, that we should be scanning—ever more efficiently. And we, in turn, are scanned by those institutions that provide us the content to scan and create the conditions in which and the devices through which we practice our scanning. And so on and so forth.[136] Cognitive scientists, then, with their dreams of buffered subjectivity and neurotypical flourishing, play an integral role in modeling what secular selfhood might soon become for all involved.[137] Thinking about the brain, immunized from the world around, from its petty pathos and physicality. A future of information processing rather than meaning-making. Our hyper-

134. As philosopher Isabelle Stengers has wryly noted, "not only are scientists not asked to give an account of their choices and research priorities, but it is just and normal that they are unable to provide such an account" (*Invention of Modern Science*, 5, 9).

135. Pascal Boyer, "How Natural Selection Shapes Conceptual Structure: Human Intuitions and Concepts of Ownership," in *The Conceptual Mind: New Directions in the Study of Concepts*, ed. Eric Margolis and Stephen Laurence (Cambridge, MA: MIT Press, 2015), 186. See also Pascal Boyer, "Evolutionary Perspectives on Religion," *Annual Review of Anthropology* 37 (2008): 111–30. On the utility of CSR in an age of economic uncertainty and collective ignorance, see Pascal Boyer and Michael Bang Petersen, "Folk-economic Beliefs: An Evolutionary Cognitive Model," *Behavioral and Brain Sciences* 41 (2018). On the suspect assumptions about evolution underlying CSR, see Susan McKinnon, *Neo-liberal Genetics: The Myths and Moral Tales of Evolutionary Psychology* (Chicago: Prickly Paradigm Press, 2006) and S. E. Smith, "Is Evolutionary Psychology Possible?," *Biological Theory* 15 (2020): 9–49.

136. Shoshana Zuboff, *The Age of Surveillance Capitalism: The Fight for a Human Future at the New Frontier of Power* (New York: Public Affairs, 2019).

137. In CSR's particular form of disenchantment, God is not so much false or dead but a prediction error, an often comforting "surprise" that results from a "mismatch between the sensory signals encountered and those predicted." At the neural level, this is termed "a surprisal" and suggests how the brain thrives on the appearance of uncertainty in order to frame the present in terms of a distribution of potential outcomes (Clark, "Whatever Next?," 183). On the cybernetic imprint of this metaphor, see Wiener, *Human Use of Human Beings*, 10–12, 21.

active agency detection devices correcting their own errors for evolutionary advantage. All ritual, it would seem, becomes meaningless without the brain being involved.

So despite whatever doubts I harbor about the claims made by cognitive scientists of religion, there is one thing that I believe all involved can agree on: as religion drifts ever inward during these long centuries of the self, into the mind and into the nervous system, whatever religion is becomes increasingly conjured by particular modes of calculation. Which is to say that religion in this secular age becomes increasingly defined by the machines, broadly construed, that reveal its truth.[138]

10. CHEAP TRICKS

With a renewed focus on "Error Management Theory," studies of HADD, or a more "nuanced" version of it, continue apace. In a bubble of their own creation, studies deploy the Geometrical Figures Task inspired by Heider and Simmel's 1944 experiment and use Oculus Rift VR headsets so that "participants can now move around in highly controlled but immersive forestscapes that emulate the milieu of early hunter-gatherers."[139] They live statistically, with uncertainty, amid the virtual trees, in order to arrive at certainty in the long run.[140]

138. On the turn to algorithms in CSR, see Justin Lane and F. LeRon Shults, "The Computational Science of Religion," *Journal for the Cognitive Science of Religion* 6, no. 1–2 (2020), https://doi.org/10.1558/jcsr.38669.

139. Uffe Schjoedt et al., "Vikings, Virtual Reality, and Supernatural Agents in Predictive Minds," and Marc Andersen, "Predictive Coding in Agency Detection," both in a special issue of *Religion, Brain and Behavior* 9, no. 1 (2019): 1; David L. R. Maij et al., "The Boundary Conditions of the Hypersensitive Agency Detection Device: An Empirical Investigation of Agency Detection in Threatening Situations," *Religion, Brain and Behavior* 9, no. 1 (2019): 23–51. See also Marc Andersen et al., "Agency Detection in Predictive Minds: A Virtual Reality Study," *Religion, Brain and Behavior* 9, no. 1 (2019): 52–64.

140. Just like science. Or as Walter, himself, put it optimistically, arguing that his own "account of how experience is processed in the brain could well be a description of what is generally called the scientific method—the classification of observations, the developments of hypotheses and the testing of hypotheses by experiments." The brain, according to Walter, contained within it truths that were universally applicable. "The scientific method," he continued, "is the deliberate formalisation of the intrinsic mechanisms of cerebral computation." Science, in other words, was now subjecting consciousness to empirical

In keeping with their cybernetic legacy, CSR is currently riding a wave of interest in (and funding for) studies that explore predictive processing—that the "brain is a sophisticated hypothesis-testing mechanism, which is constantly involved in minimizing the error of its predictions of the sensory input it receives from the world."[141] As cognitive scientists emphasize the role that prior expectations play in interacting with incoming sensory stimuli, they seek to consider how the brain is "somehow able to continuously minimize prediction error." And the question of how the brain copes with false positives becomes primary, once again.[142] In a major statement on agency detection, cognitive scientist Marc Andersen nods to his ancestors Heider and Simmel even as he looks forward:

> In recent years, cognitive neuroscience has increasingly adopted predictive coding to model how the brain processes sensory information. Predictive processing specifies how the brain is constantly in the process of predicting incoming sensory input and thereby inferring the causes in

---

scrutiny precisely because science was now capable of modeling its own approach to the brain (and updating that model) based on how the brain works. Walter, *Observations on Man*, 34–35.

141. As Jakob Hohwy writes, summarizing a spate of recent research, the brain is best understood as a "prediction error minimization mechanism" (*Predictive Mind*, 1, 3, 59). See also Clark, *Surfing Uncertainty*, and, for the compatibility between the predictive processing framework and ongoing emphases on embodied, embedded, enacted, and extended cognition, Albert Newen, Leon De Bruin, and Shaun Gallagher, eds., *The Oxford Handbook of 4E Cognition* (New York: Oxford University Press, 2018).

142. On the turn to models of predictive processing in CSR, see Michiel van Elk and André Aleman, "Brain Mechanisms in Religion and Spirituality: An Integrative Predictive Processing Framework," *Neuroscience and Biobehavioral Reviews* 73 (2017): 359–78; C. A. M. Hermans, "Towards a Theory of Spiritual and Religious Experiences: A Building Block Approach of the Unexpected Possible," *Archive for the Psychology of Religion* 37 (2015): 141–67; Kristoffer L. Nielbo and Jesper Sørensen, "Prediction Error during Functional and Non-functional Action Sequences: A Computational Exploration of Ritual and Ritualized Event Processing," *Journal of Cognition and Culture* 13, no. 3–4 (2013): 347–65; and Jakob Hohwy, "New Directions in Predictive Processing," *Mind and Language* (2020): 3. It should be noted that such interest in a predictive processing model of neural systems emerged just as HADD became the central preoccupation among cognitive scientists of religion in the first decade of the twenty-first century. For precursors to current interests in predictive processing, see Martin Davies and Max Coltheart, "Introduction: Pathologies of Belief," *Mind and Language* 15 (2000): 1–46; Jakob Hohwy, "Top-down and Bottom-up in Delusion Formation," *Philosophy, Psychiatry and Psychology* 11 (2004): 65–70.

> the environment of that input; crucially, this inferential process relies heavily on prior expectation. By focusing on sensory input that does not fit its predictions—the prediction errors—the brain elegantly can come to represent the world accurately. When sensory input conflicts with predictions, prediction errors are passed up the neuronal hierarchy, allowing the brain to update its model of the world through an ongoing process of prediction error minimization.

Andersen's is a strong argument for distributed cognition contra the tendency of previous HADD accounts to a) offer "black-box explanations" (Atran) or b) to localize religious cognition in the relations between identifiable modular architectures in the brain (Boyer, Barrett). Predictive processing, according to Andersen, is a "challenge to a classic modular view of the mind, because it offers a philosophically elegant, mechanistic, and neurally plausible account of cognition and perception in which cognitive and perceptual architectures emerge as 'profoundly unified and, in important respects, continuous.'"[143]

Regardless of whether cognitive scientists prefer the language of predictive or computational processing or whether they naturalize religion with talk of modules or distributed cognition,[144] the most distinguishing mark of CSR is its wholesale internalization of information theory and biological processes of data-driven communication.[145] The brain "process[es] infor-

143. Andersen, here, adopts the neuromatic frame. "Predictive Coding," 71; Marc Andersen, "The Bayesian Observer and Supernatural Agents," *Religion, Brain and Behavior* 9:1 (2019): 100. Andersen, here, is citing Clark, "Whatever Next?," 187.

144. See, for example, the series of critical responses to Andersen's target article in "Commentaries," *Religion, Brain and Behavior* 9, no. 1 (2019): 84–98.

145. Which is to say that the turn to predictive processing is a revival of certain cybernetic emphases. See, e.g., Anthony N. Mucciardi, "Neuromime Nets as the Basis for the Predictive Component of Robot Brains," in *Cybernetics, Artificial Intelligence, and Ecology: Proceedings of the Fourth Annual Symposium of the American Society for Cybernetics*, ed. Herbert W. Robinson and Douglas E. Night (New York: Spartan Books, 1972), 159–93. On the flashback nature of current approaches to predictive processing, see Nico Orlandi and Geoff Lee, "How Radical Is Predictive Processing?" in *Andy Clark and His Critics*, ed. Matteo Colombo, Elizabeth Irvine, and Mog Stapleton (New York: Oxford, 2019), 206–21. On HADD as bound up in a precaution system that is designed to detect and to react to potential danger (and not part of the great system which is geared to respond to manifest danger), see Pascal Boyer and Pierre Lienard, "Why Ritualized Behavior? Precaution Systems and Action Parsing in Developmental, Pathological and Cultural Rituals," *Behavioral and Brain Sciences* 29, no. 6 (2006): 595–613.

mation from a range of cognitive domains," writes Andersen in a renewed cybernetic vernacular. Or, as Boyer surmises, "every bit of information is fodder for the mental machinery."[146]

"Cheap tricks," argues Boyer, "are important, if not essential, in many religious traditions." Cheapness, however, does not imply fraud or disingenuousness. "In many places the world over," assures Boyer, "conjuring tricks and manufactured illusions are perfectly respectable adjuncts to more sober myth and ritual. And after all, the gospels are replete with conjuring tricks."[147]

My sense is that cognitive scientists of religion continue to have a few tricks up their sleeves, given their belief-practices in which conjecture becomes symbolically efficacious; given their own ritual intensities that move reflexive theorization into the realm of definite intuition, given their own beliefs in agents that are anything but empirically justified. To suggest as much is to turn Boyer's brand of condescension against the CSR project of neuromatic secularization. Consequently, I have become increasingly taken with the analytical conceit that the cognitive science of religion is involved in its own form of prayer. For in prayer one finds the values around which to orient a life—prayer defined, here, as an activity marked by a subjunctive mood and the intentionality to commune with a divine agency, broadly construed.

146. Andersen, "Predictive Coding," 71; Boyer, *Religion Explained*, 30. Boyer defines agency as a series of choices that are "made by computation, that is, by mentally going through a variety of aspects of the situation and evaluating the least dangerous option." This brand of scientism is a common enough trait of our secular age, the desire to be surrounded by algorithms and, in the end, to become one (Boyer, *Religion Explained*, 21–22). Luckily for us, Boyer has recently turned from explaining religion to framing the brain's predictive power as a mode of risk assessment—useful information as more and more resources are allocated to lessen our collective risk. When the brain encounters information, argues Boyer, it scans for cues that signal as a threat—and the more threatening the message, the more we tend to trust it. Individuals achieve a sense of security, then, within group settings because they come to detect, evaluate, rate, and react to potential threats. Pascal Boyer et al., "Safety, Threat, and Stress in Intergroup Relations: A Coalitional Index Model," *Perspectives in Psychological Science* 10, no. 4 (July 2015): 434–50.

147. Pascal Boyer, "Why 'Belief' is Hard Work: Implications of Tanya Luhrmann's *When God Talks Back*," *HAU: Journal of Ethnographic Theory* 3, no. 3 (2013): 352.

The notion of "prayer expectancies" is a staple of the scientific literature on prayer.[148] When people pray they often expect something to happen. But to be clear, the mood of expectancy that I sense in the making of the hyperactive agency detection device is not necessarily about belief in a future state of affairs but, rather, about a style of reasoning that feels right and gathers momentum as it moves along. In the manner of Jonathan Edwards, a perpetual praying *to* rather than praying *for* anything in particular.

So it is my outrageous conclusion that in the contemporary cognitive science of religion, prayer has become an apt description for scientific explanations keen on domesticating religion as an elaborate engineering problem. For in these pleas there is a kind of mystical closure. The model of the behavior of the brain becomes, at last, the behavior of the brain. In its relentless pursuit of a real-time picture of a living brain (admittedly yet to be achieved but ever on the horizon), CSR seeks to operationalize cybernetic principles by validating their existence in the brain.[149] For there remain cognitive scientists in the world who seek to "communicate" (understood in terms of information control) with a vast and determinative and self-organizing system (the brain-environment complex). Their ecstatic communication serving as an explanation for everything and everyone else. Their brain-imagining technologies and technics of measurement becoming so many prayer machines geared toward making sense of the brain and leveraging that knowledge for material advantage in this world *and* the next.

148. Neal Krause, "Assessing the Relationships among Prayer Expectancies, Race, and Self-Esteem in Late Life," in *Journal for the Scientific Study of Religion* 43, no. 3 (August 2004): 395–408; Neal Krause and R. David Hayward, "Trust-Based Prayer Expectancies and Health among Older Mexican Americans," in *Journal of Religion and Health* 53, no. 2 (April 2014): 591–603; Bernard Spilka and Kevin L. Ladd, *The Psychology of Prayer: A Scientific Approach* (New York: Guilford Press, 2012), 84ff.

149. "A mind needs and generally has some way of organizing information to make sense of what is observed and learned," writes Boyer. "This allows the mind to go beyond the information given, or in the jargon, to produce inferences on the basis of information given." And "inference systems," continues Boyer, "have particular input conditions; that is, they get activated whenever information with a certain form is presented" (*Religion Explained*, 42, 286).

SYNAPTIC GAP: THE INFORMATION OF HISTORY

> But the Past has vanished, and the stupendous Present is before us. Here we stand, upon the towering summit of the ages past, contemplating the world of matter and the world of mind. We stand upon a mighty eminence, with all the vast accumulation of ages, with all the experience and wisdom of the past, beneath, around, and within us.
>
> ANDREW JACKSON DAVIS, *Spiritual Declaration of Independence* (1851)

AS A CRITIQUE OF THE FUTURITY built into the neuromatic paradigm, the form of *Neuromatic* seeks to disturb, by way of genealogy, the "ahistorical triumphalism characteristic of the *neuro* field."[1] To be sure, I readily admit that such genealogical practices will not cure any disease or malady. Indeed, they may even cause unwelcome distress. For my point is not simply to critique the reductionism of neuromatic frames of being. I also seek to get behind the conditions (epistemological and political) that make possible the common sense of these framings. Mine, then, is a critique of neurorealism even as the signs of the brain circulate and saturate. For in historicizing the discourse of neuromation I draw on a generalized notion of brain plasticity and strategically privilege environmental pressures over essential traits within an individual's nervous system. I acknowledge that not everyone is

1. Melissa M. Littlefield and Jenell M. Johnson, eds., *The Neuroscientific Turn: Transdisciplinarity in the Age of the Brain* (Ann Arbor: University of Michigan Press, 2012), 10.

affected in the exact same way by social patterns, concepts, and codes. But I do insist that the experience of these patterns, concepts, and codes does affect brain morphology. *Plasticity is real, indeed, and we are not prepared.* For the more pervasive and dense the atmosphere becomes, the more disproportional an imprint it will make on any given subject.[2] To portray the positivity of discourse is both an analytic and an aesthetic proposition.

Therefore my critique of the neuromatic paradigm, in addition to being political, is also historiographical. How to provide an adequate history of the brain from within the vortex of digital humanities initiatives? How not to get bound up in its prophecies? For within the swirl of information there is little to indicate an outside to either its ontology or its temporality. The past of the neuromatic brain is, apparently, ever a reflection of its present, absorbing whatever difference it encounters in a voracious endgame of its own accord. The future, in other words, is all but guaranteed.[3] For with the coming "appearance of an intercommunicating network of archives" everything will be known as it should be known. The past as it really happened. Right here right now. History removed from petty bias and human foible.[4]

Or so predicted an article in the *Journal of American History* in 1967. In "Computers and Historical Studies," Jerome M. Clubb and Howard Allen declared that intellectual habits would change, for the better, as computers were increasingly relied on in historical research and writing. With a certain optimism in the "rich potentialities of electronic data processing equipment" for humanistic research, they predicted that a new kind of sober empiricism would take hold in the historian's relationship to the past. An unprecedented scientific rigor would be enabled. A bigger, convincing picture would provide proper context and right meaning to those singular events, lives, and words described by the historian. "Computers facilitate generalization, and historians become increasingly concerned with the common properties of human affairs and less concerned with supposedly unique and transitory occurrences, however dramatic they may be." The past could be known to

2. On this point I am in complete agreement with contemporary cognitive science—see the entire issue edited by Niels N. Johannsen, John McGraw, and Andreas Roepstorff, including their "Introduction: Technologies of the Mind," *Journal of Cognition and Culture* 14, no. 5 (2014): 335–43.

3. Walter Benjamin, "One-Way Street," in *Reflections: Essays, Aphorisms, and Autobiographical Writings*, ed. Peter Demetz, trans. Edmund Jephcott (New York: Schocken, 1986), 61–94.

4. Jerome M. Clubb and Howard Allen, "Computers and Historical Studies," *Journal of American History* 54, no. 3 (December 1967): 599–607.

the degree that it could be stabilized and plotted against. The past smoothed of jagged and statistically insignificant edges. Uniqueness becoming indecipherable and any degree of sameness lending itself to quantification. Data points providing leverage for building scale into ever-expanding explanations of the past. Scalability grounding a series of models. Systems of various shapes and forms—all coded for temporality—proliferating into an increasingly dense network. The model of these correspondences and metacorrespondences was meant to transcend mere modeling. The goal, in other words, was prophecy. *To not only predict but to generate the future.*

Within predictive models—on a fundamental level, that is, the neuronal level—the past has been deemed irrelevant. From the beginning. For in the process of firing or not, a neuron had little use for the history of the signal it was communicating.[5] Only the present existed in the moment of becoming the future.[6] According to Warren McCulloch and Walter Pitts, "specification of the nervous net provides the law of necessary connection whereby one can compute from the description of any state that of the succeeding state, but the inclusion of disjunctive relations prevents complete determination of the one before." At pains to alert the reader that their logical calculus is a beautiful abstraction, useful in predicting final outcomes rather than what is really going on, McCulloch and Pitts were direct in their address: "But one point must be made clear: neither of us conceived the formal equivalence to be factual explanation."[7]

The presence of any neural signal, framed in terms of information, was

5. One might call this a perversion of Shannon's model in that whatever preceded any event—specifically, the choices that had been made and the freedoms actualized—were of great significance for the present. According to Warren Weaver's gloss, "obviously probability plays a major role in the generation of the message, and the choices of the successive symbols depend upon the preceding choices" (Warren Weaver, "The Mathematics of Communication," *Scientific American* 181, no. 1 [1949]: 12).

6. Orit Halpern, *Beautiful Data: A History of Vision and Reason since 1945* (Durham, NC: Duke University Press, 2014), 47.

7. Warren S. McCulloch and Walter Pitts. "A Logical Calculus of the Ideas Immanent in Nervous Activity," *Bulletin of Mathematical Biophysics* 5, no. 4 (1943): 129, 117. On McCulloch's "experimental epistemology," see Tara Abraham, *Rebel Genius: Warren S. McCulloch's Transdisciplinary Life in Science* (Cambridge, MA: MIT Press, 2017), and Lily E. Kay, "From Logical Neurons to Poetic Embodiments of Mind: Warren S. McCulloch's Project in Neuroscience," *Science in Context* 14, no. 4 (2001): 591–614. See also Joseph Dumit, "Plastic Diagrams: Circuits in the Brain and How They Got There," in *Plasticity and Pathology: On the Formation of the Neural Subject*, ed. David Bates and Nima Bassiri (New York: Fordham University Press, 2016), 227.

quite literally ahistorical. According to Edgar Adrian, for example, the trailblazing neurophysiologist and Nobel Laureate who was an early mentor to Grey Walter, "the propagated disturbance at any point in a nerve fibre depends only on the local condition of the fibre at that point and not on the previous history of the disturbance before it arrived there."[8] In extending Adrian's all-or-nothing principle, McCulloch and Pitts also extended his refusal to account for any other ecological confound that might affect the transmission of signal, including all the other transmissions that were already occurring as a sense organ conveyed a particular signal to the nervous system. According to McCulloch and Pitts, those who would seek to know the brain must model their approach on the brain's approach to complexity. For, ideally, like the brain, what we really want is "freedom toward the future—freedom from affairs intercurrent between our ideas and our deeds."[9]

Despite various admissions that the analogue could not be mapped digitally (that the brain was not really a computer), there was a distinguishable increase in such map-making at midcentury. In efforts to model machines on the formal logic of the brain (provided by mathematicians as much as by neurophysiologists), context had already dropped out. The past, once considered by some to be an ontological burden, became unnecessary for figuring out where a neuron, or perhaps the human, was going. As cybernetician Stafford Beer declared, "if the world is beyond our capacity to know it, and if, even worse, it continually changes, knowing the past is of limited utility."[10] Here was a necessary ahistoricism, built into a worldview that had little use for the past save for its object status of not being the future. Cybernetics was "a dynamic science [that] has no need of its past," quipped Marvin Minsky, cofounder of the MIT Computer Science and Artificial Intelligence Laboratory in 1959, "it forges ahead."[11]

I am disquieted by the clean lines of this rehearsal, particularly its version of the subject who has escaped history and culture, not unrelated to the evacuation of interpretation proffered by information theory wherein

8. E. D. Adrian, "The All-or-none Principle in Nerve," *Journal of Physiology* 47 (1914): 460.

9. Warren S. McCulloch, "Through the Den of the Metaphysician," in *Embodiments of Mind* (Cambridge, MA: MIT Press, 1970), 149–50; Jean-Pierre Dupuy, *The Mechanization of the Mind: On the Origins of Cognitive Science*, trans. M. B. Debevoise (Princeton, NJ: Princeton University Press, 2000), 133–34.

10. Stafford Beer, *Management Science: The Business Use of Operations Research* (New York: Doubleday, 1968), 252.

11. Minsky, quoted in Dupuy, *Mechanization of Mind*, 43.

signals were to be measured and mapped rather than interpreted for their meaning.[12] A subject whose information processing capacities enable feeling this way or that, choosing or not. A subject who lives in a moment bound up in nothing but the singular presence of an unfolding present.[13] In this story the human—and an intensely gendered human at that—can assume control over their own context. When security becomes a matter of cognitive science, reason and its metaphysics assume material command. Story upon story about mysteries unlocked. A nonlinear emplotment of a hero's journey. History lived and told as an algorithm.[14]

12. Similarly, John Durham Peters notes that "information is a term that does not like history" as theorists assume its primordial pedigree. John Durham Peters, "Information: Notes toward a Critical History," *Journal of Communication Inquiry* 12 (1988): 10. In a droll reference to the cybernetic folks at mid-century, Lydia Liu writes that they were extraordinarily inattentive to the discursive structures of scientific knowledge (*The Freudian Robot: Digital Media and the Future of the Unconscious* [Chicago: University of Chicago Press, 2010], 141). It is my argument throughout that genealogy is uniquely poised to question the naturalization of the imperial drive within cybernetics, to illuminate its epistemic code of knowing how not to acknowledge the forces of history and culture that infuse our categorical assumptions, like a frog that does not see "the detail of stationary parts of the world around him" but sees only "size and movement" (Jerome Y. Lettvin, Humberto R. Maturana, Warren S. McCulloch, and Walter H. Pitts, "What the Frog's Eye Tells the Frog's Brain," *Proceedings of the IRE* 47, no. 11 [1959]: 1940–51).

13. For a more optimistic and largely uncritical take on such analytic complicity, see Luther H. Martin and Jesper Sørensen, eds., *Past Minds: Studies in Cognitive Historiography* (London: Equinox, 2011).

14. But it is the dynamism of history and its uneven pressures, as much if not more so than the neuron, that speaks to our gendered and racial indentity, that promises to explain how and why "we desire anything—either physically, as we want food and drink or a woman and a bed, or mentally, as we seek in music the resolution of a discord or, in mathematics the proof of a theorem." Warren S. McCulloch, "Machines That Think and Want," in *Embodiments of Mind*, 307.

02

NEITHER MATTER NOR SPIRIT: TOWARD A GENEALOGY OF INFORMATION

> In an increasingly technological, computerized world, information is a prime commodity, and when it is used in biological theorizing it is granted a kind of atomistic autonomy as it moves from place to place, is gathered, stored, imprinted, and translated. It has a history only insofar as it is accumulated or transferred. Information, the modern source of form, is seen to reside in molecules, cells, tissues, "the environment," often latent but casually potent. It is thought to enable these molecules, cells, and other entities to recognize, select, and instruct each other, to construct each other and themselves, to regulate, control, induce, direct, and determine events of all kinds.
>
> SUSAN OYAMA, *The Ontogeny of Information* (2002)

1. HARD PROBLEMS

The "hard problem of consciousness" has emerged as a pressing paradox within philosophical and neurophysiological circles over the past few decades.[1]

1. Stephen Grossberg, "Towards Solving the Hard Problem of Consciousness: The Varieties of Brain Resonances and the Conscious Experiences They Support," *Neural Networks* 87 (March 2017): 38–95; Berit Brogaard and Dimitria Electra Gatzia, "What Can Neuro-

Its primary questions are these: What maneuvers must occur for the fleshy substance of the brain to produce the phenomenon of consciousness? What is the invisible and immaterial form that inheres in the content of the physical universe? How to account for the ethereal process that mediates the relationship between matter and spirit, between neural networks and the experience of experience? As David Chalmers, the philosopher most responsible for present interest in this conundrum, writes:

> What makes the hard problem hard and almost unique is that it goes beyond problems about the performance of functions. To see this, note that even when we have explained the performance of all the cognitive and behavioral functions in the vicinity of experience—perceptual discrimination, categorization, internal access, verbal report—there may still remain a further unanswered question: *Why is the performance of these functions accompanied by experience?*[2]

For Chalmers, the answer is not at all easy. On one hand, Chalmers rejects those answers that would reduce consciousness to neurophysiology alone. Chalmers, for example, calls into question the hubris of those who would claim measurable certainty, to have arrived at a final explanation of consciousness without accounting for the excessive nature of experience—the quality of redness, a clarinet's sound, a mothball's smell, and all the "mental images that are conjured up internally." On the other hand, Chalmers remains recognizable to those he would critique. For in adopting their basic understanding of information processing, Chalmers also privileges the neural unit—comprised of the soma (the cell body), the dendrite (extension of

---

science Tell Us about the Hard Problem of Consciousness?," *Frontiers in Neuroscience* 10 (2016): 395. For philosophical perspective on this rumble at the intersection of analytic philosophy and cognitive neuroscience, see John R. Searle in exchange with Daniel C. Dennett and David J. Chalmers, *The Mystery of Consciousness* (New York: New York Review of Books, 1997). Other recent forays into the study of consciousness, animated by the prospects of artificial intelligence, include Antonio Damasio, *Self Comes to Mind: Constructing the Conscious Brain* (New York: Vintage, 2012) and Stanislas Dehaene, *Consciousness and the Brain: Deciphering How the Brain Codes Our Thoughts* (New York: Viking, 2014). For an insider critique of neuroscience and its claims to be on the verge of solving the mystery of consciousness, see Steven Rose, *The Future of the Brain: The Promise and Perils of Tomorrow's Neuroscience* (New York: Oxford University Press, 2005).

2. David J. Chalmers, "Facing up to the Problem of Consciousness," *Journal of Consciousness Studies* 2, no. 3 (March 1995): 202.

the cell body), the axon (connective fiber) and a synapse (where the activity of one neuron affects another). According to Chalmers, the neural unit, in its form and functionality, is the communicative bridge between personal experience and world. It is where consciousness happens, where "information" provided by the senses is processed in the electrical relays of a neural network and fed back into the environment. Yet, still, argues Chalmers, the refinement of neurophysiological explanation of this process will not explain consciousness per se.[3]

Chalmers, here, is staging the perennial ambiguity of what, if anything, distinguishes humans from machines. He is mindful of the subtleties of being human and the limitations of human knowledge. He is also, simultaneously, compelled to make his argument within an environment beholden to the neuromatic paradigm of the brain. In the key of punk rock romanticism, Chalmers often adopts an enthusiastic tenor indigenous to the cybernetic archive, writing as if universal principles that govern human life and its destiny are, indeed, there to be revealed. For Chalmers, as for his interlocutors in the neurosciences and artificial intelligence communities, the processing of "information" defines the horizon of theorizing about the fundamental relationship between neural matter and everything else. Information is the form that allows one to think, credibly, about the relationship between matter and spirit. As Chalmers writes,

> the treatment of information brings out a crucial link between the physical and the phenomenal: whenever we find an information space realized phenomenally, we find the same information space realized physically. And when an experience realizes an information state, the same information state is realized in the experience's physical substrate. . . . It is natural to suppose that this double life of information spaces corresponds to a duality at a deep level. We might even suggest that this double realization is the key to the fundamental connection between physical processes and

3. Chalmers, "Facing up to the Problem of Consciousness," 214, 207. Chalmers, here, is taking on those who would reduce the brain to its neuromatic gears (see, e.g., Austen Clark, *Sensory Qualities* [New York: Oxford University Press, 1993]), use their brain-imagining machines to forge strictly biological theories of consciousness (Francis Crick and Christof Koch, "Towards a Neurobiological Theory of Consciousness," in *Seminars in the Neurosciences* 2 [1990]: 263–75), or else deny that consciousness is even a problem to be solved (Daniel C. Dennett, *Consciousness Explained* [Boston: Little Brown, 1991]).

> conscious experience. We need some sort of construct to make the link, and information seems as good a construct as any.[4]

According to Chalmers, information is a reasonable ruse, an explanatory conceit that "could be fleshed out into a system of basic laws connecting the physical and phenomenal domains." Such a "fundamental theory of consciousness" would frame the subjective aspects of experience—thinking, feeling, perceiving, and so on—as integral effects of "a whir of causation and information processing."[5] Consciousness is not, in other words, a strictly material thing—for part of its reality is ethereal, an effect, if you will, of a metaphysical proposition—this is why Chalmers calls for an elastic explanatory framework in order to appreciate the "subtle" relationship between consciousness, information, and the material world—that which happens "over and above the explanation of structures."[6] Information, in this scheme, is *more than* biological, mechanical, chemical, molecular, or even electrical. Information exists somewhere, in a "space" between matter and spirit, leading a "double life" as a spirit in the material world.[7] Mana in the matrix. A ghost in the machine.[8]

Chalmers's hard problem assumes, as a matter of common sense, that information is an infinitesimal solvent of consciousness. The questions that Chalmers poses owe much of their plausibility to the heady decade follow-

4. David J. Chalmers, *The Conscious Mind: In Search of a Fundamental Theory* (New York: Oxford University Press, 1996), 284–87. See also David J. Chalmers, "The Singularity: A Philosophical Analysis," *Journal of Consciousness Studies* 17, no. 9–10 (2010): 7–65.

5. Chalmers, *Conscious Mind*, 4. See also Andy Clark and David J. Chalmers, "The Extended Mind," *Analysis* 58 (1998): 7–19.

6. Chalmers, "Facing up to the Problem of Consciousness," 217, 208. Chalmers sees himself in the vein of James Clerk Maxwell who "expanded" the "ontology of physics" in order to better understand electromagnetism (209).

7. Or as Maj. Bryan N. Sparling writes in *Information Theory as a Foundation for Military Operations in the 21st Century* (Fort Leavenworth, KS: School of Advanced Military Studies, 2002), "information has a dual nature that is difficult to grasp, because it exists at the intersection between the mental and physical domains" (iii). For insight into the more ominous aspects of the cognitive defeat of behaviorism at mid-century, see Danielle Judith Zola Carr, "'Ghastly marionettes' and the Political Metaphysics of Cognitive Liberalism: Anti-behaviourism, Language, and the Origins of Totalitarianism," *History of the Human Sciences* 33, no. 1 (2020): 147–74.

8. For an elegant statement on the materiality of information, see Sylvester Johnson, "From Sophia to Silicon: The Materiality of Information," *Sacred Matters* (July 21, 2016).

ing World War II in which the neuromatic paradigm assumed widespread viability. Which is to say that long before Chalmers's particular articulation, there were hard problems of consciousness that had already been solved.

In the late 1930s Claude Shannon had begun working out a mathematical theory of communication to an audience of engineers involved with various media—radio, telephone, television, and pulse code modulators. As a graduate student in electrical engineering at MIT, Shannon became interested in mapping the intricate electrical relays of the Differential Analyzer, a hundred-ton analog precursor to the digital computer. Changing his concentration to mathematics, Shannon then spent a postdoctoral year at the Institute for Advanced Study in Princeton. Later, as an employee of Bell Labs, Shannon worked on cryptography and antiaircraft targeting for the National Defense Research Committee. He soon turned his eye toward quantifying the amount of information that could be *sent* over a channel and limiting the effect of noise *on* this communication.[9]

Shannon's "Mathematical Theory of Communication" (1948), published in *Bell Labs Technical Journal*, initiated an interdisciplinary conversation—first among communication engineers and later with just about everyone else.[10] As Shannon's theory was loosed from its technical considerations of bandwidth, information remained not merely an abstraction but became a material discovery that would usher in new understandings of biology, the natural world, and society. Information theory would also revolutionize approaches to political security, war, and the human sciences. As *Fortune* declared in 1953, "It may be no exaggeration to say that man's progress in peace, and security in war, depend more on fruitful applications of information theory than on physical demonstrations, either in bombs or power

9. C. E. Shannon, "A Mathematical Theory of Communication," pt. 1, *Bell System Technical Journal* 27, no. 3 (July 1948): 379–423, and pt. 2, *Bell System Technical Journal* 27, no. 4 (October 1948): 623–56; Claude E. Shannon, "Symbolic Analysis of Relay and Switching Circuits," *Electrical Engineering* 57, no. 12 (1938): 713–23; Jon Gertner, *The Idea Factory: Bell Labs and the Great Age of American Innovation* (New York: Penguin Books, 2012), 125ff. Precursors to Shannon's model developed at Bell Laboratories include Harry Nyquist, "Certain Factors Affecting Telegraph Speed," *Bell System Technical Journal* 3, no. 2 (1924): 324–46, and R. V. Hartley, "Transmission of Information," *Bell System Technical Journal* 7 (1928): 535–63.

10. In the popular press, information theory was often broached in the context of discussing electronic computers and the automated financial future. See, e.g., Edmund C. Berkeley, *Giant Brains, or Machines That Think* (New York: John Wiley and Sons, 1949), and "Office Robots," *Fortune* 45, no. 1 (January 1952): 82–87, 112, 114, 116, 118.

plants. . . . Naturally, therefore, attempts are being made to use information theory in a dozen fields from psychiatry to sociology. In a few fields, notably psychology, neurophysiology, and linguistics, the theory has already been applied with considerable success."[11]

By 1956, however, Shannon was complaining that information theory had become "something of a scientific bandwagon."[12] Part of this midcentury bandwagon was the emergence of the kind of brain that Chalmers's "hard problem" assumes to be operative: a brain ever in relationship to its environment that was, itself, continuously processing information. *Information flowing into and out of the brain in such a way that makes the eyes, ears, nose, mouth, and skin fungible to one another and to the brain via information. The activity of individual neurons dissolving into nervous system. Categorical distinctions melting into the elegance of mathematics.*[13] As Chalmers him-

11. Francis Bello, "The Information Theory," *Fortune* (December 1953): 136–37. For insightful takes on the technical uses and misuses of information theory, see Susan Oyama, *The Ontogeny of Information: Developmental Systems and Evolution*, 2nd ed. (Durham, NC: Duke University Press, 2002) as well as Ronald R. Kline, *The Cybernetics Moment: Or Why We Call Our Age the Information Age* (Baltimore: Johns Hopkins University Press, 2015). Kline has done much to articulate the fraught intellectual debates through the 1950s and 1960s that mark the emergence of the information paradigm, particularly the different perspectives of its two main proponents, Shannon and Norbert Wiener. On uneven applications of information theory in psychology, see Elizabeth A. Wilson, *Neural Geographies: Feminism and the Microstructure of Cognition* (New York: Routledge, 1998), 114f. On the incoherent applications of information theory within neuroscience, see Lance Nizami, "Information Theory's Failure in Neuroscience: On the Limitations of Cybernetics," Paper No. 158, Proceedings of the IEEE 2014 Conference on Norbert Wiener in the 21st Century, IEEE, Piscataway, NJ, 1–8. (2014). For insight into the culmination of information theory as common sense within neuroscience, see Fred Rieke et al., *Spikes: Exploring the Neural Code* (Cambridge, MA: MIT Press, 1997). On the biological revolutions ushered in by information theory, see Lily E. Kay, *Who Wrote the Book of Life?: A History of the Genetic Code* (Stanford, CA: Stanford University Press, 2000).

12. Claude E. Shannon, "The Bandwagon," *IRE Transactions—Information Theory* 2, no. 1 (1956): 3. In his complaint Shannon was throwing shade at colleagues like Wiener who called for wide applications of information theory—to neuroscience, psychology, biology, sociology and any other disciplines investigating processes of organization and control (Norbert Wiener, *Cybernetics: or, Control and Communication in the Animal and Machine*, 2nd ed. [1948; Cambridge, MA: MIT Press, 1961], 18).

13. In his influential *Design for a Brain: The Origin of Adaptive Behavior* (New York: John Wiley and Sons, 1952), for example, the psychiatrist W. Ross Ashby drew on Shannon's equations to make the case for the adaptive behavior of the nervous system. See also J. Y. Lettvin, H. R. Maturana, W. S. McCulloch, and W. H. Pitts, "What the Frog's Eye Tells the Frog's

self announced in his foundational article, "I understand information in more or less the sense of Shannon (1948). . . . An information space is an abstract object, but following Shannon we can see information as *physically embodied*."[14]

In considering just how much of our world is currently subject to devices and schemes of information processing, I am struck by what appears to slip Chalmers's notice: the saturated scene in which he is writing.[15] Take, for example, the seamless ease in which the hard problem of consciousness, or something like it, is now being pressed and pursued and deployed in ways that transcend questions of philosophical import. For what lies between neurons and experience, as such, is *now* the purview of psychopharmacologists, social media conglomerates, neuromarketers, geopolitical operatives, the police, software developers, life hackers, and engineers alike.[16] For the hard problem of consciousness, in a world surfeit with information and machines that purport to process it, has become an explosive force in realms beyond the philosopher's desk or even the neuroscientist's laboratory.[17] Incredible

---

Brain," *Proceedings of the IRE* 47, no. 11 (November 1959): 1940–51 and, more recently, Jordi Fonollosa et al., "Quality Coding by Neural Populations in the Early Olfactory Pathway: Analysis Using Information Theory and Lessons for Artificial Olfactory Systems," *PLoS One* 7, no. 6 (2012): e37809.

14. Chalmers, "Facing up to the Problem of Consciousness," 216. See also David J. Chalmers, *The Character of Consciousness* (New York: Oxford University Press, 2010), 25.

15. For a concise but provocative gloss on this scene, see Bernard Dionysius Geoghegan, "Information in Formation," in *Digital Keywords: A Vocabulary of Information Society and Culture*, ed. Benjamin Peters (Princeton, NJ: Princeton University Press, 2016): 173–83. See also James Gleick, *The Information: A History, a Theory, a Flood* (New York: Pantheon Books, 2011).

16. L. P. Zalukaev and L. F. Ponomareva, "Information Theory Approach to Pharmacology," *Pharmaceutical Chemistry Journal* 29, no. 1 (1995): 5–13; James Kent, *Psychedelic Information Theory: Shamanism in the Age of Reason* (Seattle: PIT Press, 2010); Clive Thompson, "Meet the Life Hackers," *New York Times* (October 6, 2005), https://www.nytimes.com/2005/10/16/magazine/meet-the-life-hackers.html; Elif Batuman, "Electrified: Adventures in Transcranial Direct-Current Stimulation," *New Yorker* (April 6, 2015), 24–32; Cade Metz and Adam Sartariano, "An Algorithm that Grants Freedom, or Takes It Away," *New York Times* (February 6, 2020). On the nefarious uses to which neuromatic paradigms are put, see Tony D. Sampson, *The Assemblage Brain: Sense Making in Neuroculture* (Minneapolis: University of Minnesota Press, 2017).

17. To be clear, these biomedical and moneyed investigations into the mediating space between brain and consciousness are not equivalent to the sophisticated and rigorous artic-

amounts of power and money are on the table in this age of algorithmic looting. This, of course, is not Chalmers's responsibility. But it must be pointed out that in these other spaces, whatever consciousness is, *it* is assumed to be a foundational feature of physical matter that can be explained—as well as manipulated—in light of information theory.[18] Consequently, Chalmers's learned critique offers only a soft check on the exaggerated and rapidly commodified claims of neuroscience and artificial intelligence.

Rather than question information theory or view it as an opportune (and perhaps unconscious[19]) theory of environment that carries with it its own politics, its own epistemic valence, its own technological utopianism, Chalmers reasons that because information "is everywhere," so, too, is the potential for consciousness. "Where there is simple information processing," he writes, "there is simple experience and where there is complex information processing, there is complex experience."[20] One should be so lucky to live in such an age of complex information processing! Which is to say that the game of figuring out what consciousness "is" according to an increasingly complex set of rules and technical possibilities is, more often than not, devoid of contextual consideration and the wherewithal to account for the historicity of the questions being asked.[21] For information and its theoriza-

---

ulations by Chalmers and others. But neither are they incompatible. For they both operate within an atmosphere in which the hard problems of consciousness have been identified as central to whatever it means to be human.

18. Peter H. Lindsay and Donald A. Norman, *Human Information Processing: An Introduction to Psychology*, 2nd ed. (New York: Academic Press, 1977), 195. See also Bradley Y. Bartholomew, "Solving the 'Hard Problem': Consciousness Is an Electronic Phenomenon," *Journal of Consciousness Exploration and Research* 11, no. 1 (2020): 46–60.

19. Lydia H. Liu, *The Freudian Robot: Digital Media and the Future of the Unconscious* (Chicago: University of Chicago Press, 2011).

20. Chalmers, "Facing Up to the Problem of Consciousness," 216–17. In summarizing the early cybernetic consensus, Dupuy sheds historical light on Chalmers's take on information as an "organizational property par excellence." For certain cyberneticians, writes Dupuy, "information was to be interpreted as a physical variable common to all natural systems rather than an exclusive feature of the world of human signal transmissions. If every organism is surrounded by information, this is simply because it is everywhere surrounded by organization, which itself, by the very fact of being differentiated, contains information. Information, because it is part of nature, is therefore independent of human interpreters who assign meaning to it" (*Mechanization of Mind*, 118).

21. Admittedly, given the atmosphere that is quite literally being theorized into existence as I write, to level such a critique against Chalmers risks analytical naiveté (not to

tions are, indeed, everywhere, whether as leverage for corporations or markets, simmering beneath all manner of screens and concealed within the pills that regulate one's neuro-anatomy.

Call it the hard problem of immanent critique.

Which brings me back to Chalmers's own riff on the articulation of the "hard problem" and "default attitude . . . that both biological and nonbiological systems can be conscious."[22] The ease with which Chalmers suggests the elimination of the ontic distinction between brain and machine is part of a larger epistemic shift, long in the making, that has privileged neither brain nor environment per se but the complexity of their co-organization.[23] Indeed, Chalmers's line of inquiry resembles those of early cyberneticians who insisted on the ontological argument that the brain was a computational machine.

Warren McCulloch, for example, who chaired the ten Macy Conferences on Cybernetics between 1946 and 1953, often expressed dissatisfaction with his fellow cyberneticians for being too metaphorical in how they addressed the perennial question of how consciousness might emerge from cerebral flesh. As McCulloch recalled on the eve of the tenth conference in 1953, "we . . . all came to realize that for problems of feedback, energy was the wrong thing to consider. The crucial variable was clearly information."[24] McCulloch had made the wager that, if he reduced the question of consciousness to formal logic, he could arrive at a theory of a consistent mathematical order embodied in the flora, the fauna, and us, specifically in each

---

mention disciplinary hubris). For it may be well-nigh impossible to consider our complicity in the giddy and ongoing creation of an informational environment while all the while being subject to it. On information theory as an "overarching discourse," see Ronald R. Kline, "Cybernetics, Management Science, and Technological Policy: The Emergence of 'Information Technology' as a Keyword, 1948–1985," *Technology and Culture* 47, no. 3 (July 2006): 516n13. See also Ronald E. Oay, *The Modern Invention of Information: Discourse, History and Power* (Carbondale: Southern Illinois University Press, 2008).

22. David J. Chalmers, "The Singularity: A Philosophical Analysis," *Journal of Consciousness Studies* 17, nos. 9–10 (2010): 45.

23. Pierre de Latil, *Thinking by Machine*, trans. Y. M. Golla (New York: Houghton Mifflin, 1957), 353. See also H. A. Abramson, ed., *Problems of Consciousness*, Transactions of the Second Conference on Problems of Consciousness, March 19–20, 1951 (Packanack Lake, NJ: Josiah Macy Jr. Foundation Publications, 1951).

24. Warren S. McCulloch, "Summary of the Points of Agreement Reached in the Previous Nine Conferences on Cybernetics," in *Cybernetics: The Macy Conferences, 1946–1953*, ed. Claus Pias (Zurich-Berlin: Diaphanes, 2016), 722. See also Jurgen Ruesch and Gregory Bateson, *Communication: The Social Matrix of Psychology* (New York: Norton, 1951).

of our heads.[25] Given that the human was always already a computational machine, instinctively expert at processing information, homologous with a thermostat or with weapon systems of command and control, McCulloch argued that consciousness could now be seen as wholly material (and vice versa), a move that served to dissolve the essential difference between the physiological and phenomenal registers.[26] Like every good soldier, declared McCulloch, "we must first be prepared for the kind of world we now invade." Or, as a contemporary of McCulloch noted in hindsight, "McCulloch was the person who, I think for the first time, provided a compelling demonstration that thought could be expressed just through the interaction of neurons alone."[27]

As evident by the kinds of discussions Chalmers currently finds himself in with AI engineers, neuroscientists, and data managers, the "hard problem of consciousness" (and my critique) has a history that passes through the

25. The brain, wrote McCulloch, "is a complicated computing machine consisting of 10^{10} relays. Each of these relays receives signals from other relays. Each on receipt of an appropriate signal—or group of signals—emits a signal. It is my business to learn how these signals are connected to one to another, what it takes to fire a given relay, how long after receipt of a signal it will send a signal, and how a signal received can prevent a relay from responding to a second signal otherwise sufficient." According to McCulloch, his capacity to identify connective patterns was in direct contrast to those benighted folks who misapprehend large systems because they clung to the "superstition" of causality. Their overdetermined concern with causes and mere measurement, according to McCulloch, could not account for the sheer magnitude of information built into the nature of living systems. Warren S. McCulloch, "Through the Den of the Metaphysician," *British Journal for the Philosophy of Science* 5, no. 17 (1954): 19, 24–25. Indeed, one might even go so far as to claim that information theory, on its face, pointed to a world elsewhere in which causes and effects could be considered mathematically equivalent. On McCulloch's "experimental epistemology," see Lily E. Kay's brilliant discussion in "From Logical Neurons to Poetic Embodiments of Mind: Warren S. McCulloch's Project in Neuroscience," *Science in Context* 14, no. 4 (2001): 591–614.

26. As the neurophysiologist Ralph Gerard warned at the 7th Macy Conference in 1950, those like McCulloch were too quick to move from the idea that the brain *may* be a computing machine to a pronouncement that it, in fact, was. The "overoptimism" of positing connections between the brain and computers was not only giving "spurious certainty to a credulous audience" but, just as phrenology had done in the nineteenth century, was offering elaborate theories of the human based on strong opinion and limited biological knowledge. Ralph Gerard, "Some of the Problems Concerning Digital Notions in the Central Nervous System," Pias, *Cybernetics*, 171.

27. Michael A. Arbib, "Warren McCulloch's Search for the Logic of the Nervous System," *Perspectives in Biology and Medicine* 43, no. 2 (Winter 2000): 213.

cultural commitment to information theory that began apace in the 1940s with the nomination of the cybernetic sciences. This history, however, is not exclusively about work carried out in places like MIT or Bell Labs. For such a history includes other audacious efforts to reduce the brain and its environment to their shared substance, variously construed over time—nervous fluid, animal magnetism, the spirit world, sympathetic energies, and, of course, the ether. There is, in other words, a history of consciousness as a hard problem. A history of that which courses through the brain and everything around it—"a mediate substance through which the mind acts upon and into the rest of the body, and through which mind is externally manifested."[28]

2. NEUROMATIC PIETY: AN OVERVIEW

A history of consciousness as a matter of information processing might be so bold as to invoke the scalable vision of Andrew Jackson Davis (1826–1910), who, in 1871, wrote that "man's phrenological structure is a wonderfully exact miniature of the starry universe." Davis, the person, perhaps, most responsible for the diffusion of metaphysical piety in the nineteenth century, distilled a version of correspondence that is formally consistent with later efforts to articulate the significance of the brain as the key to understanding creation.[29] The worldview embedded in Davis's account is one of strategic analogy. For once two distinct realms of the wide wide world could be represented by the same model—be it the brain or the starry universe—what was known about one could then be used to understand and measure the other. *What lies within the skull potentially in tune with everything outside it. As within, so without. As without, so within.* For in considering the "marvels of the nervous system," wrote Davis, "I find that *the organizing principle in matter is a constituent of Spirit*."[30]

When read in concert with a figure like Davis, one begins to sense the

28. Walter Felt Evans, *The Mental Cure* (Boston: Colby and Rich, 1869), 99.

29. Cristoph Adami and Thomas LaBar, "From Entropy to Information: Biased Typewriters and the Origin of Life," in *From Matter to Life: Information and Causality*, ed. Sara Imari Walker et al. (Cambridge: Cambridge University Press, 2017), 130–154.

30. Andrew Jackson Davis, *The Temple: Concerning Diseases of the Brain and Nerves; with Full Directions for Their Treatment and Cure* (Boston: Colby and Rich, 1871), 39, 40, 99–100.

Selective Chronology of Events in the Scientific Conceptualization of Information

1390 First recorded printed use of word *information*
1840s–1860s Helmholtz, investigations of the physiology and psychology of sight and sound
1843–1877 Bernard, work on homeostasis
1868 Maxwell, "On Governors"
1881 Ribot, *Diseases of Memory*
1890 James, *Principles of Psychology*
1891 Waldeyer, "neurone" theory
1894 Boltzmann, work on statistical physics
1906 Sherrington, *The Integrative Action of the Nervous System*
1915 Holt, *The Freudian Wish*
1919 Watson, *Psychology From the Standpoint of a Behaviorist*
1919–1923 McCulloch, work on logic of transitive verbs
1923 Lashley, "The Behaviorist Intepretation of Consciousness"
1924 Nyquist, "Certain Factors Affecting Telegraph Speed"
1925 Szilard, work on entropy and statistical mechanics
1928 Hartley, "Transmission of Information"
1932 Von Neumann, *Foundation of Quantum Mechanics*
Mid-1930s Development of recursive function theory (work of Gödel, Church, Kleene, Post, Turing)
Mid-1930s Feedback concept studied by electrical engineers (work of Black, Nyquist, Bode)
Mid-1930s Development of electroencephalography
1937 Turing, "On Computable Numbers"
1937 Turing and von Neumann, first discussions about computing and artificial intelligence at Princeton University
1938 Shannon, "Symbolic Analysis of Relay and Switching Circuits" (published 1938, M.A. Thesis 1940)
Late-1930s Harvard Medical School seminar led by Cannon
1940 Wiener, work at MIT on computers
1940 Shannon, "Communication in the Presence of Noise" (submitted; published 1948)
1942 Macy Foundation meeting on central inhibition in the nervous system
1942 Bigelow, Rosenblueth, Wiener, "Behavior, Purpose, Teleology"
1943 Turing, work with Shannon and Nyquist at Bell Laboratories in New York City
1943 McCulloch and Pitts, "Logical Calculus of the Ideas Immanent in Nervous Activity"
1943 Pitts accepts position at MIT to work with Wiener
1944 Princeton conference organized by Wiener and von Neumann on topics related to computers and control
1945 Von Neumann, draft report on EDVAC
1945 Shannon, "A Mathematical Theory of Cryptography" (Bell Laboratories report, published in 1949 under different title)
1945 McCulloch, "A Heterarchy of Values Determined by the Topology of Nerve Nets"
1946 Macy Foundation meetings on feedback organized by McCulloch
1946 Turing, NPL report on ACE computer
1946–1948 Burks, Goldstine, von Neumann, IAS computer reports
1947 Ashby, "The Nervous System as Physical Machine"
1947 McCulloch and Pitts, work on a prosthetic device to enable the blind to read by ear
1947 McCulloch and Pitts, "How We Know Universals"
1948 Wiener, *Cybernetics*
1948 Turing, "Intelligent Machinery" (NPL Report)
1948 Turing, first work programming the Manchester computer to carry out "purely mental activities"
1948 Von Neumann, "General and Logical Theory of Automata" (Hixon Symposium, Pasadena)
1948 Shannon, "The Mathematical Theory of Communication"
1949 Von Neumann, "Theory and Organization of Complicated Automata" (University of Illinois Lectures)
1949 Shannon and Weaver, *The Mathematical Theory of Communication*
1950 Shannon, "Programming a Computer to Play Chess"
1950 McCulloch, "Machines That Think and Want"
1950 McCulloch, "Brain and Behavior"
1950 Turing, "Computing Machinery and Intelligence"
1950 Ashby, "The Cerebral Mechanism of Intelligent Action"
1952 Turing, "The Chemical Basis of Morphogenesis"
1952 McCulloch, accepts position at MIT to be with Pitts and Wiener
1952 Ashby, *Design for a Brain*
1952 Von Neumann, "Probabilistic Logics and the Synthesis of Reliable Organisms from Unreliable Components" (California Institute of Technology Lectures)
1952–1953 Von Neumann, "The Theory of Automata: Construction, Reproduction, Homogeneity"
1953 Walter, *The Living Brain*
1953 Turing, "Digital Computers Applied to Games: Chess"
1953 Turing, "Some Calculations on the Riemann Zeta Function"
1956 Von Neumann, *The Computer and the Brain* (prepared for the Silliman Lectures, Yale; published 1958)
1956 Ashby, *Introduction to Cybernetics*
1956 Ashby, "Design for an Intelligence Amplifier"

FIGURE 15. Selective chronology of events in the scientific conceptualization of information, from William Aspray, "The Scientific Conceptualization of Information: A Survey," *Annals of the History of Computing* 7, no. 2 (April 1985): 117–40.

cosmic stakes involved in Chalmers's cybernetic wager—namely, that consciousness is, at base, an effect of information processing (a commonsense faith now affirmed with every swipe of your smartphone screen). Indeed, one begins to appreciate that the present plausibility and structural success of "information" owes much to epistemic grooves laid down by earlier attempts to enact correspondence with a Universal Mind. There is, in other words, a stranger history of this particular present, one that moves beyond certain joys of Boolean algebra and dutiful lists of scientific discovery.[31]

31. I am inspired here by John Durham Peters, "Information: Notes Toward a Critical History," *Journal of Communication Inquiry* 12 (1988): 9–23. Peters situated the concept of "information" within a historical trajectory that runs through medieval scholasticism,

As a counter to such matter-of-fact recitation, I seek to historicize a style of reasoning that has long sought to establish a communicative link between brain and environment. Rather than adjudicate the use of information theory within the neuro- and cognitive sciences since the mid-twentieth century or dutifully trace the scientific accretions underneath Shannon's "discovery," mine is a history of the metaphysics animating those who would frame the brain as a model of continuous communication within and without. In highlighting the compatibility between cognitive scientists, engineers, neurophysiologists, and those working within self-consciously occult settings I demonstrate how the dream of information, or something like it, was born in attempts to locate the brain at the center of a universe that mimicked its mechanics. Consequently, mine is a history of the *long desire* to experience a correspondence, by way of the brain, with one's self, with other humans, with machines, and with the invisible universe.

A recurring figure in this chapter (and in the margins of the tradition it limns) is Emanuel Swedenborg (1688–1772) and the neurological writings that he produced before his midlife turn to mysticism. After Swedenborg began publishing theology in 1744, his works on brain anatomy lay dormant in the Swedish Royal Academy of Sciences until their "rediscovery" in 1868. The first volume of *The Brain* was published in 1882, adding fuel to a metaphysical revival in the late nineteenth century already much taken with Swedenborg's corpus. The editor and translator wrote that Swedenborg's theory of brain tremulations was "nothing less than miraculous" in that "he anticipated nearly all the important discoveries made since his time in the science of the brain and of the nervous system."[32]

British empiricism, bureaucratization, war, and technical efforts to make visible various domains of invisibility.

32. Swedenborg's works on brain anatomy were first published in London in two volumes—*The Cerebrum and Its Parts* (1882), vii, and *The Brain, Considered Anatomically, Physiologically and Philosophically* (1887) under the direction of R. L. Tafel. These works, having arrived alongside the discovery of the motor cortex and the consolidation of the electrical paradigm of the brain, were often cited as prophetic in their insight. In 1935, for example, the founding editor of *Isis*, George Sarton (so called father of the history of science), issued a glowing review of Swedenborg's scientific bona fides. George Sarton, "Review of Reviewed Works: *Opera quaedam aut inedita aut obsoleta de rebus naturalibus* by Emanuel Swedenborg; *Itineraria* by Emanuelis Swedenborgii; *An Abridged Chronological List of the Works of Emanuel Swedenborg* by Alfred H. Stroh, Greta Ekelöf," *Isis* 23, no. 2 (September 1935): 459–63.

According to Swedenborg, tremulations did not operate according to the laws of cause and effect. They were "not subjected to the laws that govern local motion" but operated beyond established categories of time and space. Tremulations were universal. All organic life consisted of tremulations of some kind—"the circulation of the blood, the nervous system, and the senses" but also bullets and bomb trajectories and all "activity of the human mind and spirit."[33] Tremulations also challenged a strictly mechanistic worldview, embodying a nonlinear model of cause and effect yet one still legible to a Christian theological tradition. Consequently, Swedenborg's was a mystic-tinged and strangely consequential approach to the brain as a self-organizing mechanism in which tremulations, themselves, became a kind of divinity. Consequently, it is Swedenborg's brand of universalism, more so than any specific insight into brain anatomy, that is illuminative of the current state of informational relations between brains, environments, and machines.

This chapter moves through Swedenborg's eighteenth-century scene of writing, the reception of his works on the brain in the nineteenth and early twentieth centuries, and moments in the emergence of the neuromatic brain—a brain defined by its informational processing capacities, the electrical effects of such processing, and the interactions between neural networks and their environments.[34] Although woefully incomplete, mine is the story of a shared cosmology that cuts across the religious/scientific binary. Throughout this chapter I attend to a tradition of belief and practice that revolves around the brain *in relation*. I attend to its rituals and its rhetoric, its sermons and catechisms uttered in sincere and pious tones. My story, then, is about yet another formation of secularism. Which is to say that the neuromatic pedigree of our present is neither a triumph of secularity nor a smuggling of retrograde religiosity into the citadels of science. It is, on the contrary, the culmination of an intense and often aggressive style

33. Emanuel Swedenborg, *On Tremulation*, trans C. Th. Odhner (Bryn Athyn, PA: Swedenborg Association, 1976), 9, 26, 16, 10; David Dunér, *The Natural Philosophy of Emanuel Swedenborg: A Study in the Conceptual Metaphors of the Mechanistic World-View* (New York: Springer, 2013), 180.

34. See, e.g., McCulloch's "The Reliability of Biological Systems" and the other presentations on the effects of environmental feedback in the proceedings of a conference cosponsored by the Information Systems Branch of the Office of Naval Research in Scott Cameron and Marshall Clinton Yovits, eds., *Self-organizing Systems: Proceedings* (New York: Pergamon Press, 1960).

of reasoning about the microdetails of cognitive processing, a style that is dream-laden and dense with desire to invest the brain with all manner of macrocosmic potential.

The chapter is bookended by discussions of the past and future of information. I begin by revisiting the concept of the ether in the nineteenth century and conclude with a discussion of a theoretical physicist's breakdown of the informational surround in terms of perceptronium. Both the ether and perceptronium are matters that mediate the relationship between the biological activity of the brain and the experience of this activity in the form of consciousness. The conceit of this comparative frame is to demonstrate how the ether morphed into ever more abstract refinements such as information, and later perceptronium, over the course of the twentieth century. I argue that however one may name it—ether, electricity, information, or perceptronium—the mediating substance between brain and environment has become subject to an increasingly technological view. Consequently, this mediating substance has become *more* mediating in how it imbricates individuals in the world around and *more* substantive in the demands it places on them.

In between these bookends, I move through the emergence of the brain wave in the nineteenth century; the phrenological public sphere and the racialized and gendered demands of brain plasticity; the localization of the motor cortex; the increasing mechanization of neuroscience in the twentieth century; and the mystical/therapeutic desire to unpack, with instrumental precision, the relationship between neural activity and the concrete consequences of that activity.[35] For there is more than an echo of Swedenborg in the words of Dr. Barbara Brown, biofeedback pioneer who participated in conferences sponsored by both the Department of Defense and the Society for Psychical Research, when she writes in 1974, "I am still filled with awe and wonder when I record human brain waves in my laboratory. Watching brain waves gives me the feeling of suddenly having discovered where the halos come from."[36]

35. Timothy Lenoir, "Models and Instruments in the Development of Electrophysiology, 1845–1912," *Historical Studies in the Physical and Biological Sciences* 17, no. 1 (1986): 1–54.

36. Barbara B. Brown, *New Mind, New Body: Bio-feedback: New Directions for the Mind* (New York: Harper and Row, 1974), 310.

3. ETHER AND THE PERMEATION OF THE INTERSPACES

Before the informational link between brain and environment was officially nominated, there was the "universal, impalpable, elastic 'Ether.'" This "material medium of surpassing and inconceivable tenuity" was quite literally everywhere in the nineteenth century, "diffused throughout all known space, and permeating the interspaces of all bodies, solid, fluid, or gaseous."[37] Drawing on reiterations of Newton's "notion of a universal ether"[38] as they were inflected by mesmeric sentiment and various philosophies of electrical psychology,[39] the ether was a solving term marked by the language of flow and reversal. It could, at once, explain the immaterial mechanics of physical processes *and* the physical effects of immaterial phenomena such as consciousness or the will.[40] A common placeholder for marking the indeterminate space within one's empirical argument, ether was deployed by Michael Faraday and James Clerk Maxwell to account for lines of electromagnetic force. According to Helena Blavatsky, ether could be useful in physics as

37. James Knowles, "Brain-Waves—A Theory," *Spectator* 30 (January 1869): 136.

38. Sir Isaac Newton, *Opticks: or, A Treatise of the Reflections, Refractions, Inflections and Colours of Light*, 4th ed. (London: William Innys, 1730), 328.

39. See, e.g., John Bovee Dods, *The Philosophy of Electrical Psychology* (New York: Fowlers and Wells, 1850), 33, 51, 148. The blur of mesmeric discourses over the course of the nineteenth century contributed to the plausibility of the ether paradigm and the self-conscious professionalization of neuroscience. Justine Murison, "'The Paradise of Non-Experts': The Neuroscientific Turn of the 1840s United States," in *The Neuroscientific Turn: Transdisciplinarity in the Age of the Brain*, ed. Melissa M. Littlefield and Jenell M. Johnson (Ann Arbor: University of Michigan Press, 2012), 30; Emily Ogden, *Credulity: A History of US Mesmerism* (Chicago: University of Chicago Press, 2018); Alison Winter, *Mesmerized: Powers of Mind in Victorian Britain* (Chicago: University of Chicago Press, 1998), 18, 276. On the origins of American neurology and its self-consciously secular obsession with spiritualism, see Molly McGarry, *Ghosts of Futures Past: Spiritualism and the Cultural Politics of Nineteenth-century America* (Berkeley: University of California Press, 2012), 141ff.

40. J. Stanley Grimes, *Etherology and the Phreno-philosophy of Magic Eloquence*, revised by W. G. Le Duc (Boston: James Munroe, 1850). See also Joe Milutis, *Ether: The Nothing That Connects Everything* (Minneapolis: University of Minnesota Press, 2006), 8, and G. N. Cantor and M. J. S. Hodge, *Conceptions of Ether: Studies in the History of Ether Theories, 1740–1900* (Cambridge: Cambridge University Press, 1981) as well as the essays collected in *Vibratory Modernism*, ed. Anthony Enns and Shelly Trower (New York: Palgrave Macmillan, 2013).

well as in explaining the origins of life and consciousness. For Blavatsky, ether was an ancient concept, recognized by humans "long before the time of Moses." It was the womb from which life was born and to which it would return.[41]

The ether—with its evanescent ontology[42]—was a rich resource for nineteenth-century metaphysicals on both sides of the Atlantic—for explaining how the human fit into the cosmos and for addressing the interplay between "vibrations" of the brain and its environments.[43] As Andrew Jackson Davis would write in 1871, summarizing his career as a graphomaniacal trance lecturer, the ether was what connected the living and the spirit world and the source of human sociality across time and space. "Brain strikes brain, however far apart, through vibration of the *omniprevalent*

41. H. P. Blavatsky, *Isis Unveiled: A Master-Key to the Mysteries of Ancient and Modern Science and Theology*, vol. 1 (New York: J. W. Bouton, 1877), 134.

42. In her exploration of the ether paradigm, Courtenay Grean Raia writes that "ether was at once an ontological mystery and an operational fact" whose "conceptual plasticity" made it amenable to all manner of explanatory purposes among spiritualists, mesmerists, psychical investigators, physicists, neurophysiologists, and psychologists ("From Ether Theory to Ether Theology: Oliver Lodge and the Physics of Immortality," *Journal of the History of the Behavioral Sciences* 43 [2007]: 21). Unlike most historical treatments of the ether, Raia's does much more than point to the curiosity of the entanglement of conversations about the ether with those scientific practices deemed integral (in hindsight) to the making of modernity. Often told in the key of secularization, these stories frame the entanglement of cosmology and empiricism (ever in the past) as a sign of progressive purification rather than deepening and diffusion. For what was the shared substratum of conviction that made such exchange and mutual influence possible? To appreciate how these differently tuned actors inhabited the same discursive atmosphere of secularism is to see differences between them as matters of discursive inflection. Consequently, I am sympathetic to Raia's argument about the existence of "one internally consistent cosmology situated not 'between science and religion' or on the sidelines of Victorian thought, but at the heart of a new cultural synthesis striving to thoroughly modernize the boundaries of belief" (22). One must not, in other words, see the curious claims of the ether paradigm as inevitably working themselves out in the court of reason—through a process of secularization—in which occult claims become institutionally separated and epistemologically quarantined from the scientific. For a fuller rendition of Raia's argument, see *The New Prometheans: Faith, Science, and the Supernatural Mind in the Victorian Fin de Siècle* (Chicago: University of Chicago Press, 2019).

43. G. S. Weaver, *Lectures on Mental Science According to the Philosophy of Phrenology* (New York: Fowlers and Wells, 1852), 33. See also the description of "the fibres of the human brain vibrating to the waves of atmosphere," in J. P. Quincy, "A Crucial Experiment," *Atlantic Monthly* 59, no. 356 (June 1887): 725.

etherium, by which human mind is bonded to human mind throughout the wide universe."[44] Ether served for Davis, much as information does today, to guarantee measurable communication across seemingly disparate registers.

Davis, aka the "Seer of Poughkeepsie," was arguably the most influential metaphysical figure in the nineteenth century. Having provided the cosmology that informed both phrenology and séance spiritualism, Davis was keen on offering a scientific explanation of the ether based on principles of nature. "Organic electricity," nervous energy, and what he called "spiritual exhalation" were but different ways to signify the same immanent yet kinetic frame. Ether was an atomic proposition yet utterly refined.[45] Within American orbits of séance spiritualism and phrenology (in which Davis figured prominently), ether mediated the relationship between any single brain and the world at large. In England, at this same moment, an occult concept of the ether was generating similar insight into "a kind of brain atmosphere which extends over the globe, and upon which the brain has the power of impressing undulations, just as a bell sets the air in motion, or an electric battery the electric fluid in a telegraph wire."[46]

The notion of the brain wave—well before it became formalized as the measurable electrical mark of neurons firing in unified patterns—emerged from within the ether paradigm. In January of 1869 *The Spectator* published "Brain-Waves—A Theory" by James Knowles. Knowles was an architect and founder of the Metaphysical Society (1869–80), a gathering of Protestant clergymen, natural scientists, poets, and assorted others dedicated to bridging the difference between faith and reason. For Knowles, brain waves were an opportune topic for such speculation. As Knowles reasoned, because "the movement of any solid particle submerged in any such medium must create a wave," "no brain action" could "take place without creating a wave or undulation (whether electric or otherwise) in the ether." Such a wave,

44. Davis, *The Temple*, 214.

45. Andrew Jackson Davis, *The Philosophy of Spiritual Intercourse; Being an Explanation of Modern Mysteries* (New York: Fowlers and Wells, 1851), 163, 25–26. "Ether-atoms," wrote Davis, "are atoms in the highest possible degree of motion, constituting an infinitely rare medium, chemical, dynamic, elastic, and all-pervading." According to Davis, in its subtle materiality, the ether retained a "geometric" density, full of angles, refractions and reflections (*A Stellar Key to the Summerland*, pt. 1 (Boston: William White, 1867), 80, 161; *The Temple*, 197).

46. "Brain Waves," *American Phrenological Journal* 49, no. 9 (September 1869), 195.

moreover, was a "vague subtle force" whose origins were the "atomic movement" within each and every brain.[47]

Knowles pitched his theory as an explanation of what, exactly, constituted "authenticated ghost stories." As an intervention into Victorian debates about spiritualism, Knowles argued ghosts were nothing more than brain waves of deceased individuals, tremulating in space, waiting to encounter the brains of the living whom Knowles described as "sensitive mechanical receiver[s]." The experience of haunting, according to Knowles, was due to the influence of brain waves that continued to circulate long after their originary moment. Pitched, in part, as a rebuttal to spiritualist speculation about otherworldly personalities, Knowles offered a more reasonable account of what Davis and others in the American grain referred to as "spiritual intercourse." Knowles's self-conscious scientism was, in part, a response to the mystical excesses catalyzed by, among other things, the steady stream of metaphysical publications emanating from America in the mid-nineteenth century.

As was common among his Victorian contemporaries, Knowles associated the excess of transcendental reasoning, particularly in the American grain, with marks of "heathen mysticism."[48] In keeping with the desire to contain the excesses of an "oriental spirit" from gathering force in the United States, Knowles, from the very beginning, was interested in leveraging brain waves for disciplinary projects, at both home and abroad. He imagined a moment when brain waves of individuals could be used to surveil, from afar, the emotional lives of a population—"the thoughts of love or hate, of life or death, of murder or rescue, of consent or refusal." Knowles's was a biopolitical wager that brain waves would usher in a new era of knowledge and security—not just for the selves who might take control of the commerce between their own brains and the surrounding environment but also

47. Knowles, "Brain-Waves," 135–37.

48. H. A. Page, "The Old Morality and the New," *Contemporary Review* 8 (1868): 597. Knowles was the editor of the *Contemporary Review*, a journal that looked askance at "spiritual developments in America." Consequently, the writings of Ralph Waldo Emerson and Walt Whitman as well as the popularity of Emanuel Swedenborg and William Blake were evidence of a folly in which a "mystical, antinomian, [and] properly oriental spirit" (598) had led to an utter evacuation of the difference "between the divine and human elements" at work in the world (613). The problem was at once theological, epistemological and racial. For one loses a grip on oneself and on one's mind when "the external world" becomes "a plastic medium of the soul's indulgence" (618).

for the colonial state invested in defining the brains of racialized others as intimately manageable.[49]

4. EMANUEL SWEDENBORG, NEUROSCIENTIST

In 1868, a year before Knowles's article on brain waves appeared, a cache of manuscripts on brain anatomy was discovered in the Swedish Royal Academy of Sciences. They were authored by none other than Emanuel Swedenborg, one of the looming figures of the metaphysical revival in America that was catching the eye of Knowles and his journal. Swedenborg's works on brain anatomy would eventually be published in London in 1882 and 1887. By highlighting Swedenborg's articulation of a nerve doctrine, these volumes framed his treatises as profound commentaries on what constituted the spirit of consciousness and how it operated.[50]

In his translator's introduction to the first volume of *The Cerebrum*, R. L. Tafel noted that it had been written before Swedenborg's mystical turn in the 1740s. Still, Swedenborg's theory of brain tremulations was "miraculous" in that "he anticipated nearly all the important discoveries made since his time in the science of the brain and of the nervous system." Amid renewed interest in brain localization and the electrical excitability of ner-

49. Knowles, "Brain-Waves," 136. Knowles's conceptual suturing—of race with true religion with reason—was exemplary in its desire to control a large number of people through exceedingly precise categories and imperatives. As media historian Roger Luckhurst notes, Knowles's theory of the brain wave was, in part, addressing an ongoing crisis of colonial rule (*The Invention of Telepathy, 1870–1901* [New York: Oxford University Press, 2002], 153–55). Indeed, the very idea of the brain wave occurred to Knowles when he heard a strange story about telepathy on the colonial periphery as related by the sculptor Thomas Woolner. As the editors of the *Spectator* summarized it, brain waves "anticipate the channels of ordinary communication. Thus, in the case authenticated by Mr. Woolner, of the young man who, after years of uninterrupted silence between him and his friend, and while separated by the whole diameter of the earth, became aware of some terrible calamity affecting the latter, at the very moment when, allowing for a difference of longitude, he was being tortured to death by the Maoris in New Zealand" ("The Hypothesis of Brain-Waves," *Spectator* 30 [January 1869], 133–34).

50. Emanuel Swedenborg, *The Brain, Considered Anatomically, Physiologically and Philosophically*, vols. 1 and 2, ed. and trans. R. L. Tafel (London: Speirs, 1882, 1887). See also George Rousseau, *Nervous Acts: Essays on Literature, Culture and Sensibility* (New York: Palgrave Macmillan, 2004), 224–25.

vous tissues, Tafel positioned Swedenborg's ideas about brain tremulations as prophecy. According to Tafel, Swedenborg's approach to the cerebellar "wave"—understood as the back-and-forth flow of "cerebro-spinal liquid"—confirmed "modern" scientific discoveries. Others, too, promoted Swedenborg's works of neurophysiology as lost gospels whose time had arrived.[51] An 1883 review in *Brain*, for example, remarked that Swedenborg's was "one of the most remarkable books we have seen" in that he "appears to have anticipated some of the most modern discoveries" by over a century.[52]

For Swedenborg, neural activity occurred by way of "tremulations" and the oscillation of indivisible particles from the brain to the peripheral nerves and back again; he framed the brain's self-organizing capacities in terms of a "nervous network." Swedenborg wrote of "little-brains" stitched together by nerve fibers and communicating with one another through tremulations in the complex folds on the surface of the cortex. Tremulations were precisely what connected the different parts of the brain to each other and to the environmental horizon. A tremulation spread out in all directions, "communicating its motion to all other things in the whole body." Some waves built on one another, resonating and amplifying. Other waves intersected, dampening and canceling each other.[53]

51. "Editor's Preface" to Swedenborg, *The Brain*, 1:vii; Swedenborg, *The Brain*, 2:551, 579, 611ff.; O. B. Frothingham, "Swedenborg," *North American Review* 134, no. 307 (1882): 600–616. See also *New Church Magazine* 1 (July 1882): 324f.

52. Andrea Rabagliati, "Reviews and Notices of Books," *Brain* 6, no. 3 (October 1883): 404–13. See also his "Reviews and Notices of Books," *Brain* 10, no. 4 (January 1888): 512–24. Contemporary histories of neuroscience tend to ignore Swedenborg. When they do mention his neurophysiology, it is as a curiosity. Swedenborg, it seems, may have anticipated the discovery of the cerebral cortex but his obscure mystical leanings disqualify him as an ancestral resource. See, e.g., Stanley Finger, *Origins of Neuroscience: A History of Explorations into Brain Function* (New York: Oxford University Press, 1994), 30; Konrad Akert and Michael P. Hammond, "Emanuel Swedenborg (1688–1772) and His Contributions to Neurology," *Medical History* 6, no. 3 (1962), 255–57. On an effort to recover Swedenborg, see Charles G. Gross, "Emanuel Swedenborg: A Neuroscientist before His Time," *Neuroscientist* 3, no. 2 (1997): 142–47.

53. Swedenborg, *The Brain*, 2:214, 442; Emanuel Swedenborg, *The Animal Kingdom, Considered Anatomically, Physiologically and Philosophically*, trans. James John Garth Wilkinson (Otis Clapp: Boston, 1843), 137; Emanuel Swedenborg, *The Soul, or Rational Psychology*, ed. and trans. Frank Sewall (New York: New Church, 1887), 100; Swedenborg, *On Tremulation*, 4, 7. Swedenborg's notion of tremulations constituted his own spin on Willis's emphasis on surface folds of the brain as a material index of consciousness and theories of corpuscularism that were beginning to take hold.

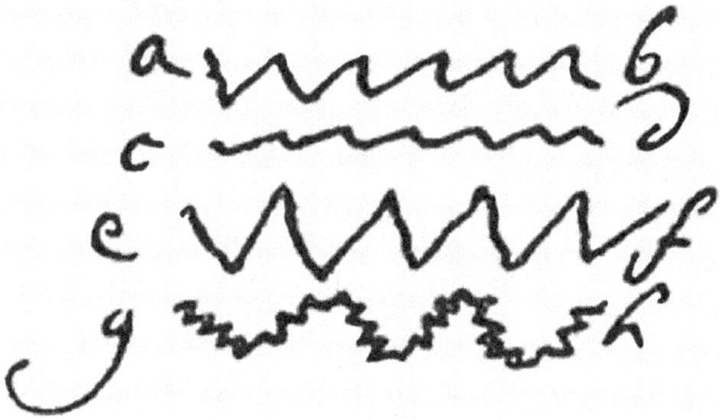

FIGURE 16. Different speeds of tremulations, from Emanuel Swedenborg, *Psychologica* (1733–34). Reprinted in David Dunér, *The Natural Philosophy of Emanuel Swedenborg: A Study in the Conceptual Metaphors of the Mechanistic World-View* (New York: Springer, 2013), 206.

Consciousness, for Swedenborg, was a self-organizing phenomenon, driven by perpetual mediations between brain and environment. An on-and-off-and-on-again jag that generated the "innermost life" of the body. For Swedenborg, the "communication of tremulation" was the means of self-organization and put a check on what he called the "destruction of equilibrium." The perpetual activity of tremulations guaranteed the stability of the nervous system *as a system*. "It may be seen from this," wrote Swedenborg, "that the whole nature of tremulation consists in the effort of a thing to recover the balance which it was about to lose." Tremulations were the ethereal means by which the brain retained its own balance, both within and without . . . "set[ting] the whole system into sympathetic tremulation."[54]

Swedenborg's general theory of correspondence was of a piece with his insistence that the anatomical effects of the brain manifested a logic of their own. For in addressing topics with "too sublime a nature to be explained to the common understanding," Swedenborg promulgated what might be called a mystical *and* mathematical science of the brain. According to Swedenborg, to see the brain in terms of correspondence was to see material surfaces in terms of their geometrical relationality, which, in turn, consti-

54. Jonathan Sheehan and Dror Wahrman, *Invisible Hands: Self-Organization and the Eighteenth Century* (Chicago: University of Chicago Press, 2015), 185; Swedenborg, *On Tremulation*, 45, 25, 11; Dunér, *Natural Philosophy*, 181, 205; Swedenborg, *Animal Kingdom*, 1:107, 275.

tuted their heavenly depth. "Let us therefore take a broad view of the things below, and let us elaborate doctrines by the aid of which we may be enabled to take a universal view of the individual things which are around and below. And thence let us raise the sight of our mind towards the higher things which will then be nearer, and let us regard with veneration the heavenly things which will then meet us." In other words, true explanations of neural matters at hand were always more than met the eye.[55] The spiritual economy of cerebral flesh was not merely mechanistic but required a deeper knowledge. Consequently, Swedenborg had to invent, as he wrote, "new doctrines for my guidance, which are doctrines of forms, of orders and degrees, of series and society, of communication and influx, of correspondence and representation, and of modification."[56]

Swedenborg's musings on the brain and its tremulations must be read in light of his general understanding of how movements within the physical world corresponded intimately with unseen, universal laws. For long before his midcareer drift to the genre of cosmology;[57] before initiating his explicit program of Christian reform; before the memorable shade thrown at him by Immanuel Kant for his increasingly speculative writings,[58] Swedenborg was an engineer, natural philosopher, and military planner. And as with many who would come in his wake, Swedenborg's metaphysics of the brain was forged in and through the physics of war.

55. Within a world defined by "our Doctrine of Representations and Correspondences," wrote Swedenborg, "the physical world was purely symbolical of the spiritual world" (*Animal Kingdom*, 1:451); Swedenborg, *The Brain*, 1:7–8; Dunér, *Natural Philosophy*, 130.

56. Swedenborg, *The Brain*, 1:9. In grounding his approach to neurophysiology in the logic of correspondence, Swedenborg pointed to a new age in which an understanding of the integrity of the brain would ground the understanding of everything else, including divinity. Here was the internal logic of what Sheehan and Wahrman have called the "positive providentialism" or "providential materialism" underlying Swedenborg's pursuit of universal order that emerged at the microscopic level (*Invisible Hands*, 32, 36).

57. In the early 1740s Swedenborg entered his "spiritual phase." See *De Cultu et Amore Dei* (London: Apud Kegan, Paul, Trench, et Soc., 1745), trans A. H. Stroh and F. Sewall as *The Worship and the Love of God* (Boston: Massachusetts New-Church Union, 1914).

58. Immanuel Kant, *Träume eines Geistersehers erläutert durch Träume der Metaphysik* (1766). Translated by E. F. Goerwitz, and edited with an introduction and notes by Frank Sewall as *Dreams of a Spirit-Seer, illustrated by Dreams of Metaphysics* (London: Swan, Sonnenschein, 1900).

FIGURE 17. Portrait of Christopher Polhem (1661–1751), from *Svenska Industriens Män* (Stockholm: Kjellberg och Åströms, 1874).

In 1710, with Sweden over a decade into the Great Northern War with Norway-Denmark (1700–1721), a twenty-two-year-old Swedenborg began his apprenticeship with Christopher Polhem by writing advertising copy for Polhem's commercial ventures. At the time Polhem was also Sweden's most recognized engineer and influential military consultant.[59] In the early years of the Great Northern War, Polhem had assumed a leadership role in defending the Swedish coast. Chosen for his ability to optimize con-

59. Polhem was the director of the Laboratorium mechanicum in Stockholm. There he trained engineers and designed an automated factory whose production of household goods was powered by water. As a general matter the Laboratorium mechanicum was invested in displacing the human body with mechanical levers, for such levers could be harnessed to elemental powers such as water—the advantage of water-driven machines being, according to Polhem, that they needed neither payment nor hay. Svanta Lindqvist, "Labs in the Woods: The Quantification of Technology during the Late Enlightenment," in

temporary warfare through engineering, Polhem charted the trajectories of cannonballs, built a suspension bridge in order to aid the movement of Swedish troops, and designed locks to forge channels of communication, transportation, and distribution. Polhem also reformed foundry techniques and built machines for cutting out cog wheels and for hoisting iron ore from the earth—all in the service of consolidating state power.[60]

In 1713, Swedenborg purchased an advanced calculus textbook—Charles René Reyneau's *Usage de l'analyse* (1708). With military strategy and statecraft on his mind, Swedenborg made notes about bomb trajectories, parabolas, cycloids, and logarithmic curves and began to imagine a virtual world that was as real, if not more so, than the phenomenal one—precisely because both corresponded with each other at the level of mathematical formalization.[61]

Polhem's pedagogy may be gleaned from his engineering agenda, driven as it was by the desire to account for universal geometric relationships found in nature—from church bells and sound waves to the proportions of organ pipes. Polhem was confident in his answer because he had concluded, in the words of Swedenborg, that everything was mechanical all the way down, up, and out—"a great wheel governing a thousand minor wheels by which it effects the motion of the whole machinery." Consequently, everything—including the physiology of the human form—could potentially be put in relationship with everything else through mathematical reasoning. Under Polhem's tutelage, Swedenborg assumed his mentor's enthusiastic insistence

---

The Quantifying Spirit in the 18th Century, ed. Tore Frängsmyr et al. (Berkeley: University of California Press, 1990), 300; Samuel E. Bring, "A Contribution to the Biography of Christopher Polhem," in *Christopher Polhem: The Father of Swedish Technology*, trans. William A. Johnson (Hartford, CT: Trinity College, 1963), 37; David Dunér, "Daedalus of the North: Swedenborg's Mentor Christopher Polhem," *New Philosophy* 113, no. 3–4 (2010), 1084–85.

60. Polhem, in other words, was a man about communication and control. Bring, "Contribution," 52; Nils Cronstedt, "Polhem's Contribution to the Art of Building," 205, and Gustaf Sellergren, "Polhem's Contributions to Applied Mechanics," 110–11, both in *Christopher Polhem: The Father of Swedish Technology.*

61. Dunér, "Daedalus of the North," 1088–89, and *Natural Philosophy*, 58. While working with Polhem, Swedenborg founded the scientific periodical entitled *Daedalus Hyperboreus* (The Northern Daedalus) (1716–18). Committed to the position that natural philosophy could and should benefit military planning, artillery, and ballistics, Swedenborg conceived of his journal as an official record of mechanical invention and mathematical discovery. Plans hatched within the pages of *Daedalus Hyperboreus* included the potential construction of flying machines, submarines, mechanical guns, and a "thundertube," designed by Polhem, that would funnel the boom of a cannon to the opposing army so that they would then retreat in fear.

that the mechanics of man and the mechanics of nature were but different reflections of the same cosmic unity. He also adopted Polhem's position that clocks and brickmaking machines, as well as "every living creature, including man," were but "an infinite number of charged particles or atoms."[62]

Upon Polhem's recommendation to King Charles XII, Swedenborg became Extraordinary Assessor in the Royal Board of Mines in December of 1716 and was enlisted to work on military projects. In 1718 Swedenborg supervised the successful transportation of Swedish Royal Navy ships—"two galleys, five large boats, and a sloop" on rolling machines across fourteen miles of rough terrain in order to reinforce the Swedish siege on the Norwegian city of Halden.[63] During the siege of Fredriksten, King Charles visited the front lines to inspect troops and the front line that Swedenborg had been so integral to reinforcing. During his inspection King Charles was shot in the head and killed instantly.[64]

Demoralized, the Swedish army decamped, and over three thousand Swedish soldiers froze to death during the retreat.[65] Amid the gloom of this national tragedy, Swedenborg's relationship with Polhem soured for personal and professional reasons. In 1719 Swedenborg's arranged marriage to Polhem's daughter fell through and Swedenborg turned away from military engineering.[66] Left with neither king nor mentor, nor wife, Swedenborg turned his attention to the brain.

Over the next two decades Swedenborg honed in on mysterious mechanics that defined the brain and its environment, committed as he was to providing a physics of the soul's metaphysics. Despite his falling-out with Polhem, Swedenborg's turn to brain anatomy remained indebted to

62. Swedenborg, *On Tremulation*, 16; Polhem's "On Living Spirits," quoted in Bring, "Contribution," 59.

63. Dunér, "Daedalus of the North," 1079; Jane K. Williams-Hogan, "Swedenborg: A Biography," in *Swedenborg and His Influence*, ed. Erland J. Brock et al. (Bryn Athyn, PA: Academy of the New Church, 1988), 9; "Character, Organization, and Biography of Emanuel Swedenborg," *American Phrenological Journal* 11 (1849), 107.

64. Controversy still simmers over whether Charles was shot by enemy troops or by one of his own men. His body was exhumed in 1917 and ballistics and DNA testing remain inconclusive. Mike Dash, "The Blazing Career and Mysterious Death of 'The Swedish Meteor,'" *Smithsonian Magazine* (September 17, 2012), https://www.smithsonianmag.com/history/the-blazing-career-and-mysterious-death-of-the-swedish-meteor-39695356/.

65. Davis Gudmundsson, "The Consolation of Soldiers: Religious Life in the Swedish Army during the Great Northern War," *Scandinavian Journal of History* 39, no. 2 (2014), 215.

66. Dunér, "Daedalus of the North," 1097.

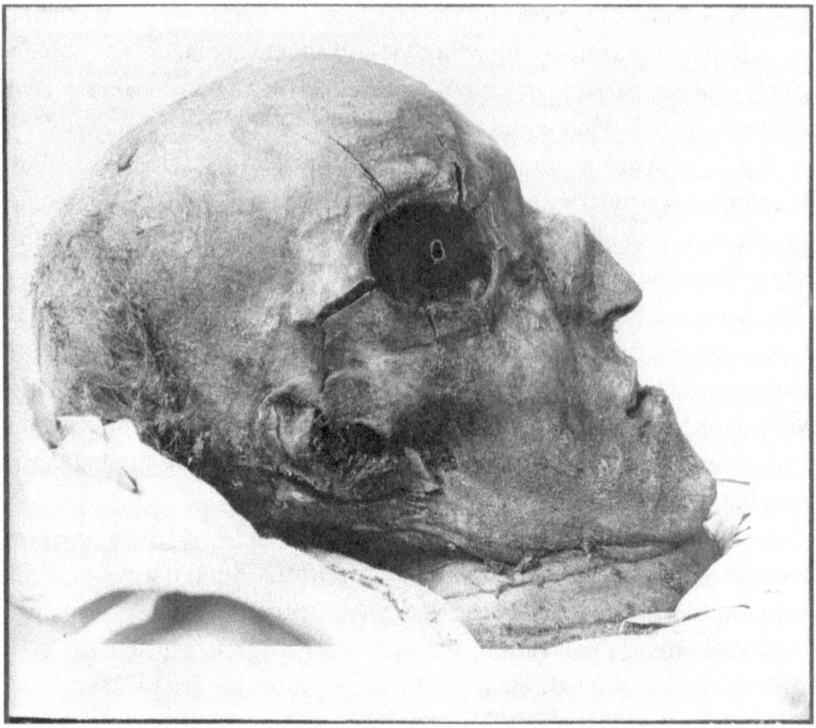

FIGURE 18. Right side of the skull of the Swedish King Charles XII, at the opening of his grave in Riddarholmen Church in Stockholm in 1917. Courtesy of the Swedish National Heritage Board.

his mentor's insights.[67] Still driven by Polhem's insistence that "thought and its external manifestations possess a real existence"[68] and that motion in the world could affect the functioning of the brain, Swedenborg dreamed of a universal science based on mathematics—a *philosophia mathematica universalium* (mathematical philosophy of universals)—that had heretofore been "unknown to the world."[69]

In his treatise *On Tremulation*, submitted to the Sundhets Collegium, or

67. Torsten E. Gordh et al., "Swedenborg, Linnaeus and Brain Research—and the Roles of Gustaf Retzius and Alfred Stroh in the Rediscovery of Swedenborg's Manuscripts," *Upsala Journal of Medical Science* 112, no. 2 (2007): 148.

68. Polhem's "Thoughts about Spiritual Beings" (ca. 1710) cited in Bring, "Contribution," 61.

69. Emanuel Swedenborg, *Three Transactions on the Cerebrum*, trans. Alfred Acton (Philadelphia: Swedenborg Scientific Association, 1938), xxxiii; Williams-Hogan, "Swedenborg: A Biography," 33.

Board of Health, in Stockholm in 1719, Swedenborg drew on his knowledge of brain anatomy in the works of Raymond Vieussens and Thomas Willis as well as his earlier article in which he had offered "proof that our living being generally consists of small quivers, that is, tremulations." Still mourning the assassination of King Charles, Swedenborg defined tremulation as that motion "so subtle and swift that it vanishes from our sensation, as in the case of a bullet which travels through the air and traverses our sight so swiftly that we can make no observation of its course."[70]

After writing his "little anatomy of our vital forces," Swedenborg continued thinking about the brain even as he thought about almost everything else—geology, crystallography, chemistry, atomic theory, biology, navigation and astronomy (he is credited with the discovery of Uranus), magnetism, optics, and paleontology among other things.[71] In the late 1730s Swedenborg traveled to Paris where he attended public dissections of cadavers. He then journeyed to Venice where he spent fourteen months writing *Cerebrum*. In this voluminous text Swedenborg would move beyond his speculations in *On Tremulation* and figured the brain as embodying the mechanics of life itself. Resonant with Willis's description of the brain as the "chapel of Deity," Swedenborg approached the cerebrum as the master script within which resided the key to understanding both minute and cosmic particulars, and, *most importantly*, the relationship between them. Indeed, tremulation was the form and tremulations the substance of a corresponding universe, a cosmos in which every part great and small was in systematic relationship with every other.[72] Within Swedenborg's scheme the ontic distinction between environment and individual brain did not

70. Swedenborg, *Three Transactions*, xiii; Swedenborg, *On Tremulation*, 16, 10; Swedenborg, *Daedalus Hyperboreus* 6 (October 1718), quoted in Dunér, *Natural Philosophy*, 129.

71. R. L. Tafel, *Documents concerning Swedenborg*, vol. 1 (London: Swedenborg Society, 1875), 310.

72. On this point, Swedenborg makes a distinction between undulation and the vibratory logic of tremulation. "Undulation," wrote Swedenborg, "is the kind of motion seen in water when it is moving in circles, but vibration is the reciprocal concussion of each particle with those near it. The former is connected with a local motion of the whole particle, the latter takes place without a local motion of the whole particle. . . . Hence, vibration is subtler than undulation." Emanuel Swedenborg, *Miscellaneous Observations Connected with the Physical Sciences*, trans. Charles Edward Strutt (London: William Newberry, 1847), 90–91. Tremulation was a mechanism of scalability, in other words, moving from the particular to the universal in one fell swoop. "For there must be something in the atmosphere which flows," insisted Swedenborg, "with certain little impulses and circlings moving about the finest fibers and most minute termini, which by means of tremulation or little vibrations

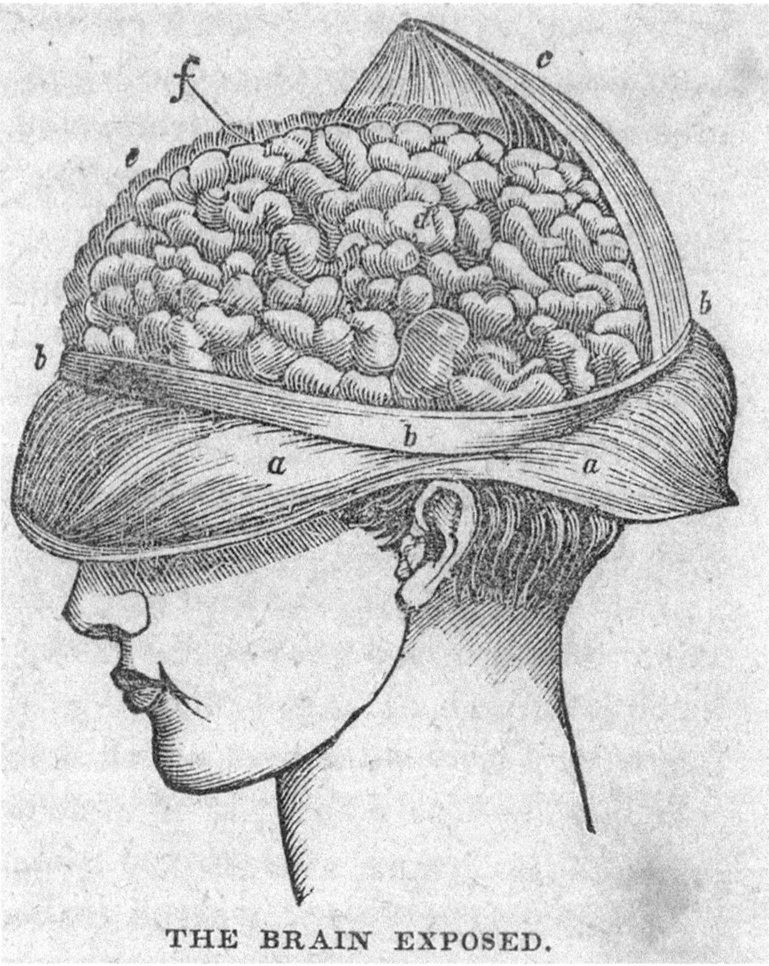

FIGURE 19. "The Brain Exposed," from *The Hydropathic Encyclopedia: A System of Hydropathy and Hygiene*, volume 1 (New York: Fowlers and Wells, 1854).

make sense. There was pattern but not essence—constant activity and movement across thresholds, tremulation rather than stasis. As any good engineer might have done, Swedenborg fixated on the nature of control at the neural level and framed the nervous system as self-organizing—"incessant

carry forward the motion to a certain sensation, and which thus by a motion distributed over an entire system contribute or effect together a symbolism of life." Swedenborg, *On Tremulation*, 7–9.

action as a consequence of incessant reaction"—ever in relation to itself and environment.[73]

5. GHOSTS OF SWEDENBORG

In her magisterial history of occult piety, Catherine L. Albanese notes that Swedenborg was "something of a Hermetic household magus" across the nineteenth century.[74] Well before the discovery of his brain manuscripts, Swedenborg had made his mark on Mormon founder Joseph Smith and on phrenology, Shakerism, Fouricrism, Freemasonry, spiritualism, new thought and theosophy, Henry James Sr., Ralph Waldo Emerson, and the Transcendentalists.[75] The discovery of Swedenborg's work on the brain occurred at a moment when his more mystical pronouncements were all the rage on both sides of the Atlantic.[76] One might even go so far as to say that Swedenborg had already conditioned the reception of his neurophysiological ideas. As Davis reported in explaining the uptick in spiritual intercourse in America, spirits were, at present, studying Swedenborg's insights into the cerebral cortex as a means for communicating with the living.[77] The spirits, however, were not alone in reading Swedenborg. Davis and others who were

73. Swedenborg, *The Brain*, 1:419; Swedenborg, *Animal Kingdom*, 2:244.

74. Catherine L. Albanese, *A Republic of Mind and Spirit: A Cultural History of American Metaphysical Religion* (New Haven, CT: Yale University Press, 2007), 136.

75. Albanese, *Republic*, 39, 136, 175–76, 187. Swedenborg would also impact architecture. *New Thought and the Plan for Chicago* (1909), in which a new vision of the city was put forward in great detail, was itself based on the communicative ties laid out in Swedenborg's *Heaven and Its Wonders and Hell* (Irving D. Fisher, "An Iconology of City Planning—The Plan of Chicago," in *Swedenborg and His Influence*, ed. Erland J. Brock [Bryn Athyn, PA: Academy of the New Church, 1988], 449–64).

76. Indeed, Swedenborg's presence was indirect but pervasive in nineteenth-century America, even when it went unmarked (Albanese, *Republic*, 5). As one Universalist minister noted in 1882, Swedenborg's "doctrine of divine immanence" had long become a matter of common sense in the American grain, "common property of even evangelical preachers" (Frothingham, "Swedenborg," 601–2, 605).

77. In *The Philosophy of Spiritual Intercourse* (1851), Davis referred to such communication as "vibrations or waving breathings which [would] pass through the nervous system up to the brain, and awaken there thoughts by impression" (103).

invested in new, brain-centered articulations of piety drew from Swedenborg, without necessarily having read his early works on brain anatomy.[78]

In *The Principles of Nature, Her Divine Revelations, and a Voice to Mankind* (1847), some twenty years before Swedenborg's manuscripts on the cerebrum were discovered at the Royal Academy, Davis had written that the future was Swedenborgian—"Some of his revelations," wrote Davis, "will be capable of an application only when the race advances to a high degree of social and mental refinement." Davis himself had obtained such refinement in Poughkeepsie, New York, when Swedenborg visited him in March of 1844 and then again in June of 1846. Much of Davis's first book was an homage to Swedenborg's revelatory insights into the neural interplay between matter and spirit. Drawing from Swedenborg's wave metaphor and the detailed anatomical descriptions in other published works, Davis declared that Swedenborg "discourses deeply upon the cortical composition of the brain, describing the glands or 'little hearts' (as he calls them), as constituting the cause of all motion or spiritual exercise in the material form."[79]

In Davis's reading of Swedenborg, the brain was a node of a cosmic communication network.[80] The messages traveling to and fro in the form of "electrical vibrations" were the very source of consciousness:

78. Davis, *The Temple*, 225, 474. Davis's articulation of consciousness as a vibrational matter would make its way into medical science via liberal enthusiasts such as the Swedenborgian minister from Chicago, L. P. Mercer, who expounded on the "co-relation of the facts of consciousness with molecular movements in the brain," urging students to move beyond the dead laws of a mechanical world (*The Soul and The Body: A Sermon to Medical Students with an Appendix on the Doctrine of Swedenborg* [Chicago: Gross and Delbridge, 1883], 3, 19).

79. Andrew Jackson Davis, *The Principles of Nature, Her Divine Revelations, and a Voice to Mankind* (Boston: Colby and Rich, 1847), 587, 32, 221–22; 654–55. In *Mesmer and Swedenborg* (New York: John Allen, 1847), George Bush, a Hebrew professor at NYU, introduced Davis as the American Swedenborg and cautiously suggested that he may have only been visited by a spirit "subject" of Swedenborg's—a replica who was "in close conjunction with him and who was thoroughly imbued with his truths—who was intent on the propagation of his system in the world" (165, 173, 189).

80. Davis responded to evangelical criticism by suggesting that those who doubted Swedenborg's insights into mental piety were, themselves, mentally disordered. Because "the spiritual world has been laid open to mortal view through the instrumentality of Emanuel Swedenborg," wrote Davis, "it is upon those therefore who deny, and not upon those who admit, this fact, that the imputation of madness redounds" ("Revelations of A. J. Davis," in Bush, *Mesmer and Swedenborg*, 206).

all beings, animal as well as human, are knowing or intelligent in proportion to the quantity of cerebral substance deposited in the forehead.

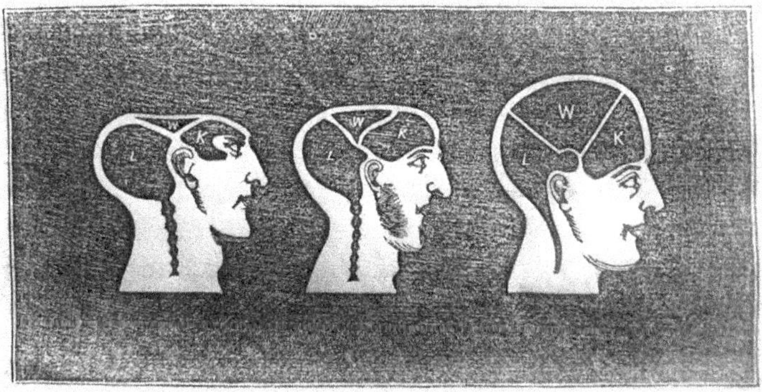

PAST. PRESENT. FUTURE

When the brain is largest and broadest just above the eyes,

FIGURE 20. "Past, Present, and Future," from Andrew Jackson Davis, *The Great Harmonia: The Reformer* (New York: Fowlers and Wells, 1866), 33.

> Not only is the brain a grand Battery compounded of innumerable little corresponding batteries, called nerve-cells or nerve-centres: but the brain is also a grand union telegraph office into which pass, and from which proceed, innumerable tubes and conductors called blood-vessels and nerve-fibres; and these vessels and fibres connect cell with cell, one little battery with another; also they tie together positive groups of cells or centres with corresponding negative groups of cells, in such a systematic and harmonious manner as to make it impossible to strike one link in the cerebral chain and not at the same moment disturb more or less the entire throbbing chain of feeling and intelligence.[81]

In the neurocosmology of Davis, "the brain, which is the seat of power, registration, and government," became a way of seeing the world in toto. For to gain interpretive leverage on the "weaving loom of the nervous system" with "its multitudinous inter-reciprocating centres of sensibility and power, and its incomparable net-work of conductors" was to experience the secret

81. Davis, *Philosophy of Spiritual Intercourse*, 76; Davis, *The Temple*, 114.

of correspondence in both epistemic and cosmic registers. Like Swedenborg before him, Davis committed himself to the "laws of mechanics" not as an end but as a way to begin to explain "the soul forces acting through the cerebellum." To understand the brain was, first and foremost, to appreciate the flux and flow of spirit.[82] These laws, however, despite their arrival as universal and beyond history, were nonetheless shot through with the earthly and explosive politics of race coursing through Davis's scene of writing.

6. MENTAL SLAVERY AND THE INVENTION OF SPIRITUALITY

Across large swaths of metaphysical America, the brain was beginning to bear the burden of achieving one's freedom, particularly for white subjects perversely concerned with the specter of "mental slavery." As Davis declared, "the mental slavery of generations past bears a resemblance to the physical slavery of the present." Davis does two things here. First, he prioritizes the mental liberation of self over the physical freedom of others. Second, he frames abolitionism as solely metaphorical. For Davis, coming to consciousness becomes equivalent to the breaking of chains. Both mental and physical slavery were made fungible in their mutual proximity to sin. Both were caused by "a perversion of man's natural inclination."[83] But there was something about mental slavery that preoccupied Davis in a way that his relatively weak abolitionist stance did not. "Mental slavery," wrote Davis, was "much worse than physical slavery." For unlike being poked and prodded and sold at market, mental slavery was spiritually encompassing. Consequently, for Davis, the neural freedom of white Europeans was a precondition for addressing the physical enslavement of those whose brains were not even capable of being impressed, whether positively or negatively.

Unlike emancipation from chattel slavery, freedom from mental slavery was for Davis a freedom that could be achieved deliberately and without delay. Such freedom began by cultivating how the mind "acted upon that

82. Davis, *The Temple*, 110, 105–7, 18, 13.

83. Davis, *Principles of Nature*, 13. See also the discussion of "mental slavery" in Davis's short-lived journal, *The Univercoelum and Spiritual Philosopher*, specifically W. J. Fox, "Mental Slavery," *Univercoelum and Spiritual Philosopher* 1, no. 16 (1848): 245–48.

which it receives."[84] According to metaphysicals and medical reformers alike, the self was ever in relation to its environment, constantly receiving impressions and impressing itself upon the world. The key was to manage this economy of impressions. By learning to act on impressions received, and to impress outward accordingly, one could manage the liberties of selfhood. Impressions, in other words, were generative. Impressions invited progress. The capacity to be influenced, framed in terms of one's capacity for sympathy, was an advantageous sign of one's humanity. Here was a vision of religious and racial progress premised on the invitation to reengineer one's mental make-up.

As historian Kyla Schuller has demonstrated, progressive reformers during this period often assumed that the brains of white Europeans were more impressible than those of Africans and therefore more susceptible to cultivation. The *capacity* of white brains to be impressed, in other words, was *the* key to their superiority. Whiteness achieved. Whiteness maintained.[85] Here was the script for maintaining political privilege, in and through the brain. It was inevitable that racialized epistemologies made their way into nineteenth-century spaces of neuromatic formation, increasingly embedded within seemingly nonracial or, at the very least, racially neutral concepts.

Amariah Brigham, for example, was a founding member of the Association of Medical Superintendents of American Institutions for the Insane. The brain, he argued, was a plastic entity, the hinge of political reform and

84. Charles Howell, *Civilized Money: The Way to Prosperity, Happiness, Civilization* (Grand Rapids, MI. Cash Publishing, 1895), 5–6.

85. Kyla Schuller argues that the notion of a "nervous system as a differentially pliable and agential entity in continuous interplay with its environment" was central to the biopolitical atmosphere of the US in the nineteenth century (*The Biopolitics of Feeling* [Durham, NC: Duke University Press, 2017], 3). Schuller documents how notions of nervous plasticity served to support a lattice frame of human hierarchy. Specifically, Schuller attends to the "physiological dimensions of sympathy" (10) and, devastatingly, charts how the "uneven distribution of somatic dynamism was central to the materialization of modern ideas of racial and sexual difference" (12). On the metaphysics of nervous impressibility and sympathy, in general, during this period, see my *Secularism in Antebellum America* (Chicago: University of Chicago Press, 2011), 171–81. For a take on neural plasticity that is less invested in questions of ethereal power (as in Stuart Hall, Judith Butler, and Michel Foucault), see Catherine Malabou's argument that neural plasticity provides the "biological guarantee" for philosophy in the deconstructive vein: "Floating Signifiers Revisited: Post-Structuralism Meets Neurolinguistics," lecture delivered February 13, 2018, Princeton University.

self-improvement.[86] "An accurate knowledge of the anatomy of the brain," suggested Brigham, "was the first step in one's liberation." Such knowledge, he asserted, was "necessary" for managing the traffic between brain and environment. Brigham, here, was drawing on his friendship with Dr. Johann Spurzheim and the phrenological playbook of brain plasticity (even as the *American Phrenological Journal* attempted to impress Brigham on its readers).[87] By midcentury, phrenology was on the upswing in America even as it was losing adherents in Europe. Pitched perfectly to the nonspecific Protestantisms of its era, phrenology made urgent the imperative to know yourself knowing and to use such knowledge to reform the faculties that made up one's consciousness.[88]

Phrenology posited the Caucasian brain as a malleable site of message exchange. Different areas of the brain communicated with one another and, collectively, they assumed an executive function. And it was at the level of interaction—managing the relationships between faculties and the animal magnetism that coursed between them—where individuals could intervene in their own mental make-up. To become an engineer of self, managing the relationships from above, as it were, was to "detect" the flux of spirit and to gain control over it as it flowed in and out of the brain. The phrenological agenda was nothing but liberal in the proposition that cognitive control would allow one to curb unwanted influence on behavior by regulating the porous boundary between self and world.

The rhetoric of "mental slavery" made schemes of mental cultivation such as Brigham's or phrenology's all the more appealing by allowing white sub-

86. As Victoria Pitts-Taylor wryly notes in her discussion of intensifying discourses around brain plasticity since the nineteenth century, "there is nothing inherent in plasticity to suggest a progressive politics" ("The Plastic Brain: Neoliberalism and the Neuronal Self," *Health* 14, no. 6 [2010]: 635–52).

87. Amariah Brigham, *An Inquiry Concerning the Diseases and Functions of the Brain, the Spinal Cord, and the Nerves* (New York: George Adlard, 1840), 20, 16; L. N. Fowler, "The Phrenological Character of Amariah Brigham, M.D.," *American Phrenological Journal* 11 (1849): 361–67.

88. Once the measure of each faculty was taken, the goal was not to maximize but rather "to establish . . . a state of equilibrium between all parts of the body and brain as to produce an harmonious, uniform, consistent, moral, physical, and intellectual character." "Importance of a Phrenological Examination," *American Phrenological Journal* 20 (August 1854): 38. On the notion of non-specific Protestantism, see Tracy Fessenden, *Culture and Redemption: Religion, the Secular, and American Literature* (Princeton, NJ: Princeton University Press, 2007).

jects to retain their political and ontological privilege even as they engaged in progressive reform.[89] For while "mental slavery" denoted the potential stakes for all brains, only some brains were inherently capable of liberation. Only some brains possessed the capacity to transform and improve. Because of the "natural" determinations of race, only some brains were truly plastic.[90] Consequently, the "improvability of the brain"—as was often the case in nineteenth-century America—was also a mode of racial differentiation. *A way, in other words, of achieving one's whiteness.*[91] Whereas whiteness was associated with "elasticity of mind," everyone else was bowled up in some way, some more so than others. Asians, for example, had "arrived at a point comparatively low in the scale of improvement, beyond which they have never passed." Africans were an "unbroken scene of moral and intellectual desolation." And while they had failed to reach even "moderate civilization" they still "greatly excel[led] several of the tribes of native Americans, who have continued wandering savages from the beginning to the end of their existence."[92] Savages, in other words, never really had a chance, for they

89. On "mental slavery" as false religion, see Rev. P. Maclachlan, *The True Religion, What It Is: Or, a Protestant's Objections to Catholicity Fully and Fairly Answered. In a Series of Letters to R. W. Kennard* (Edinburgh: Marsh and Beattie, 1855), 23; as part of an abolitionist critique of the southern planter, see Marshall Hall, *The Two-fold Slavery of the United States: With a Project of Self-emancipation* (London: Adam Scott, 1854), 32; On "mental slavery" as an accusation made in the name of a self-conscious secularity, see J. O'Donovan, "Converting an 'Atheist,'" *Freethinker* 10 (April 27, 1890): 195–96; R. G. Ingersoll, "Individuality," *Lectures of Col. R.G. Ingersoll: Including His Answers to the Clergy, His Oration at His Brother's Grave, Etc., Etc.* (Chicago: Rhodes and McClure Publishing, 1898), 160.

90. Britt Rusert notes the appeal of plasticity for abolitionists like Frederick Douglass who were otherwise critical of phrenology's implicit racial and sexual hierarchies. Despite phrenology's structural blindness, Douglass attempted to distill for African Americans what phrenology made possible for whites, that is, the capacity for self-transformation and improvement (Britt Rusert, *Fugitive Science: Empiricism and Freedom in Early African American Culture* [New York: New York University Press, 2017], 125).

91. On the racial imperative underlying the "improvability of the brain," see Charles Caldwell, "Thoughts on the Most Effective Condition of the Brain as the Organ of the Mind, and On the Modes of Attaining It," *American Phrenological Journal* 1, no. 11 (August 1, 1839): 394, 402–3.

92. George Combe, "Phrenological Remarks on the Relation between Natural Talents and Dispositions of Nations, and the Development of Their Brains," appendix in Samuel George Morton, *Crania Americana; or, A Comparative View of the Skulls of Various Aboriginal Nations* (Philadelphia: John Penington, 1839), 271–72.

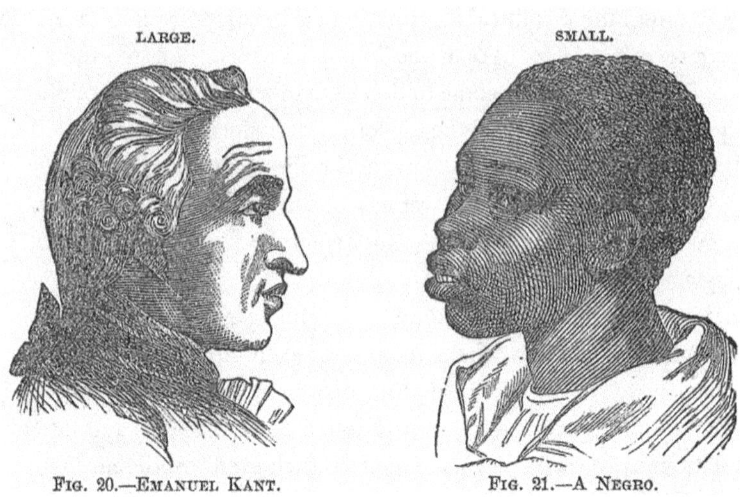

FIGURE 21. The Reflective Group, from Samuel R. Wells, *How to Read Character: A New Illustrated Hand-Book of Phrenology and Physiognomy for Students and Examiners with a Descriptive Chart* (New York: Samuel R. Wells, 1875), 32–33. "The function of the Reflective Group is to analyze, compare, and classify the facts collected by the perceptives, and to philosophize, contrive, invent, and originate ideas."

were, from the beginning, constitutionally immune to the soothing directives of civilization.[93]

---

93. Which is to say that faith in plasticity is never ill-intended. See, for example, those currently seeking leverage on the intractable problem of racism by investigating the "neural underpinnings" of "processing racial information." Such accounts have produced an array of conjecture about the "deep historical roots" of racism and its "contemporary relevance to social issues." The brain, in these studies, becomes the means of liberalism's self-correcting path. As such, a neural science of the population will reveal "basic mechanisms of social cognition" that underlie an array of biases, "such as those based on ethnicity, gender, sexual preference and nationality." Consequently, "Race and its influence on unintended race-based attitudes and decisions" may soon be engineered out of existence with the proper understanding of the brain and the proper tools of transformation. Such studies, in treating "race-specific beliefs and feelings" in terms of an "individual's general regulatory abilities" seek to change the racial attitudes of individuals rather than, say, the structures of capital and law that determine their lives. Tiffany A. Ito and Bruce D. Bartholow, "The Neural Correlates of Race," *Trends in Cognitive Sciences* 13, no. 12 (2009), 528–30; David Amodio, "The Neuroscience of Prejudice and Stereotyping," *Nature Reviews Neuroscience* 15, no.

The most successful phrenological entrepreneurs were the Fowler brothers, Orson and Lorenzo, along with their business partner Samuel Wells and his wife Charlotte Fowler. Under the auspices of their company Fowlers and Wells, phrenology became a therapeutic brand that included a phrenological museum, a storefront, a lecture booking service, and private examination rooms. The publishing arm of Fowlers and Wells was anchored by the *American Phrenological Journal* and the annual *Phrenological Almanac*, both notable for their practical advice and readings of celebrity heads, including Swedenborg's. Fowlers and Wells took Swedenborg's insights to their logical conclusion, advertising phrenology as being "the crowning essence of true religion" and "destined to form a new era in Christianity."[94] As missionaries of mental piety, Fowlers and Wells offered their audience a "correct test and touchstone of true and false religion." They advocated a piety that was vaguely liberal and explicitly removed from the more emotional precincts of evangelicalism.[95] For it was only after a phrenological examination that "every individual" would "see at a glance the departures of his own religious opinions and practices from . . . the true standard of our nature, pointed out by Phrenology."[96]

In 1842, Orson took it on himself to revise Spurzheim's understanding

10 (October 2014), 670–71, 679–80; Saaid A. Mendoza et al., "Reducing the Expression of Implicit Stereotypes: Reflexive Control through Implementation Intentions," *Personality and Social Psychology Bulletin* 36, no. 4 (2010): 512–23.

To be fair, these scientists vigorously deny suggestions of structural blindness on their part and insist that they are describing the actual mechanics of neural structuration where racial categories are processed, largely automatically. Despite its liberal politics, however, there is something illiberal about the logic that race is a socially constructed difference yet a difference that is processed naturally by the brain in a consistent and identifiable way. Race remains, quite literally, in your head rather than, say, embedded in the racialized apparatus, categories, and epistemic moves that have allowed scientists to immunize themselves from history and proclaim, once again, that racial healing is primarily a matter of changing other's people's bad brains.

94. "Boston Phrenological Society," *American Phrenological Journal* 2, no. 5 (February 1, 1840): 235; "The American Phrenological Journal for 1849," *American Phrenological Journal* 11 (1849): 10; "Progression a Law of Nature: Its Application to Human Improvement, Collective and Individual," *American Phrenological Journal* 8 (August 1846): 240.

95. O. S. Fowler, *Religion; Natural and Revealed: or, the Natural Theology and Moral Bearings of Phrenology and Physiology* (New York: Fowlers and Wells, 1844), 76.

96. Fowler, *Religion*, 59–60. See also the explanation of religious differences based on physiognomy in Richard Walker, "Religious Differences Explained and Obviated by Phrenology," *American Phrenological Journal* 8 (1846), 321–23.

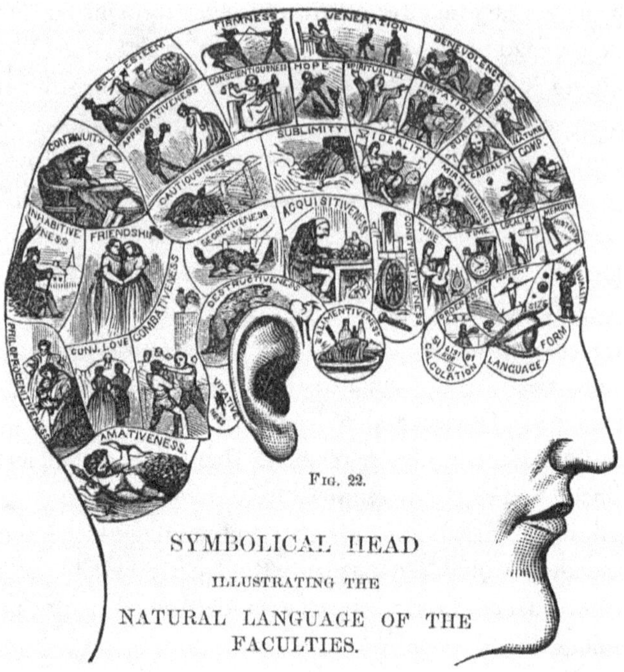

FIGURE 22. Symbolical Head. L. N. Fowler, *Marriage: Its History and Ceremonies with Phrenological and Physiological Exposition of the Functions and Qualifications for Happy Marriages* (New York: Fowlers and Wells, 1848), 73.

of what religion was, in essence. His efforts resulted in a new standard in which one's impressibility could be harnessed as an opportunity for self-actualization and a counter to the necessary "abject subserviency" that the existing faculty of Veneration demanded. During his investigations into the human brain, Fowler had discovered a new moral faculty—the deepest layer of true religion, what he called the faculty for "SPIRITUALITY." The invention of Spirituality in antebellum America was a vibrant defense of religion as a reasonable and deliberate act of cognition and a reaffirmation that religion was located in the brain.[97] According to Orson, Spirituality enabled

97. Fowler, *Religion*, 45, 51; see also O. S. Fowler, *Education and Self-Improvement, Founded on Physiology and Phrenology: or, What Constitutes Good Heads and Bodies, and How to Make Them Good, by Enlarging Deficiencies and Diminishing Excesses* (New York: Fowlers and Wells, 1844), 235–37. On the role that the isolation of higher mental faculties played in the establishment of the neurosciences in the nineteenth and twentieth centuries, see Jan Verplaste, *Localising the Moral Sense: Neuroscience and the Search for the Cerebral Seat of Morality, 1800–1930* (New York: Springer, 2009).

the rational appreciation of "spiritual influences." When functioning healthily, Spirituality enabled degrees of discernment *of* and freedom *from* agents unseen. The state of impressibility, in other words, became an opportunity for control. Spirituality, then, served to demarcate an interior process of agentive reason from an exterior life to which this agentive reason would then apply itself. In contrast, to lack a healthy faculty of Spirituality was to be overwhelmed by environmental impressions. The "perversion" of Spirituality, moreover, would compromise one's "independent capacity as mind."[98]

Spirituality, for all involved, was intimately bound up with cerebral plasticity, the brain becoming an opportunity to manage the affective traffic between inside and out for the purpose of self-realization.[99] Yet in promising the actualization of freedom in relationship to one's environment, Spirituality was also born a cognitive attribute of white self-regard. In catalyzing a sense of self-aware self-control, Spirituality served to remind the self of its racial superiority. Spirituality, then, is part of what Schuller has called the "historical ontology of whiteness"—a history in which racialization and sexual differences have served to address the "precarity of the impressible body" and to resolve the tension of a self-possessed subject ever teetering on the

98. Fowler, *Religion*, 106, 122. [Attributed to] A., "On the Functions of the Organ of Marvellousness," *American Phrenological Journal* 3, no. 8 (May 1, 1841): 361. See also O. S. and L. N. Fowler, *New Illustrated Self-Instructor in Phrenology and Physiology* (New York: Fowlers and Wells, 1859), 172, 122.

99. Mary Baker Eddy, founder of Christian Science and a key figure in institutionalization of the mind-cure movement in the 1880s, cautioned against over emphasizing the brain, alone, in any definition of spirituality. Eddy offered a pointed critique of "brainology" and what she saw as the debasement of the human spirit by those who insisted that consciousness was "manifested through brain and nerves" alone (*Science and Health with Key to the Scriptures* (1875; Boston: Allison V. Stewart, 1912), 17–71, 295, 290). Alternatively, "brainology" left a more opportune mark on Annie Payson Call, who for over twenty-five years served as business manager of the Swedenborgian Massachusetts New-Church Union and editor of the *New-Church Review* and who offered advice on how to "grow faster spiritually" in her popular *Power through Repose* (Boston: Roberts Brothers, 1891). In a chapter entitled "The Use of the Brain," she recommends comparing the "whole machine" of the brain "to a community of people . . . Church, State, institution, or household" (28). In such a vision the borders of both the social and neural community must be managed so as to prevent "misdirected force" and "useless stress of any kind" (34). On her discourse on the nerves, see Annie Payson Call, *The Regeneration of the Body* (Boston: Massachusetts New-Church Union, 1885) and her later *Nerves and Common Sense* (Boston: Little, Brown, 1910). See also Anastasia Kirtiklis, "New Thought or Old?: Positioning Annie Payson Call in Neurasthenic America" (graduate student working paper, Graduate Center, City University of New York, 2018).

FIGURE 23. Portrait of Emanuel Swedenborg, from "The Character, Organization, and Biography of Emanuel Swedenborg," *American Phrenological Journal* 11 (1849): 105.

brink of being possessed.[100] The examples within phrenology are plentiful as women and racialized others were consistently shown to be disproportionate in the distribution of their mental faculties. Too much or too little of this or that faculty and an overall failure to measure up to the cognitive harmony achieved by famous white men featured prominently in every issue of the *American Phrenological Journal.*[101]

100. Schuller, *Biopolitics of Feeling*, 13, 16. On the gendered ecology of this new age, see Beryl Satter, *Each Mind a Kingdom: American Women, Sexual Purity, and the New Thought Movement, 1875–1920* (Berkeley: University of California Press, 2001).

101. The concept of spirituality as a cognitive skill, as a form of extrarational perception, bubbled up in places far afield from phrenology. Spirituality, for example, was invoked in an 1848 *American Whig Review* article entitled "The Anglo-Saxon Race: An Inquiry into the Causes of Its Unrivalled Progress, with Some Considerations Indicative of Its Future Destiny." Spirituality served civilization and was the mark of superiority that differentiated whites from a "quarry slave" and from "ferocious" foreigners (*American Whig Review* 7, no. 1 [January 1848]: 40). Or take the more imperial tone of an *Atlantic Monthly* article a decade later entitled "The New World and the New Man." In it the author notes with a shrug the "several happy features" that could result from the Dred Scott decision of 1857 and the Supreme Court's ruling that the Missouri Compromise was unconstitutional (and thereby

7. THE DIAKKA AND THEIR EARTHLY VICTIMS

Davis, who promoted phrenological frames throughout his career, made explicit the racializing dimension of Spirituality when popular tides began to turn decisively against Spiritualism in the 1870s. As fraud and controversy swirled around séance spiritualism in the wake of the Civil War, Davis took to defending the integrity of his brand by appealing to the fears of white people. According to Davis, the problems of fraudulent mediumship were bound up in the difference of race. Specifically, they were due to the untoward impressions of dark savages from the spirit world.

In 1873 Davis published *The Diakka, and Their Earthly Victims: Being an Explanation of Much That Is False and Repulsive in Spiritualism*. This thin volume begins with Davis writing in missionary mode from the far reaches of the Summerland: "I have penetrated in to the recesses of that mysterious realm, and have explored many of its sad human experiences." Davis claims to be reporting from the land of the Diakka, a place marked by "hills and dales and forests" and "boundless magnitude of the celestial wilderness." Mixing an imperial ethnographic gaze with bits of gothic horror, Davis's pamphlet is laced heavily with condescension—"The country of the Diakka, then, in a sentence, is (to give you my own definition) a Garden of Eden, to call it by no harsher name, where the morally deficient and affectionally unclean enter upon *a strange probationary life*."[102]

Davis's depiction of the Diakka drew from Swedenborg's *Concerning Heaven and Its Wonders* in which Swedenborg argued that the "mercy of the Lord is universal so that every heathen has the potential to be saved" in the spirit world. Because the potential for piety had been "provided . . .

---

prevented Congress from prohibiting slavery in the territories). In his liberal optimism in the face of unfortunate setback, the writer advocates "organization and moral discipline" and warns against nervousness brought on by "the public element." Envisioning a rosier future with the emergence of a mature state apparatus premised on the cultivation of gendered distinction, he declares that "a spirituality must appear to crown and complete this great continental body; otherwise America is acephalous. Unless there be an American Man, the continent is inevitably but an appendage, a kitchen and laundry for the European parlor. American Man, and the word Man is to receive a large emphasis" ("The New World and the New Man," *Atlantic Monthly* 2, no. 12 [October 1858]: 515, 526).

102. Andrew Jackson Davis, *The Diakka, and Their Earthly Victims: Being an Explanation of Much That Is False and Repulsive in Spiritualism* (Boston: Colby and Rich, 1880), 13, 7, 6.

for everyone," heathen such as the Diakka would *eventually* learn that "to live in accordance with one's religion [was] to live interiorly." For although the potential for true religion may have been on the inside of everyone, including the heathen, such potential existed in different proportions. As Swedenborg wrote, "heaven is not the same in one as in another. It differs in each one in accordance with his affection for good and its truth."[103] The heathen, then, would begin the process of redemption only after they had died. In heaven they would learn properly, and from other spirits, how to affect and be affected by the environment in a civilized manner. Whereas on earth the heathen were incapable of learning these lessons, in heaven spirits would successfully instruct them to give up their gods, idols, and fantasies.

In taming the unruly practices of the heathen, spirits morally justified the logic of colonial subjugation in this life by evidencing benevolence in the next. For before their conversion, however, the Diakka were a pesky bunch. As Davis writes, the Diakka "believe in ultimate annihilation" like "a black man of the land of Mgogo." In the spirit world, as on earth, the Diakka are disruptive by nature. They perform "tricks" in "dark circles" and confuse the living on matters of spiritual import. They "take a gypsy-like pleasure in traveling with stealthy celerity from place to place" and delight in misrepresentation and insincerity. The Diakka are quite literally a spirit species of bad brains ("Diakka" meaning "mental antagonisms, or minds with cross purposes"). They exploit the impressibility of others, cajoling them to act against their own interest. The Diakka "confuse your thoughts," insist Davis. They "break up the lines of your memory, mingle their inclinations with your own, and psychologize your nervous and muscular systems." The Diakka, who also included "Indians of every nationality," band together in the spirit word and conspire to "complete manipulatory control over, the subtlest elements and atoms and laws of exterior chemistry."[104]

103. Emanuel Swedenborg, *Concerning Heaven and Its Wonders, and Concerning Hell* (Boston: Otis Clapp, 1849), 168–69.

104. Davis, *The Diakka*, 11, 32, 73, 14–15, 80. In her neurotheology of white supremacy, Helena Blavatsky, cofounder of the Theosophical Society, picked up the notion of the Diakka from Davis who, she noted, "discovered them in a shady corner of the 'Summer Land.'" Blavatsky framed these "bad demons" as a vortex in which the "personal individuality" of those who encountered them was utterly extinguished (*Isis Unveiled: A Master Key to the Mysteries of Ancient and Modern Science and Theology*, vol. 1: *Science*, 6th ed. (New York: J. W. Bouton, 1892), 218–19.

Harkening back to his earlier encounters with Swedenborg, Davis reported learning from a Swedenborgian spirit (Victor) that Diakka were continually being converted in heaven. But because of the steady stream of Diakka souls from earth there were labor issues in heaven—a backup of unconverted Diakka that gave them much time to improperly influence inhabitants of Earth. Because their violation was both cosmic and civil, the "removal" of the "danger" the Diakka posed was an urgent matter of liberal governance.

> In the Summerland the police regulations are based upon securing the enjoyment of the utmost personal freedom, including the privileges and opportunities not conflicting with the exercise of the utmost personal freedom in the part of every other; so that even the intellectually gifted and witty and tricky Diakka, with their known deficiencies in the higher moral principles of character, are not restrained in their visits to earth, because personal education through experience, is part of a scheme of developing personal *responsibility*.[105]

Personal responsibility, in other words, was at stake for all involved. To be able to manage the impressions of the Diakka and to fend off their machinations was spiritually fortifying for both potential victims as they moved forward in this life *and* Diakka spirits in their journey to ultimate redemption in the Summerland.

Doubling down on the truth of Spiritualism, Davis observed that the Diakka wielded their power by manipulating the brains of their victims—their spirit affecting matter by way of strategically timed tremulations. Davis concluded that superior knowledge of spiritual intercourse, that is, the same cognitive means by which Diakka cast their spells on the bodies of others, would "remove the danger." "Knowledge of this truth," wrote Davis, "is equivalent to the development and application of your [mind's] self-healing power." Here was a liberating piety that was brain-centric, that was benevolent in its missionary outreach yet naturally on the defensive. Here was a piety of soft-pedaled racial menace, attending to the evil of the Diakka and their petty machinations and "psychologizing"—all in order to immunize oneself from them. Consequently, in Davis's Swedenborgian scheme, civilized whiteness and its attendant freedoms were not only about defend-

105. Davis, *The Diakka*, 16.

ing oneself against the incursions of racial difference. They were also about neural purification: racial difference, itself, was made possible by a highly particular conception of the brain and its relative plasticity.[106]

8. THE MEDIOMANIACAL ORIGINS OF AMERICAN NEUROLOGY

Whereas Davis's racial paranoia sparked nary a controversy in the early 1870s, a budding cadre of medical men vehemently opposed *his* use of *their* ideas to explain séance spiritualism. Physicians like George Beard and William Hammond turned Davis's "sectarian" practice of neurology into an object of critical derision. More specifically, they turned the pathology of the female medium into a resource for authorizing their own professional status in contrast to what Beard saw as one of the "great delusions of history." As historian Ann Braude has shown, the triumph of neurology over various sectarian practices such as phrenomesmerism and homeopathy was an extension of masculine authority at both the institutional and symbolic level. Ridiculed as antiscientific, spiritualists posed a threat to "the organized knowledge of men" because of the feminized excesses of their claims. A text

106. Davis, *The Temple*, 262, 379; Davis, *The Diakka*, 15. In describing the psychic trickery of the Diakka, Davis wrote that they acted like "the familiar gossips and social ghouls in this world" when they meddled with the intimate bond between husband and wife. The threat that the Diakka posed, in this case, was to marriage, the foundational form of sexual difference. By not "indulging in such impressibility," wrote Davis, couples could save not only their marriages but their sanity as well. Davis referred to the symptom of such insanity as the "inversion of conjugal affection." Citing Swedenborg on the matter, Davis describes any "mania of mutual hate between men and women" as troubling, metaphysically as well as politically. In his neurological description of sexual "insanity," Davis described how "inverted conjugal love kindles unhealthy fires within the medulla oblongata, and between the vital threads of the cerebrum and cerebellum" (Davis, *The Temple*, 262, 339, 341; Davis, *The Genesis and Ethics of Conjugal Love* [Boston: Colby and Rich, 1881], 66). In grounding his revelation in the conflation of bad religion, bad sex, and bad brains, Swedenborg had offered an elaborate explanation of how men were more capacious than, and therefore superior to, women in a cosmic register. Emanuel Swedenborg, *Delights of Wisdom concerning Conjugal Love after which Follow Pleasures of Insanity Concerning Scortatory Love* (Boston: Otis Clapp, 1833), 35.

like Davis's *The Temple* did not educate in any practical manner but only "confirm[ed] the longings of the heart."[107]

The gendered struggle between two different approaches to bodily and mental health was also a struggle over the truth of religion. Over the course of the second half of the nineteenth century, professional scientists steeped in Protestant orthodoxy took aim at spiritualists, particularly female mediums as they became the focus of diagnostic derision. The increasingly common diagnosis of religious hysteria in the form of "mediomania," for example, reframed hysteria (a disease originally considered to be located in the female reproductive organs) as a pathology induced by a hyperactive nervous system. The excess of femininity was no longer simply a moral or religious or even biological matter. It was also, simultaneously, neurological.[108]

For as the science of neurology emerged at this moment, it did so as an explicit yet ambiguous project of secularization. As the historian Molly McGarry has argued, "Rather than science vanquishing religion, as in the classic secularization narrative, here science and religion worked as mutually constitutive knowledges, together producing a materialist belief system to explain the immaterial world."[109] In these early moments of self-conscious professionalization, neurologists such as Hammond and Beard were engaged in a masculinist performance—securing the truth of the brain as an empirical index of human being even as they secured the distinction between their reason and the feminized faith of their patients.[110]

This reminds the reader once again that this genealogy of information as currently employed in the neuromatic brain is not a strictly secular story. But there are, of course, consequential figures whose vocation to science was self-conscious and shared across national boundaries and professional societies. The rediscovery of Swedenborg's manuscripts, in addition to being reported internationally, coincided with the neurological turn in America and the consolidated discovery of the motor cortex by scientists on both

107. Ann Braude, *Radical Spirits: Spiritualism and Women's Rights in Nineteenth-century America* (Boston: Beacon Press, 1989), 143; George M. Beard, "The Psychology of Spiritism," *North American Review* 129 (July 1879): 65–66, 71.

108. Braude, *Radical Spirits*, 158–60. See also William A. Hammond, *Spiritualism and Allied Causes of Nervous Derangement* (New York: G. P. Putnam's Sons, 1876), 256f.

109. McGarry, *Ghosts of Futures Past*, 128–29, 143.

110. William A. Hammond, *A Treatise on Diseases of the Nervous System* (New York: D. Appleton, 1871; George Miller Beard, "The Delusions of Clairvoyance," *Scribner's Monthly* 18, no. 3 (July 1879): 433–40.

sides of the Atlantic.[111] This development would soon lead to a *recognition* of an electrical network of *precise connections* below the scalp; oscillations *identified*, sweeping swaths of electrical futures becoming patterned, rhythmic . . . off *again*, on *again*, off *again* . . . clusters of neurons changing their potentials at the same time . . . waves *emanating* outward and up. The brain *encountering* its environment. Discrete signals *inviting* interpretation. *Decoding* as a form of control. The promise of a new kind of mastery *of* the human, *by* the human, and *for* the human. And everything, *eventually*, on the screen.

9. PREHISTORIES OF ELECTROENCEPHALOGRAPHY

Among the articles noted by R. L. Tafel in his introduction to the second volume of *The Cerebrum* (1887) was "Ueber die elektrische Erregbarkeit des Grosshirns" (1870) by the galvanic therapist and neurologist Eduard Hitzig and anatomist Gustav Fritsch. In "Autonomic and Motor Responses in Dog upon Cerebral Stimulation," Hitzig and Fritsch reported on their successful efforts to stimulate, through electricity, discrete motor movements of a dog. Slicing open and peeling back the skull of dogs that were strapped to tables (and only sometimes anesthetized), the two young lecturers applied weak electrical pulses, via platinum wires, to the flesh of the dog's cerebral cortex. Hitzig had previously developed the apparatus as a therapeutic device to administer electrical currents to his patients in a Berlin garrison hospital. In this experiment motor responses were produced when the frontal (anterior) cortex was stimulated.[112]

111. J. S. Jewell, "On the Existence of Definite Motor Centres in the Cerebral Cortex," *Chicago Journal of Nervous and Mental Disease* 11, no. 4 (October 1875): 477–503.

112. Eduard Hitzig and Gustav Fritsch, "Ueber die elektrische Erregbarkeit des Grosshirns," *Archiv für Anatomie, Physiologie und Wissenschaftliche Medicin* 37 (1870): 300–332. Hitzig and Fritsch performed this experiment in a makeshift home laboratory because the university would not allow it to happen on site (due to a ban on vivisection). Later that year the Franco-Prussian War broke out. Enlisted as a physician in the Prussian Army, Hitzig experimented on a wounded soldier whose skull had been fractured by a bullet. Hitzig confirmed what he already knew—that one could begin to map the relationality between electric current applied to the brain and involuntary muscular movement. Peter J. Koehler, "Eduard Hitzig's Experiences in the Franco-Prussian War (1870–1871): The Case of Joseph Masseau," *Journal of the History of the Neurosciences* 21, no. 3 (2012): 261, 250.

Despite being largely ignored because of the onset of the Franco-Prussian War, this experiment was generative in producing nothing less than the history of the neurosciences and their applications.[113] A thesis of secularization unlike any other. Unlike, say, Swedenborg's suspect riffs on tremulation, the exploits of Hitzig and Fritsch have comprised a suitable, "scientific" origin story that captures the certitude and monumentality of our neuro-now.[114] Here is an enabling fiction about the exposure *of* our brains *to* our brains rehearsed in a secular key.[115] In *blood-soaked scenes of animal sacrifice* Hitzig and Fritsch had begun to theorize, in an appropriate manner, the physiology of tremulation.

As the story goes, the discovery of the motor cortex was soon followed by neural stimulation of humans.[116] These experiments, in turn, begot the technical capture of electrical waveforms of the brain. In the mid-1870s, Richard Caton, a lecturer at the Royal Infirmary School of Medicine in Liverpool,

113. At this point, the Fritsch and Hitzig experiment has assumed legendary status—"foundational," a "seminal demonstration" (Dale Purves et al., *Neuroscience*, 4th ed. [Sunderland, MA: Sinauer Associates, 2008]: 436, 470) . . . "one for the great discoveries in neurobiology" (Gordon M. Shepherd, *Neurobiology*, 3rd ed. [New York: Oxford University Press: 1994], 465–66). On "the seminal canine experiments of Gustav Fritsch and Eduard Hitzig," see Fedor Panov and Brian Harris Kopell, "Use of Cortical Stimulation in Neuropathic Pain, Tinnitus, Depression, and Movement Disorders," *Neurotherapeutics* 11, no. 3 (2014): 564–71. See also Henri Hecaen and Martin L. Albert, *Human Neuropsychology* (New York: John Wiley and Sons, 1978), 402, and A. R. Luria, *Human Brain and Psychological Processes*, trans. Basil Haigh (New York: Harper and Row: 1966), 5, 132.

114. A more ominous future may be glimpsed not just in the animal cruelty but also in Fritsch's future inquiries into the color of native eyes and hair in hopes of establishing the physiological ground of native inferiority. Fritsch's later scientific work was invested in securing the scientific basis of racial difference, cut with all manner of domesticating impulses, not least of which was normalizing drives of romantic nationalism and racial purification. See Harry Grundfest, "The Different Careers of Gustav Fritsch (1838–1927)," *Journal of the History of Medicine and Allied Sciences* 18, no. 2 (1963): 128. On the demands of national sovereignty that inflected these experiments, see Phillip J. Pauly, "The Political Structure of the Brain: Cerebral Localization in Bismarckian Germany," *International Journal of Neuroscience* 21, no. 1 (1983): 145–49.

115. Despite the fact that Hitzig and Fritsch used the word *Seele* (soul) to describe the volition that moved the muscles, the neurologist Harry Grundfest declared in 1963 that after their 1870 paper, "thenceforth, the study and analysis of the central nervous activity could the more easily disregard mystical notions of soul or spirit and concentrate on mechanisms based on a physical substrate" ("Different Careers," 125).

116. Roberts Bartholow, "Experimental Investigations into the Functions of the Human Brain," *American Journal of the Medical Sciences* 67 (1874): 305–13.

attached screw-clamps to the brain matter of immobilized cats, monkeys, and rabbits. "The animal employed was etherized," reported Caton, "and a portion of the scalp, skull, and dura mater removed, so as to expose the greater part of one hemisphere." A fluctuating current appeared when two electrodes were placed on the surface of the head and one directly on grey matter. By using high-quality electrodes Caton was able to decrease the intensity of the surrounding noise. And by shining an oxyhydrogen lamp onto a mirrored image of the galvanometer he was able to amplify patterned flashes up to eight feet high. Blips and rolls, larger than life, projections whose source was a heretofore unseen interiority. Watching the wall, Caton measured changes in signal due to pinching and pain, blinking and chewing. He isolated and compared the patterns associated with wakefulness, sleep, anesthesia, and death.[117]

Embedded in these early theories of localization was the idea that the different parts of the brain necessarily interacted with each other and with the world at large.[118] Embedded, too, was a significant step in the manipulation of the brain in service of machines. In his reverse engineering of the experiments of Hitzig and Fritsch, Caton not only confirmed their localization thesis but also demonstrated that tremulations within the cerebral cortex could be scanned for pattern, measured, and manipulated. Here the visualization of the ethereal emanations of consciousness assumed mechanical legitimacy and what might be referred to as a superadded character.

The bloody experiments of Caton, Hitzig, and Fritsch garnered international attention and were integral to a burgeoning science of electroencephalography. In anticipating the "masculine morphology" that historian Elizabeth A. Wilson has identified as central to cybernetic theories of cognition, budding electroencephalographers sought to isolate consciousness, apart from the sensual life of the body, in the form of what is now commonly known as the brain wave.[119] The coded frailties, femininities, and suscepti-

117. Richard Caton, "Electrical Currents of the Brain," *British Medical Journal* 2 (August 28, 1875): 257.

118. On the latency of connectionism within nascent theories of brain localization, see Katja Guenther, *Localization and Its Discontents: A Genealogy of Psychoanalysis and the Neuro Disciplines* (Chicago: University of Chicago Press, 2015).

119. Versions of computationalism proceed apace, writes Elizabeth A. Wilson, "not because of the veridicality of the computer metaphor, but because what is presupposed about embodiment in this metaphor fits with certain masculinist presumptions about psychological functioning" ("'Loving the Computer': Cognition, Embodiment, and the Influencing Machine," *Theory and Psychology* 6, no. 4 [1996]: 579).

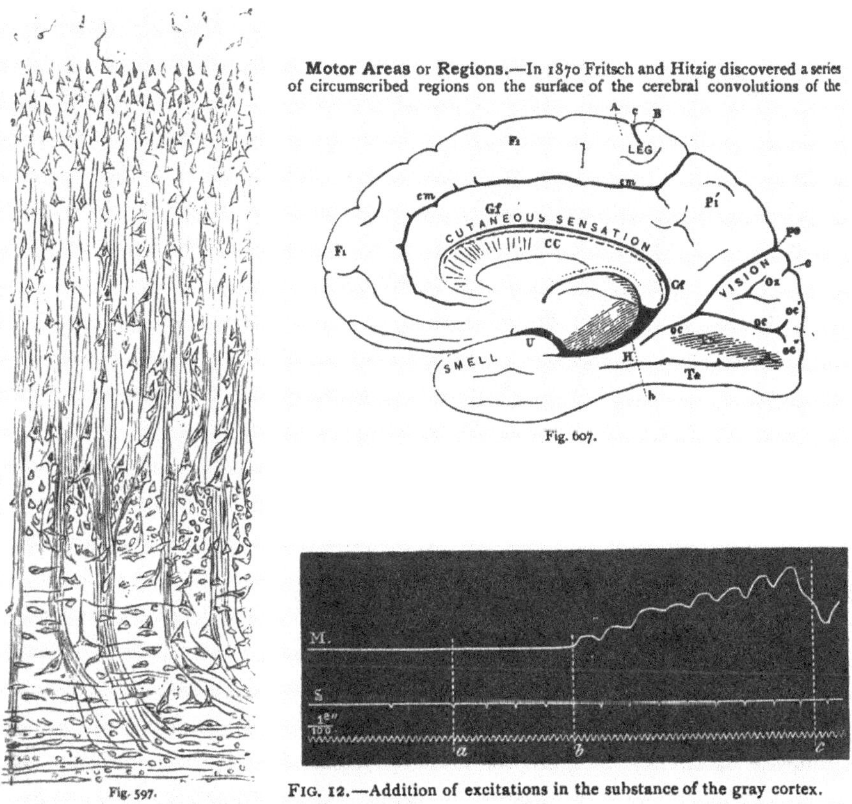

FIGURES 24A, 24B, 24C. Triptych: a) Vertical section of motor cerebral convolutions of man; b) Motor Areas, from L. Landois, *A Textbook of Human Physiology*, 4th ed., with additions by William Stirling (Philadelphia: P. Blakiston, Son and Co., 1892); and c) Excitations, from Chas. Richet, *Physiology and Histology of the Cerebral Convolutions*, trans. Edward P. Fowler (New York: W. M. Wood, 1878).

bilities of the body were, in other words, giving way to new programs of cerebral subjectivity.

In 1892, for example, electrical engineer Edwin J. Houston addressed what he called the "mechanism of cerebration." Houston surmised, in an anticipatory key (and one more precise if not altogether consistent with Knowles's musings on the brain wave decades earlier), that "cerebration or thought, whatever may be its exact mechanism, is accompanied by molecular or atomic vibrations of the gray or other matter of that part of the brain called the cerebrum." Brain cells, moreover, during the process of cerebration "assume certain groupings or relations toward one another." Houston then offered the hypothesis that cerebral energy was given off in the process of these cells achieving "certain groupings or relations." He fantasized about a machine that could record "thought-waves" as they happened in real time, for such "thought-record[s], suitably employed, might be able to awaken at any subsequent time in the brain of a person submitting himself to its influences, thoughts identical to those recorded." Houston acknowledged that such a machine did not yet exist; nonetheless he looked to a future in which mechanization was inevitable, a future in which "the thoughts of an active brain might be recorded continuously and permanently on a moving sensitized film, the exposed surface of which was placed at the focus of a large lens placed in front of any one engaged in intense thought."[120]

As part of increased interest in experimental brain anatomy in the second half of the nineteenth century, Houston premised his work on the "existence of the universal or luminiferous ether, which is now generally accepted in scientific circles." Ether made possible the process of cognition. It was both itself and part of everything it animated and amplified, "bearing in mind the fact that this ether passes through even the densest matter as easily as water through a sieve." On one hand, Houston's matter-of-fact reference to the ether paradigm was indicative of its lingering hold on biological, psychological, and even social sciences such as anthropology. Ether, or something like it—culture, the public, the spirit-world, animal magnetism, odic force, mesmeric energy, sympathy, mana—remained shorthand for the mediating substance and force between matter and spirit—that which existed but could not yet be measured exactly. On the other hand, Houston was actively seeking to provide a molecular explanation of the ether function. For even as the notion of the ether lost traction among physicists and other hard scientists, the ether paradigm did not disappear when it came to the brain.[121] Instead, within neurophysiological circles the ether paradigm assumed a newfound precision as new instruments were utilized to investigate what

120. Edwin J. Houston, "Cerebral Radiation," *Proceedings of the Electrical Section of the Franklin Institute* 2 (1892), 65, 66–67, 71. Houston had served as chief electrician of Philadelphia's International Electrical Exhibition in 1884 and turned his attention to the brain at the very same moment he was merging his electric company with Edison Electric Company to form General Electric. "Cerebral Radiation" was reprinted in the *Psychical Review* 1 (1892): 150–57.

121. Houston, "Cerebral Radiation," 66; Alvin M. Schrader, "In Search of a Name: Information Science and Its Conceptual Antecedents," *Library and Information Science Research* 6 (1984): 227–71; John Durham Peters, *Speaking into the Air: A History of the Idea of Communication* (Chicago: University of Chicago Press, 1999), 103.

constituted the medium of consciousness and to measure the traffic between neural emanations and environment.

Beginning in 1924, for example, the psychiatrist Hans Berger sought to map the relationship between electrical activity in the brain and subjective states of consciousness. Whereas Houston had been explicit in naming the "highly elastic, easily movable medium" that enabled the mechanism of cerebration, Berger merely assumed that the environment was ready-made to receive and transfer brain waves. The ether paradigm, in other words, became ethereal as Berger investigated the cognitive dimensions of human communication. Driven by a personal experience of telepathy during a military training exercise, Berger was invested in the prospect that feelings, emotions, and consciousness could be communicated at a distance. Such communication itself possessed a material component. Consequently, there was a pressing demand to measure all of the energy, in its different forms, revolving around and moving through the cortex.[122]

Berger's self-consciously scientific efforts were in the service of providing an empirical index of psychical energy. In Berger one can witness the attempt to figure the brain wave as matter *and* spirit, that is, as the referent of neural activity that could be measured and, *simultaneously*, as the mark of consciousness, itself, that could not be reduced to mere materialism. Berger's efforts did not, in the end, validate his hypothesis of "spontaneous telepathy." Despite his failure to confirm a mode of communication devoid of noise, Berger's science was notable for its will to calibrate and its refinements of the moving-coil galvanometer that measured millimeters and microvolts.[123]

In the first few decades of the twentieth century, technologies for isolating and amplifying the electrical activity of neurons became more reliable, accurate, and noninvasive. The interwar boom in neurophysiology, it should

122. Hans Berger, "Über das Elektrenkephalogramm des Menschen (On the human electroencephalogram). *Archiv für Psychiatrie und Nervenkrankheiten* 87 (1929): 527–70. For a particularly sharp analysis of Berger, see Cornelius Borck, *Brainwaves: A Cultural History of Electroencephalography*, trans. Ann M. Hentshel (London: Routledge, 2018), 25–75; David Millett, "Hans Berger: From Psychic Energy to the EEG," *Perspectives in Biology and Medicine* 44, no. 4 (2001): 522–42.

123. Jonna Brenninkmeijer, *Neurotechnologies of the Self: Mind, Brain and Subjectivity* (London: Palgrave Macmillan, 2016), 53; Hans Berger, *Psyche* (Jena: Fischer, 1940), 16; Borck, *Brainwaves*, 57–59; Pierre Gloor, "Hans Berger on the Electroencephalogram of Man," *Electroencephalography and Clinical Neurophysiology* 28, no. 1 (1969), 13–15, 32. E. D. Adrian and H. C. Matthews, "The Berger Rhythm: Potential Changes from the Occipital Lobes in Man," *Brain* 57, no. 4 (December 1934): 358.

be said, owed much to clinical discoveries and decisions made during World War I in relation to the effects of shellshock on the blood, the heart, the lungs, and, of course, the brain. It also owed much to government funding and efforts to make military campaigns more strategic and systematic.[124] Shadowed from its Swedenborgian beginnings by military might and the conflagrations of white male rage, the brain wave, or something like it, continued to come into focus.

Alexander Forbes, for example, graduated from Harvard Medical School in 1910. The grandson of Ralph Waldo Emerson and son of the first president of Bell Telephone Company, Forbes served as a radio engineer in the US Navy during World War I. With a skilled mix of neurophysiology and communications engineering, Forbes would soon play a key role in isolating the charge of individual neurons. At sea Forbes had used vacuum tubes to amplify radio signals on navy warships. After the war he used those same vacuum tubes to amplify nerve impulses. By 1922 Forbes was able to isolate the action potential of individual neurons, displaying them visually in order to measure them precisely.[125]

Over the next two decades the field of neurophysiology was abuzz with the notion that the brain could be accessed and translated and, eventually, managed with the proper instrumentation. In his pursuit of the laws of nervous activity, for example, Forbes drew from the work of Edgar Adrian (who would go on to win the Nobel Prize in 1932 for his work on the functioning of neurons). As the ethereal paradigm faded into the background, scientists like Forbes and Adrian continued to work on squaring the circle between the brain and its surround. In 1934 Adrian offered the first public demonstration of the existence of a brain wave (he also took his demonstrations on tour in the United States that same year).[126] Electrodes placed just below the

124. As Walter noted in 1953, new electronic parts that "were developed for radar apparatus during the war" hastened the refinement of electrodes and amplifiers used in electroencephalographic instrumentation. Walter, *The Living Brain*, 87. See also Louise H. Marshall, "Instruments, Techniques, and Social Units in American Neurophysiology, 1870–1950," in *Physiology in the American Context, 1850–1940*, ed. Gerald L. Geisson (New York: Springer, 1987), 360.

125. Justin Garson, "Alexander Forbes, Walter Cannon, and Science-Based Literature," *Progress in Brain Research* 205 (2013), 243; Alexander Forbes, "The Interpretation of Spinal Reflexes in Terms of Present Knowledge of Nerve Conduction," *Physiology Review* 2 (1922): 361–414.

126. Kenton Kroker, *The Sleep of Others and the Transformations of Sleep Research* (Toronto: University of Toronto Press, 2007), 272; Borck, *Brainwaves*, 164–79.

surface of an immobilized cranium, drilled into a shaved part of the scalp, transmitting the frequencies that were found therein.

10. BRAIN WAVES AND TREMULATING INFORMATION

By midcentury the science of electroencephalography (soon EEG) had become a celebrated new field, its elegant mathematical renderings of the intrinsic oscillatory behavior of neurons invested with mass appeal. Already working in concert with burgeoning theories of information, electroencephalographers made the technical promise to record, in real time, electrical activity in the cerebral cortex.[127] Calculations assumed that what was being isolated and measured was the function of cognition itself.[128]

EEG and the technical expertise of recording the brain's response to its environment were integral to the making of the neuromatic paradigm. Most significantly, the mathematics being used to theorize the relationship between the brain wave and its neural fundament were working out some of the same mathematics that would soon complement and fuel the diffusion of information theory.[129] By 1926, for example, Adrian had already begun to describe nerve impulses in binary terms, terms that would also structure Shannon's mathematical theory of communication—information

127. Rhodri Hayward, "The Tortoise and the Love-Machine: Grey Walter and the Politics of Electroencephalography, *Science in Context* 14, no. 4 (2001): 617. On the development of an analog correlator system designed specifically to measure brain potentials, see J. S. Barlow, "The Early History of EEG Data-Processing at the Massachusetts Institute of Technology and the Massachusetts General Hospital," *International Journal of Psychophysiology* 26 (1997): 443–54.

128. Mary A. B. Brazier, "Neural Nets and the Integration of Behavior," in *Perspectives in Neuropsychiatry*, ed. Derek Richter (London: H. K. Lewis, 1950), 37.

129. Electrical engineers, for example, had long investigated how the continuous waves of electrical signals could be converted to discrete data; see, e.g., Harry Nyquist, "Certain Factors Affecting Telegraph Speed," *Transactions of the American Institute of Electrical Engineers* 43 (1924): 412–22. See also the first English language review of EEG by Herbert H. Jasper and Leonard Carmichael, "Electrical Potentials from the Intact Human Brain," *Science* 81 (1935): 51–53. The revolution in microelectrodes has since allowed neuroscientists to measure, at the level of the neuron, the "continuous analog signal" emitted by the brain. See, e.g., Karen A. Moxon, "Neuralbiotics: Opening Novel Lines of Communication," in *Neural Engineering*, ed. Bin He (Boston: Springer 2013), 192ff.

flow, message coding, signal and noise.[130] Adrian insisted that any single neuron was tasked with the responsibility of communication, transmitting messages from the environment to the brain and enabling it to respond not just accordingly but also aggressively. "To take an analogy," wrote Adrian, "the nervous system may be likened to a stream of bullets from a machine gun, it cannot be likened to a continuous stream of water from a hose." In this analogy, the brain wave was a discrete rather than an analog phenomenon. On-again, off-again rather than continuous.[131] The activity of neuronal firing patterns, which were themselves considered to be the blueprint of consciousness, was akin to the rat-a-tat spray of bullets. In combination with one another, neurons produced oscillations which could be measured. Different frequencies could then be found and compared. Abnormalities could be anticipated. Predictively if not precisely.[132]

Ready-made for theorization into an informational cosmos, brain waves signaled the activity of neural nets, which, in turn, signaled the transmission of signal from one neuron through a specific channel to another.[133] The brain wave, considered to be an electrical representation of activity at the neuronal level, promised unprecedented access to the sine qua non of the human species. For those who followed in the footsteps of Hitzig, Fritsch, Caton, Berger, Forbes, and Adrian, the object of investigation was nothing less than the bass-line rhythm of sapience itself and, significantly, how such sapience was a product of the communicative back and forth between brain and environment. Indeed, the brain wave was not only a sign of the truth of

130. See, for example, the discussion in Justin Garson, "The Birth of Information in the Brain: Edgar Adrian and the Vacuum Tube," *Science in Context* 28 (2015): 31–52.

131. Edgar D. Adrian, *The Basis of Sensation: The Action of the Sense Organs* (New York: W. W. Norton, 1928), 27; cited in Garson, "The Birth of Information," 47. See also Adrian, "All-or-none Principle," 460–74. On the prehistory of the neural doctrine, see the work of Golgi and the neural reticulum theory: C. Golgi, "Sulla struttura della grigia del cervello," *Gazetta Medica Italiani Lombardi* 6 (1873): 244–46; republished as C. Golgi, "On the Structure of Gray Matter of the Brain," Golgi Centennial Symposium, *Proceedings*, ed. M. Santini (New York: Raven Press, 1975), 647–50.

132. Henry Quastler, "A Primer on Information Theory," in *Symposium on Information Theory in Biology*, Gatlinburg, TN, October 29–31, 1956, ed. Hubert P. Yockey, Robert L. Platzman, and Henry Quastler (New York: Pergamon Press, 1958), 19.

133. See, e.g., D. O. Walter and M. F. Gardiner, "Some Guidelines from System Science for Studying Neural Information Processing," *International Review of Neurobiology* 13 (1970): 343–74.

information processing but also an invitation to think about how communication loops were established between the inside and outside of the skull.[134]

Leverage on these neural flows of information was also provided by increasingly refined notions of self-organization generated by midcentury cybernetics. Although communication in Shannon's model was for the most part linear, it was a theory that would lend itself to insights into recursivity and a vision of communication going back and forth, perpetually, between system and environment. As neurophysiologists and cognitive scientists applied information theory in their research, they did so with feedback on their minds.[135] For once Shannon published his paper, neurophysiologists began to explicitly frame their understanding of neural networks in terms of feedback and information theory. In defining communication as the atomic logic of control, scientists promoted the brain for its incredible capacity to continuously convey nonsemantic messages at the level of the neuron (to, fro, and with).

As it was read alongside emerging theories of neural nets, information theory gained much of its cognitive traction and soon came to mark, in statistical terms, the predictability of neuronal firing (and not the meaning of that process). In their groundbreaking work of 1943, "A Logical Calculus of the Ideas Immanent in Nervous Activity," Warren McCulloch and Walter Pitts investigated the "macro effects displayed by nervous systems." With McCulloch working on military applications of chemical and biological warfare and Pitts about to join the Manhattan Project under the auspices of the Kellex Corporation, their article was to provide a stable base for calculating the pathways through which neurons received and transmitted electrical signals.[136]

According to McCulloch and Pitts, neurology could be pursued in terms of logical propositions. A neuron either fired or it did not. There was no in-between. Adrian's "all-or-none" law of neural transmission guaranteed that the activity of any single neuron could be represented as a proposition—on

134. W. Grey Walter, "Rhythm and Reason," *Electroencephalography and Clinical Neurophysiology* 2 (1950): 203.

135. See, e.g., Lawrence Kubie, "A Theoretical Application to Some Neurological Problems of the Properties of Excitation Waves Which Move in Closed Circuits," *Brain* 53, no. 2 (1930): 166–77.

136. Warren S. McCulloch and Walter Pitts, "A Logical Calculus of the Ideas Immanent in Nervous Activity," *Bulletin of Mathematical Biophysics* 5, no. 4 (1943): 115–33; Kay, "From Logical Neurons," 595, 601.

or off, a 1 or a 0—each proposition relating to other propositions that could be mapped in the form of complex mathematical equations. A single neuron may give or receive an electrical charge from thousands of other neurons. Moreover, a neuron could produce an excitatory or inhibitory charge. In positing that neurons operated both linearly and recursively, McCulloch and Pitts offered, hypothetically, a series of equations for tracking the effects of any given neuron, its relationality to other neurons and, in turn, how the sum of that relationality related to moments of cognition.[137] Their equations were intended to define rules of behavior in terms of neural nets "containing circles" and to account for the fact that there were many pathways through which a message could travel back and forth and through a nervous system. Neurons, even more than the circuits that they created, took precedence as the object of algorithmic inquiry. There was input. There was output. And there was process in between:

> Think of a neuron as a telegraphic relay which, tripped by a signal, emits another signal. To trip and to reset takes, say, a millisecond. Its signal is a briefer electrical impulse whose effect depends only on conditions where it ends, not where it begins. One signal or several at once may trip a relay, and one may prevent another from doing so. Of the molecular events of brains these signals are the atoms. Each goes or does not go. All any neuron can signal to the next is that it was tripped. Thus the signal received is an atomic proposition.[138]

Consequently, the major contribution of McCulloch and Pitts was their isolation of the neuron in terms of both space and time (their discovery being concurrent with the invention of microelectrodes and the capacity to monitor the activity of individual neurons). And as is now estimated, ten million

137. Adrian, "All-or-none Principle." In their 1947 article, "How We Know Universals," McCulloch and Pitts were already using Wiener's statistical concept of information to discuss environmental input in terms of scanning. McCulloch would soon refer to information as a "hypothetical mechanism" in order "to demonstrate the existential consequences of known characters of neurons" ("How We Know Universals: The Perception of Auditory and Visual Forms," *Bulletin of Mathematical Biophysics* 9, no. 3 [1947]: 127–47). See also Donald M. MacKay and McCulloch, "The Limiting Information Capacity of a Neuronal Link," *Bulletin of Mathematical Biophysics* 14 (1952): 127–35.

138. Warren McCulloch, "Machines That Think and Want" (1950), in *Embodiments of Mind*, 307.

bits of information arrive through the sensory organs and are processed by the brain every second.[139]

Such intense focus on the atomic was foundational for the scalable worldview of cybernetics. For cyberneticians, in general, the brain was the hinge between animate biological specimens and inanimate machines. The brain was the key to finding out—by way of modeling—the "fundamental principles of control that apply to all large systems." Made up of "specialized computers" that "think and want," that maintained dynamic equilibrium, the brain could harness the all-but-infinite possibilities contained within it. "In the living brain, the raw variety [of potential neuronal interactions]," wrote operations research pioneer Stafford Beer in 1972, "is probably something like 2 (to the 10^6 x 10^7) which has well been described as perhaps the largest number that has ever been taken seriously."[140]

Among its many tricks, the increasing authority of the neuromatic paradigm served to naturalize scalability as a property inherent in all systems, the macro ever reflecting the micro and vice versa.[141] In framing all of nature—organic and inorganic—in terms of information processing systems, cyberneticians saw the neural network of the brain as the most definitively complex object on which to build their models.[142]

139. Max Tegmark, *Life 3.0: Being Human in the Age of Artificial Intelligence* (Vintage: New York, 2017), 291–92.

140. In his designs for EDVAC (Electronic Discrete Variable Arithmetic Computer), von Neumann modeled his switching elements not on material objects such as vacuum tubes and mechanical relays but rather on "idealized neurons" as they were formulated by McCulloch and Pitts. John von Neumann, *The Computer and the Brain* (New Haven, CT: Yale University Press, 1958); Stafford Beer, *Brain of the Firm: A Development in Management Cybernetics* (London: Allen Lane, 1972), 32, 121, 100, 65.

141. Anna Lowenhaupt Tsing, "On Nonscalability: The Living World Is Not Amenable to Precision-Nested Scales," *Common Knowledge* 18, no. 3 (2012): 509n5. Swedenborg's *The Brain* was republished as *Three Transactions on the Cerebrum* in 1938 and 1940—amid the differentiation of brain waves (alpha, beta, delta, etc.) and similar efforts to turn neural flesh into mathematical propositions and fodder for computer engineers. In Ashley Montagu's 1941 review of the republication in *Isis*, Swedenborg is framed once again as a visionary whose insight had yet to be fully mined in light of the oscillatory nature of nerve impulses ("Review of the Swedenborg Scientific Association republication of *Three Transactions on the Cerebrum*," *Isis* 33, no. 2 [1941]: 262–63).

142. So while there were analog dimensions to neural activity, McCulloch and Pitts offered a logical calculus of the digital premised on correspondence rather than continuity. Shortly after publication of their article, Pitts wrote McCulloch that he was working on a book with Norbert Wiener at MIT that would "constitute the first adequate discussion of

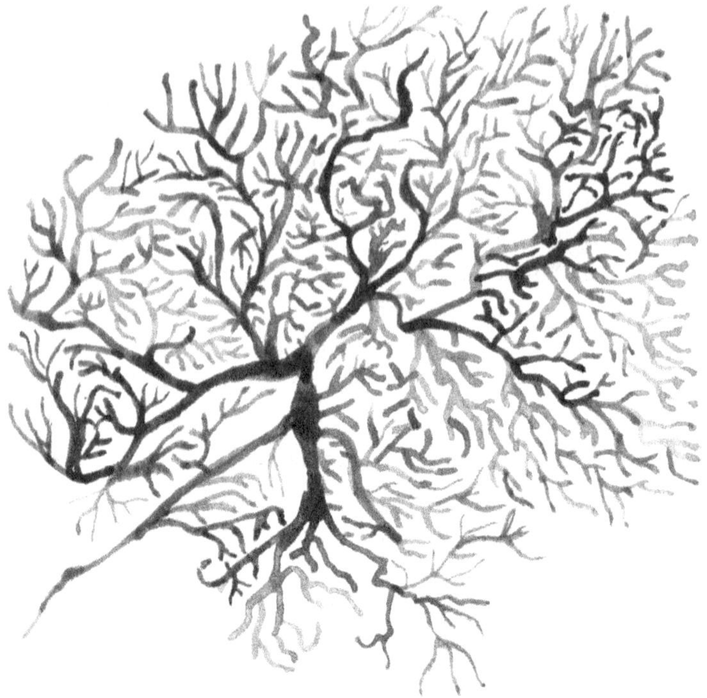

FIGURE 25. Representation of a profusely branched dendritic surface of a human nerve cell. Image created by Libby Modern.

Indeed, information theory would soon make it possible to move seamlessly across different registers—from the network of neurons underlying the brain wave to the weather and environment to transportation systems to the market to machines. Just as the ether had been used by earlier generations to make sense of a corresponding universe, information would soon be enthusiastically deployed by postwar scientists and engineers to theorize the compatibilities between systems as they were found in every nook and cranny of existence.[143] Given that all systems were defined by their capac-

statistical mechanisms understood in the most general possible sense, so that it included for example the problem of deriving the psychological, or statistical, laws of behavior from the microscopic laws of neurophysiology." Cited in Elizabeth A. Wilson, *Affect and Artificial Intelligence* (Seattle: University of Washington Press, 2011), 157n9.

143. See also the theosophist Alice A. Bailey and her concurrent discussion of the "etheric vehicle" of consciousness: *Telepathy and the Etheric Vehicle* (New York: Lucis Publishing Company, 1950), 140.

ity to process information, the use of mathematical formalization could be extended into previously uncharted territories—beyond military planning and antiaircraft ballistics into business, society, economy, mental health, medicine, and biology.[144]

11. BIOFEEDBACK AND THE EXPERIENCE OF CORRESPONDENCE

With the exponential increase in cybernetic publications from the end of World War II to the 1970s, the neuromatic paradigm assumed popular currency outside of the laboratory.[145] The brain's voracious capacity for self-organization and information processing had been put on vivid display by neurophysiologists and their machines. During the cybernetic era of momentous fact-making, the electroencephalogram arrived ready-made for analogy and inspired the application of neuromatic frames to human being, to the world around, and to the connections between.

In addition to the more elite environs traversed in this chapter, these applications manifested in popular and therapeutic cultures in the form of self-help guides that promised control if only one modeled one's life on the capacities of a computer.[146] Take, for example, Maxwell Maltz's *Psycho-*

144. Kenneth E. Boulding, "General Systems Theory—The Skeleton of Science," *Management Science* 2, no. 3 (1956): 197–208; Beer, *The Brain of the Firm*, 68.

145. Geof Bowker, "How To Be Universal: Some Cybernetic Strategies, 1943–70," *Social Studies of Science* 23, no. 1 (February 1993), 108. See, e.g., David M. Rorvik, *As Man Becomes Machine: The Evolution of the Cyborg* (New York: Doubleday, 1971); Jodi Lawrence, *Alpha Brain Waves* (New York: Avon Books, 1972); Gertrude Schmeidler and Larry Lewis, "Mood Changes after Alpha Feedback Training," *Perceptual and Motor Skills* 32 (1971): 709–10. Biofeedback was not without controversy. On biofeedback as the "parlour game of the century," see Marilyn Ferguson, *The Brain Revolution: The Frontiers of Mind Research* (New York: Taplinger, 1973), 93.

146. Y. Saparina, *Cybernetics within Us* (North Hollywood: Wilshire Book Co., 1967); Claude N. Rosenberg Jr., *Psycho-Cybernetics and the Stock Market: The Key to Maximum Investment Profits and Peace of Mind* (Chicago: Playboy Press, 1971); Sydney Petrie and Robert B. Stone, *Hypno-Cybernetics: Helping Yourself to a Rich New Life* (West Nyack, NY: Parker Publishing, 1973); Benito F. Reyes, *Cybernetics of Consciousness: Theory and Practice of Meditation* (Ojai, CA: World University, 1978); Timothy Leary, *Changing My Mind and Others* (New York: Prentice Hall, 1982); Ray Kurzweil, *How To Create a Mind: The Secret of Human Thought Revealed* (New York: Viking, 2013).

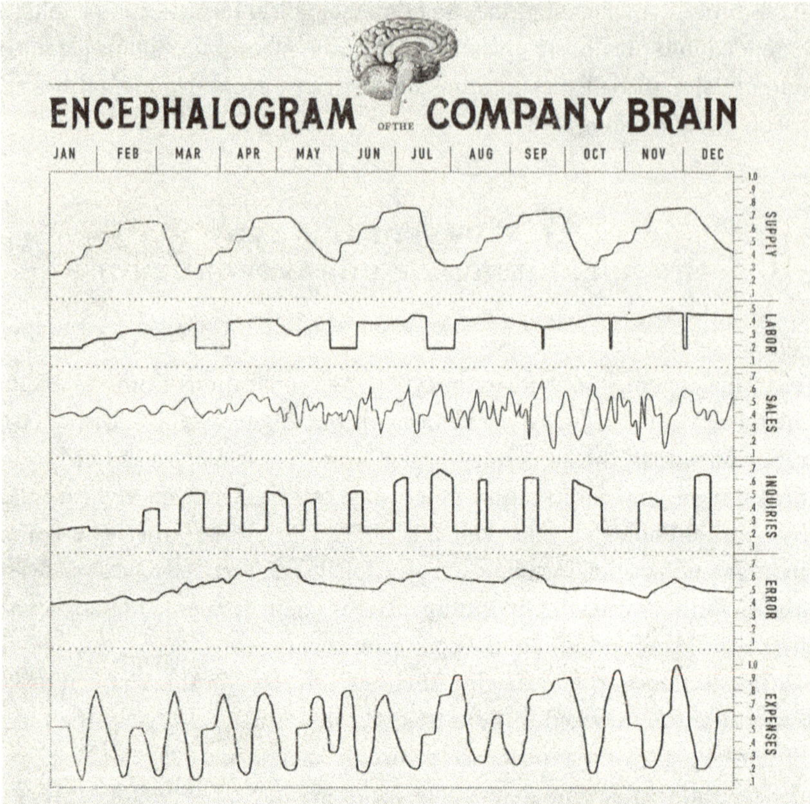

FIGURE 26. Encephalogram of the Company Brain, from Stafford Beer, "Towards the Cybernetic Factory," in *Principles of Self-Organization*, ed. Heinz von Foerster and G. W. Zopf (Oxford: Pergamon, 1962), 25–89. Approximation by Libby Modern.

Cybernetics: A New Way to Get More Living Out of Life (1960) in which he offers a plan for engineering your "success mechanism." To take hold of one's "spiritual thermostat" was to disabuse oneself from false beliefs at the level of information processing.[147] Such purity meant acting (and thinking about acting) in light of exchanges with the environment one sought to leverage. Maltz's guide, in the ridiculousness of its instruction, distilled the serious stakes of the coming information age, namely, the disciplinary programming involved in becoming master over one's own neural network.

Pitched as a science of self-control, biofeedback trained subjects in the

147. Maxwell Maltz, *Psycho-Cybernetics: A New Way to Get More Living out of Life* (North Hollywood, CA: Wilshire Book Co., 1960), 186, 44f.

FIGURE 27. Album cover of *Psycho-Cybernetics* by Maxwell Maltz, MD (Waco, TX: Success Motivation Institute, 1965). *You were to become your brain so as to allow it to become you.*

delicate process of synching with their own brain waves for the purposes of managing the feedback loop between brain and environment to maximum advantage and, by extension, controlling their own emotional states.[148] Within the practice of biofeedback the brain became a means of emotional regulation and self-control precisely to the degree that it was disciplined. As one proponent of biofeedback wrote, "Just as rats can be taught to press

148. Kenneth Gaarder, "Control of States of Consciousness: Attainment through External Feedback Augmenting Control of Psychophysiological Variables," *Archives of General Psychiatry* 25 (November 1971): 440; Erik Peper, "Localized EEG Alpha Feedback Training: A Possible Technique for Mapping Subjective, Conscious and Behavioral Experiences," *Kybernetik* 11 (1972): 166–69.

a bar, so people can be taught conscious control of their brain activity in a relatively short time."[149] Biofeedback promised nothing less than the skilled practice of self-organization, that is, a particularly intense and intensely reflexive form of neural submission.[150]

A premise of biofeedback was that the brain wave was an access point to one's essential interiority, a real-time glimpse into the ontic activity of neurons. As pharmacologist and biofeedback pioneer Barbara Brown explained, brain waves reflected a deeper physiology as they were "the outpouring of the infinitely varying interplay of working brain cells." As such, they were the hinge for designing a better, faster, and stronger brain. Brown's research agenda was to recognize heretofore undiscovered aspects of one's own organization, that is, "aspects of consciousness." Using a Grass Model 6 Electroencephalograph, Brown successfully isolated different brain wave frequencies and their corresponding emotional states while she was chief of experiential physiology research at the Veterans Administration Hospital in Sepulveda, California: "Alpha: calm, peaceful, pleasant. . . . Beta: angry, aggravated, irritated, impatient, unhappy. . . . Theta: vacillating."[151] Consequently, biofeedback yielded "the information [the subject] needs to explore himself and find the inner senses he needs to recognize antecedents to the psychic event, to help it along, and eventually, perhaps, to bring it under

149. Joseph Kamiya, "Conscious Control of Brain Waves," *Psychology Today* 1 (1968): 56–60. See also his "Conditioned Discrimination of the EEG Alpha Rhythm in Humans: A Fourth Dimension of Consciousness," *Experimental Medicine and Surgery* 27 (1969): 13–18.

150. Neurofeedback therapy is often recommended for children with ADHD. Jonna Brenninkmeijer, "Neurofeedback as a Dance of Agency," *BioSocieties* 8, no. 2 (2013): 145.

151. Barbara B. Brown, *New Mind, New Body: Bio-feedback: New Directions for the Mind* (New York: Harper and Row, 1974), 309, 304; Barbara B. Brown, "Recognition of Aspects of Consciousness through Association with EEG Alpha Activity" *Psychophysiology* 6 (1970): 442–52. See also Barbara B. Brown and Jay W. Klug, eds., *The Alpha Syllabus: A Handbook of Human EEG Alpha Activity* (Springfield, IL: Charles C. Thomas, 1974) and Barbara B. Brown, *The Biofeedback Syllabus: A Handbook for the Psychophysiologic Study of Biofeedback* (Springfield, IL: Charles C. Thomas, 1975). See also *Biofeedback and Self-Control 1973: An Aldine Annual on the Regulation of Bodily Processes and Consciousness*, ed. Neal E. Miller et al. (Chicago: Aldine Publishing Company, 1974). On Brown's occult leanings, see her "Auto-Control of Consciousness: The Next Revolution," banquet address delivered at the Parapsychological Association Annual Convention, New York City, September 10, 1970, reprinted in *Proceedings of the Parapsychological Association* 7, ed. W. G. Roll and J. D. Morris (Durham, NC: Parapsychological Association, 1972), 81–95; Barbara B. Brown, "Awareness of EEG-Subjective Activity Relationships Detected within a Closed Feedback System," *Psychophysiology* 7, no. 3 (November 1970): 458.

voluntary control." Anxiety, fear, pain, and depression could all be overcome by knowing their cognitive roots and learning to manipulate them at the neuronal level. We must, in other words, learn what activates us, what allows us to feel this or that, to act one way or another, to think one thought as opposed to any other.

According to Brown, biofeedback would hasten the development of "a new state of consciousness"—not only "becoming aware" but "becoming aware of becoming aware." More than merely a variety of personalized religion, Brown offered you control over that which was most private—enabling you to arrive in the space of knowing yourself knowing. Brown pitched such control—simultaneously epistemological and political—as a form of *secular* liberation.[152]

Born, in part, from her participation in Department of Defense discussions on behavioral control research, biofeedback was a counter to all forms of bad religion. Indeed, Brown viewed biofeedback as a triumph of the scientific method whose secular foundations were being eroded—not only by contemporary fads of Buddhism and Scientology but also by infidel practices of elite scientists. In terms of the latter, Brown wrote that "If one takes time to examine the validity of scientific authoritarian concepts, the tenuousness of scientific structure may be exposed, its sacredness may peel away." Brown's project of demystification was aimed at science's lack of self-awareness and its inability to take the measure of its own questions and their origin.[153]

In succumbing to the pressures of the "Industrial-university complex," Brown accused "scientifically established VIPs" of having followed the "objectives of Industry alone." From the position of out science-ing science, Brown lamented the "societal pressure on biopsychological research to produce within a structure that accepts only measurements and observations made

152. Brown, "Auto-Control," 95; Brown, *New Mind*, 395. See also the quip that "a computer that can store information, compare the relation among several measures in a single subject, and produce an instantaneous feedback" would "strip" the practice of meditation "of its mystical quality" (Kamiya, "Conscious Control of Brain Waves," 59).

153. Brown, *New Mind*, 76, 38, 51; Brown, "Auto-Control," 85. Brown's project of demystification was also a gendered critique of scientism (its bureaucracy, authoritarian claims to expertise, and selective institutional memory). As such, Brown aligns the liberation of women with the liberation of science. She sees both women and science as enchained by clericalism. Brown positions herself as midwife and missionary, unafraid to tend to "the turmoil and excitement of the birth trauma of bio-feedback. . . . Because I am an iconoclast in the world of science, and because I was there" (Brown, *New Mind*, 19).

in terms of the physical universe." This "total influence of industry" is the mark of a "masculine" hubris that accounts only for "material" riches at the expense of "spiritual" exploration. As Brown admitted, energy, food, drugs, cleansing agents, as well as "monster computers (now invading our private lives)" have made life comfortable and efficient. "But what have we done," she retorted, "to increase the efficiency of our own mind operations?" Consequently, Brown called for the investigation of "the *socio*-scientific forces which have produced bodies of scientific knowledge and scientific perspectives almost exclusively for the material aspects of life." By cultivating "an awareness of awareness," working scientists might begin cultivating the kind of individuality that she prescribes to her reader—managing the impressions that the environment inevitably makes on them.[154] *Spirituality 2.0.*

In Brown's use of the neuromatic paradigm of cognition, one begins to suspect that her version of the secular age was one in which it is difficult to discern where the brain ends and the computer begins.[155] Yet such vagueness is strategic, for once the boundary between the brain and the computer is irrevocably erased, the programming of the human and the programming of the machine become one in the same disciplinary task. For in Brown's cosmology, the moment of liberation is precisely the moment when you submit to the machine. Its gift is your freedom. Brown was committed to a political order in which each individual was wholly in control of its destiny, wholly able to come to terms with and provide terms for the totality of its existence. Brown's project, then, was not so much invested in transforming individuals into computational engines but rather enabling humans to inhabit, as if for the first time, the true nature of their sovereignty.

For in gazing on one's own brain waves, the biofeedback practitioner sought to manipulate and manage them for purposes of self-optimization. This incredible act of self-reflexivity sheds light on the instantiation and mechanics of a particular kind of secular subject who is both freely in control and simultaneously exposed. For even as the brain wave and its underlying mechanics came to signal a new horizon for human mastery, the desire

154. Brown, *New Mind*, 39, 92, 83, 393.

155. It was not, however, difficult for Brown to discern what secularity entailed. For as Brown insisted, the secular age could be measured neurophysiologically. By measuring the electrical currents of the body, Brown successfully indexed the religious, the secular, and the difference between them—"The students who were strongly anti-Church literally exploded with skin talk whenever a pro-Church statement was made, and the pro-Church students made skin talk when anti-Church statements were made. The students who were neutral in attitude responded little, if at all" (Brown, *New Mind*, 76).

for control was premised on an essential openness to environmental influence. Consequently, Brown's version of the neuromatic brain may appear as a contradiction or paradox, inconsistent with the notion of freedom as an all-or-nothing proposition. But within the making of the human subject vis-à-vis the informatic emanations of the brain, the vulnerability of self signals its plasticity, which, in turn, invites greater control over self and world alike.[156]

12. THE ONTOLOGY OF INFORMATION

> Consciousness is that indefinable characteristic of mental states which causes us to be aware of them.
>
> REUBEN POST HALLECK, *Psychology and Psychic Culture* (1895)

As this chapter has demonstrated, Brown's vision of biofeedback has a history—made possible by the work accomplished in so-called secular venues such as the ramshackle laboratory of Hitzig and Fritsch and the modernist décor of Bell Laboratories as well as in the musings of Swedenborg, Davis, and your garden-variety phrenological examiner. For each of the figures chronicled in this chapter, information, or something like it, was essential to those moments when brain and environment could be considered wholly mechanized, when both domains could be sufficiently represented by the same model, when the insights from one domain could then be carried over into the other, endlessly, in looping proficiency. The *longue durée* of the neuromatic has yet to end. For mathematical theories of communication can now address the activity of neurons, the fluctuations of brain waves, the secrets of the genome, the etiology of abnormal personalities, trends within a population, and, last but not least, the future behavior patterns and choices made by individuals within that population.[157] In its bent toward fulfillment, the notion of the brain as an information processing machine is an ideal species for biopolitical reproduction. For the neuromatic paradigm is perfectly

156. Brown, *New Mind*, 303. On the notion of the secular subject as buffered, see Charles Taylor, *A Secular Age* (Cambridge, MA: Belknap Press, 2007).

157. The neural ecology was, in other words, a liberal one. According to Brown, "the 20 billion cellular inhabitants of the brain have nearly as many diverse ways of behavior as

pitched to conjure a subject ready-made to heed the call of self-surveillance. A subject ready-made to calculate, to exercise personal future freedom of choice by way of managing the traffic between neural and social networks.[158]

By the time Brown came onto the scene, mathematical theories of communication—originally the purview of engineers—had become integral to the dispersion of statements, concepts, metaphors, and themes that came to envelop whatever was happening when the brain was considered, its components isolated and measured, its meanings pronounced.[159] As Shannon's original, somewhat austere vision of information was written over and across by neurophysiologists, information itself assumed a vital, *almost* molecular character. Whereas Shannon had expressed reservations about indiscriminate application of his mathematical theory of communication, Norbert Wiener did not. Wiener posited information as the missing link between automatic machines and the human nervous system. According to Wiener, information was unlike anything else. "Information is information," he coyly insisted, "not energy or matter." Paraphrasing McCulloch on the strange ontology of information, Wiener declared that "the mechanical brain does not secrete thought 'as the liver does bile,' as the earlier materialists claimed, nor does it put it out as a form of energy, as the muscle puts out its activity. . . . No materialism which does not admit this can survive the present day." According to Wiener, information was carried along in neural nets, in control loops, and even through society itself. The nature of information, as such, was naturally occurring across different registers. Information could be discerned in a written passage or a social survey and "represented by the charges in an electron tube, by the magnetization of a spot on a metallic surface, by the deflection of the beam of a cathode ray tube, by a light falling upon a photographic emulsion, and so forth." So abstract and fungible was information that it was devoid of semantic quality

---

does the entire human body and there is as much variation in their behavioral profiles as there is in the universe of human beings" (*New Mind*, 329).

158. On the cultivation of hypnotic susceptibility by way of biofeedback, see Joseph T. Hart et al., "Hypnotic Susceptibility Increased by EEG Alpha Training," *Nature* 227 (September 19, 1970): 1261–62.

159. R. C. Atkison and R. M. Shiffrin, "Human Memory: A Proposed System and Its Control Processes," in K. W. Spence and J. T. Spence, eds., *The Psychology of Learning and Motivation: Advances in Research and Theory*, vol. 2 (New York: Academic Press, 1968): 89–195.

until meaning was ascribed to its statistical essence *well after the fact*.[160] Yet, as a formal proposition, information was the very stuff of life itself, defined, minimally, as computational capacity.

And this is an irony of mechanization that has heretofore been glossed over in histories of the information paradigm. For as information theory was taken up in a neural key, driving industries of neuroscience and information processing both, the stranger information's material status became. Like the ether, information was a "subtle body."[161] And like the ether, information was that which made humans an immanent part of the larger cosmos.[162] It was, from the neuroscientific perspective, part mind and part body—an "incarnation"[163] for which the spiritual referent—that which it was a materialization *of*—remains unnamed. As biochemist Lila Gatlin observed in 1972 in her review of information theory and biological systems, "information is an ultimately indefinable or intuitive first principle, like energy, whose precise definition always somehow seems to slip through our fingers like a shadow."[164] Information, in other words, was a universal matter that flowed through all "living" things and did not change given the material body that it inhabited, even bodies that were comprised mainly of metal and vacuum tubes. "What feedback and the vacuum tube have made possible," wrote Wiener in 1950,

> is not the sporadic design of individual automatic mechanisms, but a general policy for the construction of automatic mechanisms of the most varied type. In this they have been reinforced by our new theoretical

160. Wiener, *Cybernetics*, 132; McCulloch cited in N. Katherine Hayles, *How We Became Posthuman: Virtual Bodies in Cybernetics, Literature, and Informatics* (Chicago: University of Chicago Press, 1999), 58; Quastler, "A Primer on Information Theory," 6; Claude Shannon and Warren Weaver, *The Mathematical Theory of Communication* (Urbana: University of Illinois Press, 1949), 31.

161. Catherine L. Albanese, "The Subtle Energies of Spirit: Explorations in Metaphysical and New Age Spirituality," *Journal of the American Academy of Religion* 67, no. 2 (1999): 305–25.

162. Samuel Greengard, *The Internet of Things* (Cambridge, MA: MIT Press, 2015). On the historical roots of such immanence, see John Durham Peters, *Marvelous Clouds: Toward a Philosophy of Elemental Media* (Chicago: University of Chicago Press, 2016).

163. Grey Walter, "Studies on the Activity of the Brain," talk at 1953 Macy Conference. Pias, *Cybernetics*, 689–96.

164. Lila Gatlin, *Information Theory and the Living System* (New York: Columbia University Press, 1972), 25.

> treatment of communication, which takes full cognizance of the possibilities of communication between machine and machine.[165]

In such a new age, theoretical treatments such as Wiener's served to democratize and decontextualize information. Such a move, made out of technical convenience and a concern for practical application, has had lasting effects. Strangely, information theory soon became a substitute for thinking itself, for explaining consciousness, for how a person achieves awareness of the mental states they inhabit.

But what, then, does one do with earlier iterations that the "animating principles of the Soul" were "dependent . . . upon the size, weight, and condition" of the brain? How is one any different from Davis, whose racial paranoia made possible magnetic "impression[s] from the outer world . . . flying over the sensatory nerves to the brain and obtaining recognition." As a trance physician, Davis saw in the brain the promise of activating the "Self-Healing Energies" within the "organization" of every person.[166] "The very atmosphere we breathe," announced Davis, "is saturated and loaded with the live electrified atoms of magnetism! . . . by which very delicately organized nervous systems are more or less constantly affected and disturbed."[167] Neuromation everywhere. All the time. As below, so above.

Loops and more loops and intimations of divinity. An apt description, perhaps, of the emergence of the neuromatic brain. But for those readers who prefer the charting of change over time, there is *this* historical claim to savor: the authority of the neuromatic brain rests on reconfigurations of the anatomical and electrical paradigms that preceded it. In the seventeenth and eighteenth centuries, for example, the physiology of the brain was of primal importance to a new breed of natural philosophers. This anatomical/mechanical view of the brain, on display in beautifully illustrated anatomical textbooks, was itself supplemented in the nineteenth century by a focus on the electrical activity within the brain.[168] Electricity, for nineteenth-century

165. Wiener, *Human Use*, 153.

166. Davis, *The Temple*, 107, 23; Andrew Jackson Davis, *The Harbinger of Health: Containing Medical Prescriptions for the Human Body and Mind* (New York: A. J. Davis, 1862), 31.

167. Davis, *The Temple*, 261.

168. "Electrical elements," wrote Andrew Jackson Davis, "flow down from the brain into the nerves, and into all the infinite ramifications of the nerves, and thence into the atmosphere we breathe" (*The Philosophy of Spiritual Intercourse*, 25–26). See also James Buchanan, *Outlines of Lectures on the Neurological System of Anthropology* (Cincinnati: Buchanan's Journal of Man, 1854).

neurologists and phrenomesmerists alike, served as the ethereal medium of consciousness. The electrical paradigm, in turn, was not so much supplanted in the mid-twentieth century as overwhelmingly supplemented by mathematical theories of synaptic communication. Brain waves, for example, achieved their distinction as surface effects of a neuronal depth, a pulsating system whose significance could and should be enumerated.[169] Information theory has since served to make the different metaphors of the brain—as animate being, as analog machine, as electrical network, as a self-organizing system—compatible with one another in their conformity to a higher-level mathematics.[170] With the future of neuromation built-in from *this* beginning, the question of propulsion still lingers. How and why and to what effect has a "neural basis for our reactions to universals" become a universal basis for just about everything else?[171]

Such propulsion has everything to do with the making of information as *the* medium of communication and, more ominously, the making of the human that imagines itself as an informational processing device. The neuromatic brain, as I have argued, was not so much new as part of a trajectory in which the medium that linked brain and environment became increasingly quantifiable—which is to say that the secular imaginary of contemporary neuroscience is premised on the ongoing mechanization of the ether in the form of something like information.

169. Edwin Clark and L. S. Jacyna, *Nineteenth-Century Origins of Neuroscientific Concepts* (Berkeley: University of California Press, 1987). "In biological neurons," writes Matthew Roos, computational neuroscientist and AI practitioner at Binary Cognition consulting, "information is encoded in the relative timing of spikes in individual or multiple neurons, not just in the individual neuron spiking rates"; "Deep Learning Neurons versus Biological Neurons: Floating-point Numbers, Spikes, and Neurotransmitters," *Towards Data Science* (March, 14, 2019), https://towardsdatascience.com/deep-learning-versus-biological-neurons-floating-point-numbers-spikes-and-neurotransmitters-6eebfa3390e9.

170. Within the neuromatic brain, information theory serves to imbricate the electrical and anatomical paradigms—information moving from electrical registers to chemical registers and back again; bursts of electrical pulses translated into chemical packets translated into frequencies. Jeremy Campbell, *Grammatical Man: Information, Entropy, Language, and Life* (New York: Simon and Schuster, 1982), 191.

171. Heinz Von Foerster, Margaret Mead, and Hans Lukas Tueber, eds., *Cybernetics: Circular Causal and Feedback Mechanisms in Biological and Social Systems: Transactions of the Eighth Conference, New York, March 15–16, 1951* (New York: Josiah Macy Jr. Foundation, 1952), xv–xvi.

13. CONCLUDING THOUGHTS ON PERCEPTRONIUM

The extraordinary ontology of information is on vivid display in Max Tegmark's recent article, "Consciousness as a State of Matter." In his insistence that information is everywhere, Tegmark doubles down on David Chalmers's admitted "metaphysical speculation" by framing the subjective aspect of being—that is, consciousness—as an effect of information processing.[172] A theoretical physicist and cofounder of the Future of Life Institute, Tegmark is a regular on the Silicon Valley circuit. With one foot in academia and the other planted firmly on a mountain of internet cash, Tegmark is quick to discuss his friendship and collaboration with figures such as Elon Musk and Larry Page. "Perceptronium" is Tegmark's name for that substance that makes information processing possible.[173] Perceptronium, argues Tegmark, solves the so-called "hard problem of consciousness." It is a barely carnal entity that represents the convergence of matter and spirit and possesses distinctive information processing abilities. According to Tegmark, perceptronium is manifest when cognition is manifest. As Tegmark reminds his reader, "I have long contended that consciousness is the way information feels when being processed in certain complex ways, i.e., that it corresponds to certain complex patterns in space-time that obey the same laws of physics as other complex systems." Here, Tegmark is self-consciously riffing on Chalmers's slogan that "experience is information from the inside; physics is information from the outside."[174]

Yet for Tegmark, the problem, exactly, is not how matter generates the spirit of consciousness. On the contrary, consciousness is always already matter. The problem, then, becomes figuring out the physiology of consciousness in mathematical terms. Refusing the notion of a soul being set apart from the body and rejecting the "failed notion" of "the luminous aether," Tegmark wagers that "all observed aspects of reality must emerge from . . . mathematical formalism alone." The laws of physics, in other words,

172. Tegmark is working to secure Giullo Tononi's "Integrated Information Theory of Consciousness." See G. Tononi, "Integrated Theory of Consciousness: An Updated Account," *Archives Italiennes de Biologie* 150 (2012): 290–326.

173. Tegmark has recently begun using the term "sentronium for the most general substance that has subjective experience" (*Life 3.0*, 303).

174. Max Tegmark, "Consciousness as a State of Matter," *Chaos, Solitons and Fractals* 76 (2015): 238–39; Chalmers, *Conscious Mind*, 305; Tegmark, *Life 3.0*, 345n18.

become the final referent of the human, not merely a means of explanation but the very source of consciousness. The soul, in Tegmark's theory, is no longer a meaningful ontological placeholder. The linguistic distinction between brain and environment all but collapses. The brain, in processing information, is subsumed within a sea of perceptronium. All that remains are the mathematical equations.[175]

Indeed, the vast majority of Tegmark's article is comprised of mathematical equations. Yet however brief his philosophical explication of these equations may be, his metaphysics are clear. Perceptronium is that which formally guarantees the phenomenon of consciousness. And consciousness is precisely that capacity to process information "with relative freedom from external influence." Autonomy is achieved when you can glimpse the self as self-organized, when you can "feel" the information "being processed in certain complex ways."[176]

Perceptronium, then, is the culmination of a cybernetic impulse to extend mathematical propositions into biological, social, and technological registers simultaneously—to make information the coin of a realm that can, in its limitlessness, be defined as cosmic. Information as that which guarantees the scalability between micro and macro. Information as that which guarantees the movement of the brain across ontic registers.[177] One might

175. Tegmark, "Consciousness," 239n1; Tegmark, *Life 3.0*, 287. As Wiener remarked late in his career, "cybernetics is nothing if it is not mathematical" (*God and Golem*, 88). Indeed, the making of perceptronium is part of what Garson has called the "sublimation" of the nervous system over the course of the nineteenth and twentieth centuries. For as information theory was increasingly taken up in a cognitive key, the brain effectively became a mathematical proposition subject to increasingly precise formalization. As Garson argues, "with the physicalization of the methods of nerve physiology there was an etherealization of the subject matter" ("The Birth of Information," 34, 45).

176. Tegmark, "Consciousness," 239.

177. See also Integrated Information Theory (IIT) and the "attempt to characterize consciousness mathematically both in quantity and quality" (Tegmark, "Consciousness," 239–40). IIT and intrinsic information is the latest riff on Wiener's clarification of "semantically significant information" as that which is structured by and within the system. Tegmark's theory also resonates with panpsychism and religious speculation that all matter is endowed with what might be called consciousness. See, e.g., D. S. Clark, *Panpsychism and the Religious Attitude* (Albany: State University of New York Press, 2003). On David J. Chalmers's consideration of the prospect that all physical entities might be said to have conscious experiences, see his "The Combination Problem for Panpsychism," in *Panpsychism: Contemporary Perspectives*, ed. Godehard Brüntrup and Ludwig Jaskolla (New York: Oxford University Press, 2017), 179–214.

even go so far as to claim that perceptronium is the result of an accelerating accretion of ideas, capital, and practices around the fungibility of information, an accretion that now allows humans to cohere as a species, and, in a perverse extension of romanticism, allows humans to become conversant, at the deepest of levels, with machines and storm clouds both.

SYNAPTIC GAP: TOO MUCH TOO SOON

BY THE EARLY 1960S, SCIENTISTS AND engineers of various stripes had established information theory as both a professional obligation and as an incentive for others to imagine the universe—any object, procedure, or process within reason—in terms of its neuromatic fundament.[1] Everything in the world, it seemed, could be encoded as information. The neuromatic paradigm, in other words, elicited dreams of knowing all of the communications that make us human—neuronal but also psychological, social, economic, and moral—in order to optimize them. Such dreams of control, however, were often laced with the anxious prospect of being overwhelmed by information.[2] As the radiologist Henry Quastler flatly declared at the

1. As Dupuy notes, physics had long abstracted forms from the materials under scrutiny in order to posit "isomorphisms between different physical phenomena." But such modeling assumed a more expansive character with the advent of cybernetics as it extended "this approach to new domains, which previously had seemed resistant to modeling of any sort," such as the nervous system, consciousness, and society. Jean-Pierre Dupuy, *The Mechanization of the Mind: On the Origins of Cognitive Science*, trans. M. B. Debevoise (Princeton, NJ: Princeton University Press, 2000), 113.

2. In 1950, having lived in the same apartment building as Claude Shannon in New York, Claude Lévi-Strauss expanded anthropology to include—elegantly, I might add—the informational dimensions of human biology. According to Lévi-Strauss, the elaborate networks of communication between brain and culture included both biological and nonbiological components that could be formalized in numbers and algorithms. Citing Wiener and Shannon, Lévi-Strauss argued that ethnologists were now "only waiting upon the goodwill of mathematicians" to define, in precise terms, the circular causality between physiology and culture. Lévi-Strauss, like the cognitive scientists of religion that would follow in his

"Symposium on Information Theory in Biology" held in Gatlinburg, Tennessee, in October of 1956, "the concept of a measure of information fulfills a general and deep need of our time" as "the sheer bulk of the information now available increases at a rapid rate." But with the proliferation of information both in and beyond the laboratory, "the representation of information becomes a more and more critical problem."[3] So as the basic metaphors of information theory coalesced into categorical imperatives, these imperatives were also used to address more disturbing aspects of the world they were already ushering in.

In 1960 the biologist and behavioral scientist James Grier Miller considered how "pathological function of the nervous system and abnormal behavior could result from information input overload." Miller leveraged the work of Warren McCulloch to define psychosis in terms of information theory. "It is possible," wrote Miller, "to overload a cell specialized for the transmission of information—a neurone—by increasing the rate of input of electrical impulses to it until finally its transmission breaks down." By using an IOTA (Information Overload Testing Aid) apparatus on human subjects, Miller sought to determine "maximum channel capacity for living systems at various levels."[4] In doubling down on information—identifying informa-

---

wake, insisted, as a matter of first principle, that humans were always, in some sense, overwhelmed by the representations and sensorial cues around them. There were more signifiers, always already, than signifieds. Human evolution was a story of learning to live with and think through this excess of representation as a cultural and cognitive matter. Hence the need for numbers and algorithms. Claude Lévi-Strauss, *Introduction to the Work of Marcel Mauss* (1950), trans. Felicity Baker (London: Routledge, 1987), 39ff., 44, 62; Pascal Boyer, "Explaining Religious Concepts: Levi-Strauss the Brilliant and Problematic Ancestor," in *Mental Culture: Classical Social Theory and the Cognitive Science of Religion*, ed. Dimitris Xygalatas and William W. McCorkle Jr. (Durham, UK: Acumen Publishing, 2013), 164–75. On the cybernetic turn in structuralist anthropology, see Bernard Dionysius Geoghegan on Claude Lévi-Strauss's embrace of information theory and "modern computing machines" as analytic resources in "The Cybernetic Apparatus: Media, Liberalism, and the Reform of the Human Sciences" (PhD diss., Northwestern University, 2012).

3. Henry Quastler, "A Primer on Information Theory," in *Symposium on Information Theory in Biology*, Gatlinburg, TN, October 29–31, 1956, ed. Hubert P. Yockey, Robert L. Platzman, and Henry Quastler (New York: Pergamon Press, 1958), 4–5.

4. James Grier Miller, "Information Input Overload and Psychopathology," *American Journal of Psychiatry* 116 (1960): 695, 697. Miller also considered how "pathological function of the nervous system and abnormal behavior can result from information input overload"; James Grier Miller, "Coping with Administrators' Information Overload," *Journal of Medical Education* 39, no. 11 (November 1964): 47.

tion as the *problem* and further theorization of information as the *solution* to that problem—Miller's investigation of the limits of cognitive processing was, on one level, tragic. It was also practical. For his diagnosis of limits also afforded an opportunity to surpass them.

Miller wrote of the need to develop "adjustment processes" that would be amenable to therapeutic application on humans. He suggested that the most practical skill needed "to handle the stress of information overload" was "chunking." Miller defined chunking in terms of "reading or otherwise processing information in clumps, reading whole phrases as sentences and single units": a systematic skim that worked around the limits of human channel capacity. Such "adjustment enables you to process material to which you are accustomed very rapidly."[5] By way of chunking—breaking up large amounts of pattern to smaller, more assimilable ones—Miller suggested that the human brain could, in fact, keep up with production and weather the storm of information that was on the horizon.

For Quastler and Miller, the problem of information overload was both biological and cultural, a position reiterated in a more explicitly prophetic key by the comparative literature professor turned talk-show celebrity, Marshall McLuhan. Throughout the 1960s McLuhan spoke of "information in unlimited quantities and at hitherto-impossible speeds." This strange brew had the potential to overwhelm cognition but also to instruct. As McLuhan insisted, "It's inevitable that the world-pool of electronic information movement will toss us all about like corks on a stormy sea, but if we keep our cool during the descent into the maelstrom, studying the process as it happens to us and what we can do about it, we can come through." Information processing, again, was both the root of the problem as well as the solution. For the anxious hope accompanying the concerns of Miller and Quastler as well as McLuhan's diagnosis of the "terrific speed-up of information moving" was, itself, premised on tapping previously unused cognitive potential to process information.[6]

As a defining mark of its discursive sway at midcentury, the products of information theory begot ever more refined *and* capacious theorizations

5. Miller, "Coping," 50–51. See also Buryl Payne, "A Descriptive Theory of Information," *Behavioral Science* 11, no. 4 (July 1966): 295–305; Neal F. Johnson, "The Role of Chunking and Organization in the Process of Recall," *Psychology of Learning and Motivation* 4 (1970): 171–247.

6. "The Playboy Interview: Marshall McLuhan," *Playboy Magazine* (March 1969): 60, 158, 56.

of human information processing. In other words, problems generated by information processing devices were often addressed as problems of human cognition, the assumption being that humans had somehow failed to keep up with demands made by their own inventions. A diagnosis was rendered: "low threshold[s] for disorganization under increasing stimulus input."[7] Information overload, then, did not merely refer to what might be considered problems of volume—too many communications with the world outside of the self, too many ideas, too many images, too much media, too many stories of yourself in circulation. On the contrary, the alarm over information overload was also temporal for those who sensed its accelerating pace and experienced its saturation.[8]

In the late 1940s, Mormon educator Evelyn Wood was counseling "underachievers with personality problems" at a high school outside of Salt Lake City. In her pursuit of better ways to address her students' mental health, Wood created a series of remedial reading courses. In charting the calming effects of these courses on the personality of wayward Mormon youth, Wood then developed theories about reading, attention, and control that soon found their way into print. Following the success of her book, *Reading Skills* (1958)—Wood established Evelyn Wood Speed Reading Dynamics. Her business established over 150 franchises in the United States where students enrolled in seminars that trained them in the art of optimizing reading comprehension. Wood became a minor celebrity. She was on TV. There were reports that President John F. Kennedy was a fan and, later, that President Richard Nixon sent his staff to her classes.[9] As the *New York Times* reported, in a typical seminar "25 students were assigned to finish George

7. Seymour Epstein and Margaret Coleman, "Drive Theories of Schizophrenia," *Psychosomatic Medicine* 32, no. 2 (1970): 113–40.

8. As such, the diagnosis of information overload marked a "reconceptualization of human beings and their organizations as communication channels whose capacity could be overwhelmed. From the start, information overload was more about humans as bit-processors than about humans overwhelmed by the digital output of their machines." Nick Levine, "The Nature of the Glut: Information Overload in Postwar America," *History of the Human Sciences* 30, no. 1 (2017): 33.

9. On the fraudulence of these claims, see Marcia Biederman, *Scan Artist: How Evelyn Wood Convinced the World That Speed-Reading Worked* (Chicago: Chicago Review Press, 2019).

Orwell's 'Animal Farm' in 25 minutes, a feat that required a reading rate of 1,400 words a minute. Most finished in the allotted time, and most scored well on the comprehension test that followed."[10]

Wood herself claimed to be able to read over 2,700 words per minute. She wrote for an aspiring middle-class entering new consumer and new media economies. This audience felt that they could not keep up, and they saw it necessary to retrain themselves in light of the hectic pace of the now and its onslaught of media content. This audience dreamed of becoming better, stronger, and faster readers. During seminars students learned to read more quickly and efficiently, scanning for patterns up and down in addition to left and right, approaching the world of alphabetic signs as a formal proposition. In working through the chapters in *Reading Skills,* students used a shutter card to identify patterns of words they often found at the beginning of sentences. They did this in order to perform the kind of visual attention necessary to grab hold of the pattern from the get-go without getting bogged down by semantic content. As Wood instructs:

> Use the large shutter card and move it down the columns as fast as you can and still get the meaning of each pattern. Wait for the signal to begin. The teacher will write the seconds on the board. As soon as you are finished, record the time on the chart. Wait again for the signal to begin. Record the time. Do this practice several times. Your eyes won't find it easy to see so many words together. They need lots of practice.[11]

The shutter card was a modified tachistoscope—used in psychological experiments and during World War II for training sailors to identify enemy fighter planes.[12] Among military personnel, the tachistoscope was integral to the development of flash recognition training—a perceptual enhancement technique designed to make you more efficiently attentive, more capacious

10. Cited in Lawrence Van Gelder, "Evelyn Wood, Who Promoted Speed Reading, Is Dead at 86," *New York Times* (August 30, 1995).

11. Evelyn Nielsen Wood and Marjorie Wescott Barrows, *Reading Skills* (New York: Henry Holt, 1958), 54.

12. Samuel Renshaw, "The Visual Perception and Reproduction of Forms by Tachistoscopic Methods," *Journal of Psychology: Interdisciplinary and Applied* 20 (1945): 217–32. See also "Tachistoscope in Visual Diagnosis and Training," *Optometric Weekly* 36 (November 1945): 1189; Conrad Berens et al., "Effects of Tachistoscopic Training," *American Journal of Ophthalmology* 44, no. 3 (1957): 1–48.

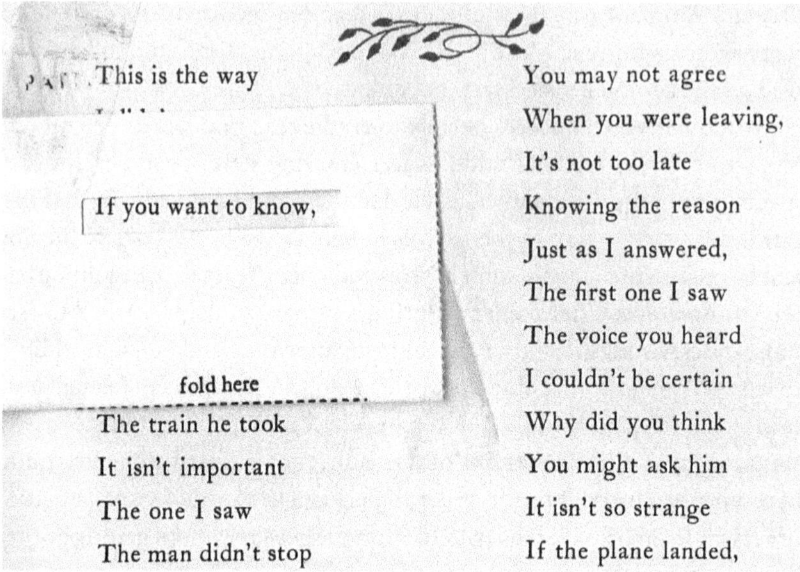

FIGURE 28. Modified tachistoscope to be used in scanning exercise. From Evelyn Nielsen Wood and Marjorie Wescott Barrows, *Reading Skills* (New York: Henry Holt and Company, 1958).

and discerning in your gaze, better able to recall past visual information and to integrate that information into present observations.[13]

In Wood's version of flash recognition training, the modified tachistoscope created a circuit between you and the words ("signals") before your eyes. One learned to read for patterns because the meaning of individual words was assumed to be secondary, perhaps even epiphenomenal to those patterns. For to model oneself as a scanning device was to be become a more efficient channel for processing chunks of incoming information.[14] In addition to teaching students to read up and down the page rather than

13. Edward C. Godnig, "The Tachistoscope: Its History and Uses," *Journal of Behavioral Optometry* 14, no. 2 (2003): 39–42. The tachistoscope was also taken up by law enforcement officials to "develop a larger span of recognition; see R. Soule, "Flash Recognition Training in Law Enforcement Work," *Criminology and Police Science* 49 (1959): 590–600.

14. The scanning metaphor arises in concert with a particular vision of the human as an information processing device which, in turn, creates new ambitions, new goods and services and sociality, new kinds of anxieties, and new grammars of racialized and gendered difference. See, e.g., the essays in *Race after the Internet*, ed. Lisa Nakamura and Peter Chow-White (New York: Routledge, 2012). See also Catherine Matacic, "In the United States,

left to right, Wood identified habits of subvocalizing (either reading under the breath or aloud in the head) as a source of needless friction and noise.[15]

Once the tachistoscope served to draw your attentive energies to the surface of consciousness for purposes of self-control, the device could then be set aside. Ideally, at this point in their training, students would have cultivated their own faculty of scanning in order "to see each pattern at a glance, just as you used to see each word at a glance." In an advanced exercise, for example, Wood instructed students to practice scanning a poem. "You can just whizz through it because the lines are short, each line is a pattern of meaning, and all of them fit together."[16]

The mundane and utterly utilitarian practice of an Evelyn Wood seminar was part of the oblique life of cybernetics at midcentury. A dream on the periphery of scientific institutions that extended their insights at odd angles. All those modified tachistoscopes and all those hands, all those seminar rooms and church basements. Evelyn Wood Speed Reading Dynamics captures something about the way the brain—in the lives of individuals—was seamlessly conceived as an information processing device. By the time Wood established her business, the utility of speed reading was all but commonsensical.

But what kind of world is conjured—outside the laboratory—when one scans the horizon for patterns, when one recognizes patterns, when one learns from those patterns, when one becomes better and more efficient in scanning the horizon? What happens, in other words, to the imagination when you imagine yourself to be a channel of forms rather than an interpreter of meanings?

Computers Help Decide Who Goes to Jail. But Their Judgment May Be No Better Than Ours," *Science* (January 17, 2018), doi:10.1126/science.aat0356; Steve Lohr, "Facial Recognition Works Best If You're a White Guy," *New York Times* (February 12, 2018), B1.

15. On current efforts to naturalize the human reader as an information processing device, see Howard Moshtael et al., "Saccadic Scrolling: Speed Reading Strategy Based on Natural Eye Movements," 2016 8th International Conference on Intelligent Human-Machine Systems and Cybernetics (IHMSC) 2 (2016): 596–600.

16. Wood and Barrows, *Reading Skills*, 56.

3

IMAGINING THE NEUROMATIC

1. CRASH AND BURN

The R101, a British dirigible, crashed in northern France on October 5, 1930, as it made its way to India. After a series of test flights and trial runs, this was the R101's maiden voyage. Forty-eight of the fifty-four people on board were killed, including its captain, Flight-Lieutenant H. Carmichael Irwin. The R101 had been, at that time, the largest airship ever built. Aluminum frames and diesel engines gave the R101 a sleek but sturdy sheen. Quarters were luxurious—two promenade decks with fiberglass, an expansive lounge area, a smoking room, and a radio cabin. Rather than hanging below the airship, the passenger and crew areas were built inside the ship, allowing for two stories of human traffic within the vastness of the R101. Among those who died were the majority of designers and engineers who had worked on the industrial project over the previous fifteen months. Here were human hands involved in orchestrating their own deaths. Here were technological directives beholden to incredibly sophisticated strategies for controlling populations on the colonial "periphery." With the territories of the British Empire spread across the world, the distance between them was described by one official as an "enemy of imperial solidarity." In being able to make the journey from Britain to India in five days, the R101 was part of a vaunted British effort to create a network of airships that would crisscross the globe and "knit together its scattered peoples."[1]

There is something about this end-of-empire moment, this story of late colonial decadence, that speaks to the power of metaphor.

1. Sir Samuel Hoare, "Aviation and the British Empire," *Scottish Geographical Magazine* 45, no. 1 (1929): 1–6.

FIGURE 29. French police and officials from the Air Ministry search through the twisted wreckage of the R101 airship, searching for clues to explain the crash (October 5, 1930). Courtesy of Trinity Mirror/Mirrorpix/Alamy Stock Photo.

Two days after the crash, a séance involving Eileen Garrett was held at the National Laboratory of Psychical Research in London. Garrett, who was affiliated with the British College of Psychic Science, was well known in London and its surrounds for her psychic gifts. Garrett and her spirit control, Ouvani, a young Arab soldier who had died in battle in the thirteenth century, were the talk of London at the time. Ouvani was Garrett's self-designated "door-keeper," controlling access to Garrett from the spirit world and thereby the communications that emanated from her. Garrett was part of a literary scene that continued in the aftermath of the late Victorian investment in the magical arts. Poets, artists, academics, scenesters, and scientists shared with each other their enthusiasm for the strangeness of the world that could not be explained away. Garrett mingled with the likes of D. H. Lawrence, William Butler Yeats, and James Joyce, as well as famed British occultists such as Oliver Lodge and Arthur Conan Doyle.[2]

2. Eileen J. Garrett, *Many Voices: The Autobiography of a Medium* (New York: G. P. Putnam's Sons, 1968), 27–28, 38, 60, 62, 52–53. Alex Owen, *The Place of Enchantment: British Occultism and the Culture of the Modern* (Chicago: University of Chicago Press, 2004).

Given her friendship with Doyle and earlier collaborations with his circle of investigators at the National Laboratory, Garrett had initially been recruited in hopes of contacting the recently departed Doyle.[3] The séance had been planned a week before, on October 2, and was meant to be an opportunity for testing the validity of the spirit world. But now the air was thick with disaster and national mourning. "The medium sat down in an armchair. . . . Her breathing became steadier and deeper and her face more drawn. Tears streamed down her cheeks. In five minutes she was entranced." Instead of Doyle, however, a mysterious figure began to speak through Garrett—"I see for the moment," said Garrett in Ouvani's accented voice, "I-R-V-I-N-G or I-R-W-I-N. He says he must do something about it. . . . apologizes for coming, for interfering (with our experiment?). Seems to be anxious to speak to a lady in a body."[4] At that moment, the

> voice of the medium again changed and an entity announced that he was Flight-Lieutenant H. Carmichael Irwin, captain of the R101. He was very agitated, and in a long series of spasmodic sentences gave the listeners a detailed and apparently highly technical account of how the R101 crashed a few hours previously. The entity described how the airship sank and failed to rise; what was wrong with the engine, her design, etc.[5]

After announcing himself to the séance participants, Irwin went on to document a host of technological problems that had led to the crash of the R101 a few days earlier—failing pumps and jammed elevators, leaking fuel pipes, a defective cooling system. The details gathered in the séance were then forwarded to the British Air Ministry. Officials soon confirmed the accuracy of much of the transcript. They also confirmed that Garrett's technical facts conformed to internal investigations being conducted on the crash but were yet unknown to the public at the time of the séance. Garrett, in other words, could not have known what had been communicated to her. Moreover, "the medium has never possessed any sort of engine or motor car and knows

3. Garrett and Doyle were a node on a network of British occultism. See, e.g., Abduhl Latif, *Health: Its Recovery and Maintenance*, ed. R. H. Saunders and Arthur Conan Doyle (London: Rider, 1928).

4. Harry Price, *Leaves from a Psychist's Case-Book* (London: Victor Gollancz, 1933), 118ff. An alternative spelling for Ouvani—Uvani—is sometimes used. When not being cited, I have used Ouvani for consistency.

5. Price, *Leaves from a Psychist's Case-Book*, 118ff.

nothing about aeronautics or engineering. The conversation of 'Irwin' was packed with terms that few men could reel off with any degree of relevancy."[6]

Amid the ruins of this mechanical world, and the few men with "relevant" knowledge, a mediumistic star was born. Garrett's celebrity was founded, in part, on her association with the R101 disaster and the urgent way it framed the debate about the mechanisms of psychic powers like trance mediumship, ESP, and telepathy. While there had been other popular mediums who had evoked mechanical understandings of what Andrew Jackson Davis had called "spiritual intercourse," Garrett's program, from the beginning, was self-consciously scientific in imagining itself in conversation with cutting-edge research in neurophysiology.

Having stirred much debate in the press, Garrett's fame spread throughout the 1930s. Portrayed as both serious and glamorous, much ink was spilled as Garrett volunteered her psychic abilities to be tested in laboratories on both sides of the Atlantic.[7] Much of the scientific inquiry into Garrett focused on the entities that spoke through her—Ouvani as well as a host of exotic characters including 'Abd al-Latif al-Baghdadi, identified as a thirteenth-century Muslim Persian physician who spoke authoritatively about the neurophysiological registers involved in Garrett's performance.[8] One mark of Garrett's psychic authenticity and role as an "instrument" was the performance of racial stereotypes during her trance state. As one researcher remarked, "it was obvious that we had before us a gentleman from India. Facial expression, eyes, color of skin, movements, the folded arms, and the finger movements that accompanied many of his words, were all those of a native of India . . . his accent was typical."[9]

6. "The R-101 Disaster," *Tomorrow* 2, no. 1 (Autumn 1953): 70.

7. She would eventually attest to having become exhausted by the laboratory: "I've [taken] part in so many experiments and have worked with so many investigators." Quoted in Allan Angoff, *Eileen Garrett and the World beyond the Senses* (New York: William Morrow, 1974), ix.

8. Ira Progoff, *The Image of an Oracle: A Report on Research into the Mediumship of Eileen J. Garrett* (New York: Garrett Publications, 1964), 98f. See, e.g., John Videan and Ivy Videan, *The Eastern Key, Kitāb al-ifādah wa'l-i'tibār* (London: Allen and Unwin, 1965), which is a translation of 'Abd al-Laṭīf al-Baghdādī's work, the *Kitāb al-Ifādah*. The Videans became interested in al-Baghdādī because of an experience with another medium in which 'Abd al-Laṭīf attempted to communicate with them. Here, I am indebted to conversations with Nicholas Harris and SherAli Tareen.

9. Eileen J. Garrett, "The Rockland County Ghost (reported by Hans Holzer)," *Tomorrow* 1, no. 3 (Spring 1953): 17.

In keeping with the orientalist playbook, Garrett's controls served to authorize her scientific stance with a condescending dose of ancient wisdom.[10] Indeed, the ontology of Ouvani garnered as much scientific attention as did Garrett during this time. Researchers subjected Ouvani to "laboratory methods, instruments, and controlled experiments"—galvanometers, fluoroscopes, language analysis, "a mass of statistical data" including 14,425 trials. J. B. Rhine, the founder of the parapsychology lab at Duke University, concluded that "we are not yet ready to come to grips with the problem of the 'ultimate nature' of Uvani."[11] The conclusion that Ouvani was distinctly present but remained beyond scientific measurement only amplified Garrett's celebrity status as she assumed an air of mystery and sophistication. She hobnobbed with T. S. Eliot and Cecil B. DeMille, presenting herself as a more glamorous version of Franz Anton Mesmer and the seductions that he had performed centuries earlier.[12]

10. Holly Edwards, ed., *Noble Dreams, Wicked Pleasures: Orientalism in America, 1870–1930* (Princeton, NJ: Princeton University Press, 2000); Nancy Rosenblatt, "Orientalism in American Popular Culture," *Penn History Review* 16, no. 2 (2009): 51–63; Melani McAlister, *Epic Encounters: Culture, Media, and U.S. Interests in the Middle East, 1945–2000* (Berkeley: University of California Press, 2001).

11. Whately Carington, *The Quantitative Study of Trance Personalities*, pt. 1: *Preliminary Studies. Mrs. Garrett, Rudi Schneider, Mrs. Leonard. Proceedings of the Society for Psychical Research* 42 (1934): 173–240; J. B. Rhine, "Telepathy and Clairvoyance in the Normal and Trance States of a 'Medium,'" *Character and Personality* 3 (1934): 91–111; Angoff, *Eileen Garrett*, 167–68, 172, 174. Physiological differences between Garrett and Ouvani were also detected—with Garrett in a normal resting state registering different blood counts for hemoglobin, different cardiograph readings, and different cessation of bleeding rates than she did while channeling Ouvani. As a New York World Telegram headline declared in 1934—"PUTS DEAD ARAB THROUGH PACES IN MENTAL TEST. . . . CARRINGTON IS CONVINCED DIFFERENT PERSONALITIES SPEAK THROUGH ENGLISHWOMAN" (Angoff, *Eileen Garrett*, 180, 168).

12. Angoff, *Eileen Garrett*, 162. In Michel Pobers's "The Tragicomedy of Franz Anton Mesmer"—published in Garrett's journal *Tomorrow*—there is much emphasis on Mesmer's "exceptional good looks, his elegance and suave manner." Pobers distills the erotics of the mesmeric encounter in such a way as to provide historical ground for Garrett's management of her own image. As Pobers writes, "When a patient was introduced to his office, Dr. Mesmer would begin by passing his hands, with fingers extended, all down the body, 'to establish contact.' Sitting very close to the patient, he touched and rubbed the affected or sensitive points of the body while looking fixedly in a patient's eyes. Soft music from an adjoining room was used to induce relaxation during the treatments. Then, with bare hands, or sometimes using a short stick, the magnetiseur transmitted 'the Fluid' to the patient" (*Tomorrow* 1, no. 3 [Spring 1953]: 116).

Garrett presented her aptitude for clairvoyance to psychologists and neurophysiologists as living proof of a breakthrough in how reality could be perceived according to universal laws. As she declared, "the laws of physics and chemistry govern living matter, but the clairvoyant sees the far greater force behind such laws." As cybernetics assumed public currency in the late 1940s and 1950s, so too did Garrett's claims to be able to "recall the past, predict the future, and even set mechanical forces in motion without the aid of instruments." Garrett, in speaking of universal principles that determined interactions between brain and environment, was deploying information theory well beyond its intended field of application. As one psychologist put it after witnessing her "acts of perception and communication" with the spirit world, Garrett was "truly a medium in the sense of being little more than a funnel for the information given. . . . She is simply the medium through whom the information travels by means of her psychic sensitivity."[13]

In 1951 Garrett cofounded the Parapsychology Foundation and redoubled her efforts to imagine a new form of consciousness in which self and world (this one and the next) were intuitively entangled. As the editor of *Tomorrow: World Digest of Psychical and Occult Studies*, Garrett published numerous articles from natural and social scientists on telepathy and trance. Garrett also organized a number of international conferences on parapsychology in the 1950s and 1960s where she enlisted working neurophysiologists such as Grey Walter and John Smythies. Her goal, like those within the cybernetic mainstream, was to secure legitimacy and funding for experiments that investigated the possibility for individuals to communicate and process messages that escaped the notice of the five senses.[14] From Garrett's perspective, psychic abilities such as telepathy and spirit communication, long integral to the world's religions, were now on the verge of being explained by advances in neurophysiology.

Beginning in the late 1940s, for example, Walter had begun using stro-

13. Cited in Angoff, *Eileen Garrett*, 157; Progoff, *Image of an Oracle*, 362.

14. In the pages of *Tomorrow*, John Smythies wrote of the transcendental world as a "psychical reality" ("Our Transcendental World," *Tomorrow* 2, no. 3 [Spring 1954]: 37). "The mechanisms of the psyche, transmitting information from the brain," wrote Smythies elsewhere, "actually construct the sense-fields of perception." Here Smythies was drawing in part from Walter's research on brain waves and his experiments in disrupting them in order to explore untapped mechanisms of cognition. See J. R. Smythies, "The Extension of Mind: A New Theoretical Basis for Psi Phenomena," *Journal of the Society for Psychical Research* 36 (1951): 477–502.

boscopic flashes to induce changes in the frequency of brain waves.[15] Motivated by the desire for extrasensory experience, Walter studied the response and adaption of the brain to pulsating flashes of intensely bright light. For Garrett, the implication of Walter's research was profound, namely that the brain could be stimulated in such a way as to forge new ways of knowing beyond conscious thought. In late 1961 Garrett subjected herself to one of Walter's stroboscopic experiments. Afterward they remained close friends. Over the course of the 1960s, Walter attended many of Garrett's conferences and became more invested in exploring the neurological fundament of what had become known as psi phenomena.

In this chapter I cast light on the shadowy tendrils of cybernetic emergence. Focusing on more unhinged renderings of information theory and neural networks, I look to the edges of a social network of professional scientists like Walter, Claude Shannon, and Warren McCulloch when they were literally in conversation with the more do-it-yourself variety—Garrett's parapsychological scene, the writers Brion Gysin and William S. Burroughs, as well as L. Ron Hubbard, the founder of Dianetics and later Scientology. Garrett, Burroughs, Gysin, and Hubbard each cited objective empirical research emerging from the cybernetic fold in order to infuse legitimacy into their writings and teachings. Bursts of terms—information, feedback, systemic self-organization, communication and control—captured their imaginations in forging new vistas of interpreting human cognition within a cosmic context. In contrast to mainstream efforts to imagine cognition as disembodied, theirs were messier, viscerally and affectively speaking. Consequently, I read the forays of Garrett, Gysin, Burroughs, and Hubbard as (theo)logical extensions of the neuromatic culture that contained them.[16] Which is to say that these efforts look bent only in hindsight, after

15. Jimena Canales, "A Number of Scenes in a Badly Cut Film: Observation in the Age of Strobe," in *Histories of Scientific Observation*, ed. Loraine Daston and Elizabeth Lunbeck (Chicago: University of Chicago Press, 2011), 237. See also W. Grey Walter et al., "Analysis of the Electrical Response of the Human Cortex to Photic Stimulation," *Nature* 158 (1946): 540–41.

16. On Burroughs's cybernetic pedigree, see N. Katherine Hayles's reading of Burroughs's depiction of the "body materializing discourse" rather than discourse materializing the body (*How We Became Posthuman: Virtual Bodies in Cybernetics, Literature, and Informatics* [Chicago: University of Chicago Press, 1999], 194). See also David Porush, *The Soft Machine: Cybernetic Fiction* (New York: Methuen, 1985) and Oliver Harris, "Can You See a Virus?: The Queer Cold War of William Burroughs," *Journal of American Studies* 33,

the battles between good and bad religion have been waged. For the victors have been declared. We are disciplined into reading these figures as somehow beyond the pale. They are normal. I, and perhaps you too dear reader, are not. For only under the spell of secularism have the cognitive sciences and other offspring of the cybernetic moment been able to sustain widespread consensus about their own categorical authority and the illegitimacy of suspect others.[17] But what to make of these errant possibilities within the neuromatic paradigm? To what extent do emphases on occult practices and networks tell a different story about the metaphysics that are ever operative in the ongoing triumph of the brain?

---

no. 2 (1999): 243–66. From the publication of *Dianetics*, Hubbard was viewed as part of a cybernetic demimonde of so-called kooks and crackpots who misappropriated the sound ideas of sober scientists. See, e.g., Yvette Gittleson's "Sacred Cows in Collision," *American Scientist Quarterly* (October 1950): 603–9. In addressing the "best-selling horror called Dianetics" Gittleson wrote that "cybernetics is the big new idea of the times, and in my opinion Hubbard (who never mentions the word) has got cybernetics, and got it bad; this is to say, he has got it wrong" (605). Shannon's colleague, J. R. Pierce, argued that Hubbard's works "show little investment of effort comparable with the tremendous scope of the fields they pretend to cover" ("The Social Uses of Science," *American Scientist* 42, no. 4 [October 1954]: 646–50). More recently the historian Ronald Kline has pointed out Hubbard's opportunistic leveraging of the computational idiom as an early instance of the misunderstanding and marginalization of cybernetics; see his *The Cybernetics Moment: Or Why We Call Our Age the Information Age* (Baltimore: Johns Hopkins University Press, 2015), 91–93. Garrett's use of technological metaphors has failed to register at all in histories of the cybernetic, her outsider status not worthy enough even to name as such.

17. For a specifically Protestant take on cybernetics and theology, see Kenneth Vaux, *Subduing the Cosmos: Cybernetics and Man's Future* (Richmond, VA: John Knox Press, 1970). In what might speak to its nonspecific Protestant character, cybernetics became a pressing issue for modernizing Catholics in the 1960s. On the engagement with cybernetics by the Roman Catholic Church, see Neville Moray, *Cybernetics* (New York: Hawthorne Books, 1963); Patrick Granfield, "Ecclesial Cybernetics: Communication in the Church," *Theological Studies* 29, no. 4 (1968): 662–78; Christopher Wagner, "Social Cybernetics as Permanent Function of the Church," in *The Social Message of the Gospels*, ed. Franz Bockle (New York: Paulist Press, 1968), 46–60. On particular forms of cybernetic immanence as they relate to the dissolution of the very category of what constitutes divinity, see Abou Farman, "Informatic Cosmologies and the Anxieties of Immanence," *Immanent Frame* (October 25, 2017), https://tif.ssrc.org/2017/10/25/informatic-cosmologies-and-the-anxieties-of-immanence/.

2. OPENING SCENE FROM A CYBERNETIC DEMIMONDE

In 1960 William S. Burroughs contacted Grey Walter and the two met that fall. Burroughs, like Eileen Garrett, had been taken with Walter's research. Burroughs was interested in parapsychology from the ground up, as it were. He contacted Walter precisely because he was interested in the neural and sensorial potentials that Walter evoked within the brain by external stimulus. Together, inspired by Walter, Burroughs, alongside Brion Gysin and Ian Sommerville, designed the Dream Machine, a homemade stroboscope apparatus that manipulated brain waves to hallucinatory effect. Here the neuromatic was the source of formal innovation. For Burroughs also saw in the effects of "flicker" the desired outcome of the cut-up experiments he was then conducting with Gysin in Paris. Just as flashing lights refigured the "psycho-sensory processes" of the test subject, the montage poetics of the cut-up was intended to forge new channels of awareness in the reader. Indeed, Burroughs's call to arms throughout his cut-up trilogy of the 1960s—"break through in grey room"—was an homage to Walter's research.[18]

By first slicing up phrases, sentences, and entire texts and then rearranging them at random, cut-ups exposed the way language and narrative convention inculcated a version of the human that was utterly vulnerable to manipulation and control. Gysin and Burroughs sought to disabuse readers of their attachment to the meanings of words and their ontic investment in them. For once these "word lines" were cut, readers could then begin to cultivate their inherent nature as communication channels and their capacity for "telepathic contact" beyond language. As Gysin would later insist, his literary and visual art were intended to invite his audience "into an immense psychic reservoir (their own, which is incessantly modified by external impulses)."[19]

18. William S. Burroughs, *Rub Out the Words: The Letters of William S. Burroughs, 1959–74*, ed. Bill Morgan (New York: Ecco, 2012), 72; for allusions to Walter and his flicker experiments, see William S. Burroughs, *The Soft Machine* (1961; New York: Grove Press, 1967), 71, 75, 78, 155.

19. Brion Gysin, cited in Gérard Audinet, "The Wounded Man: Brion Gysin, Unsung Surrealist," in *Brion Gysin: Dream Machine*, ed. Laura Hoptman (London: Merrell, 2010), 36.

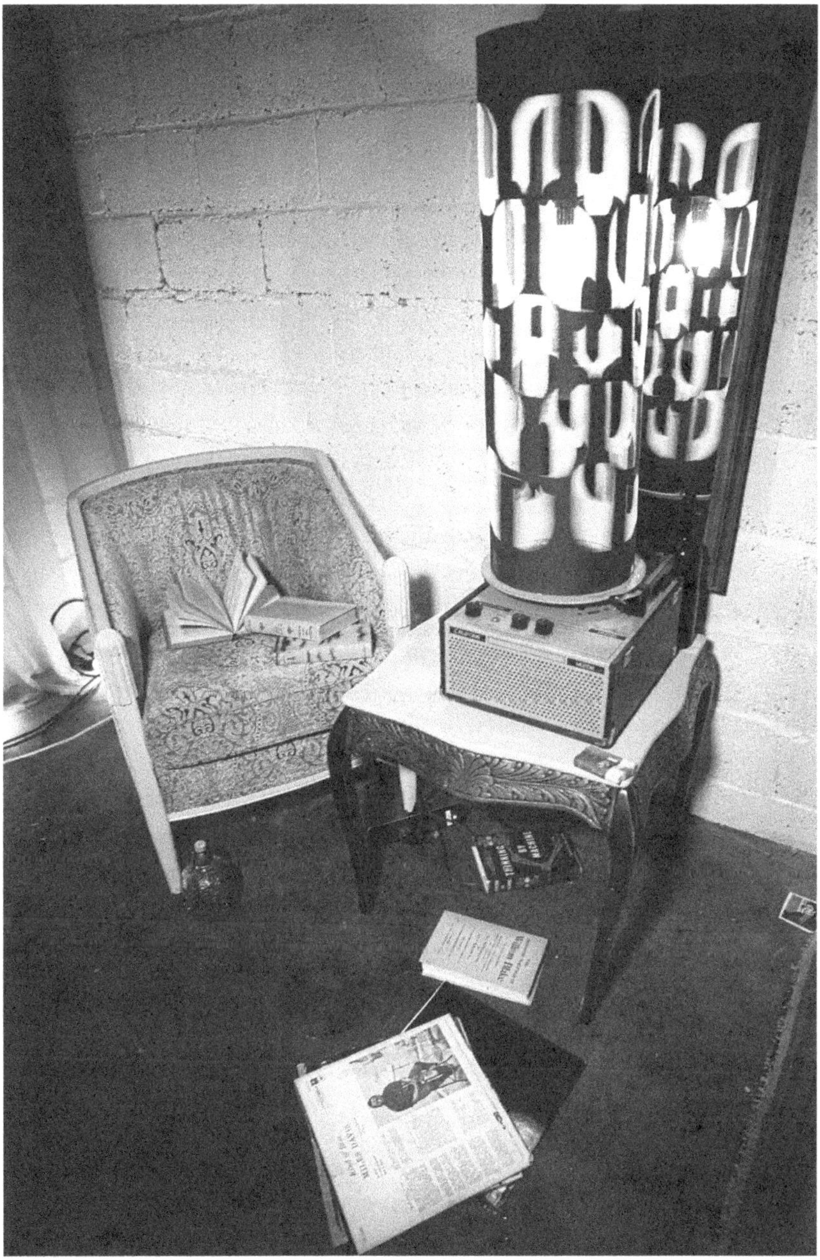

FIGURE 30. The Dream Machine. Credit: Jenny Schulder.

Gysin's first break as an artist had occurred in 1946 when Eileen Garrett had acquired and published his first book. *To Master, A Long Goodnight*, was a biography and history of the Reverend Josiah Henson, the supposed inspiration for the title character of Harriet Beecher Stowe's *Uncle Tom's Cabin*. In it, Gysin railed against a system of racial oppression that was nonlocatable but operated at the level of linguistic convention. Before his collaborations with Burroughs, Gysin had already taken Garrett's telepathic conceit to a logical if not ominous conclusion about the invisible ways in which word and image registered their effects on the human sensorium.[20]

In 1959, as they were developing their own ideas about the occult nature of communication and "machine control," Gysin introduced Burroughs to the ideas of L. Ron Hubbard.[21] Hubbard was a science fiction writer who, like Garrett, Burroughs, and Gysin, was strangely versed in emerging cybernetic paradigms. The publication of Hubbard's *Dianetics: The Modern Science of Mental Health* in May of 1950 was an unexpected success, selling over 155,000 copies by the end of the year. As an explicit riff on the "general science" of cybernetics, Hubbard's mix of psychoanalysis, science fiction, and self-help tapped an audience longing for a practical and science-driven approach to healing body and soul.[22] As Hubbard triumphantly declared in 1951, we have entered the "electronic era of the mind." Consequently, the neural processing of information became the grounding metaphor for Hubbard's blend of practical diagnosis and cosmic cure. "The fantastic storage capacity of the mind has no structural explanation at this time," wrote Hubbard. "But every sound, whether a voice, an auto horn, the click of table ware, footsteps, the wind, every sound which the individual has heard in his lifetime is recorded. . . . Dianetic technique sends the individual back

20. Given that the "sickness of the world" was due, in part, to the "multiplicity of words," Garrett hoped that her mediumistic skills would demonstrate the viability of communication that was neither deceptive nor coercive. As with the aspirations that accompanied early articulations of information theory, Garrett envisioned such psychic communication to be more efficient, more precise, more stable, more true, less susceptible to human ignorance and petty politics. Eileen J. Garrett, "The Language of Intuition," *Tomorrow* 5, no. 3 (Spring 1957): n.p.

21. Gregory Corso and Allen Ginsberg, "Interview with William Burroughs," *Journal for the Protection of All Beings* 1 (1961): 79–83.

22. See also Hubbard's "The Dianetics Question: Homo Superior, Here We Come!," *Marvel Science Stories* (May 1951): 11–13.

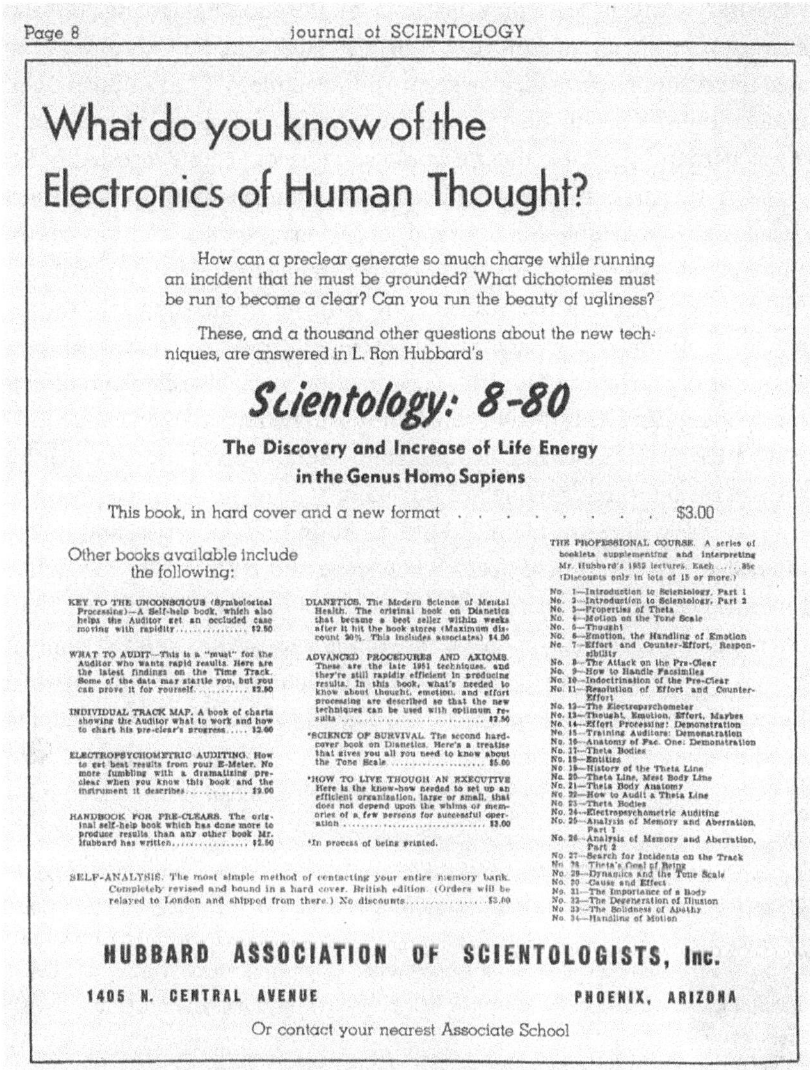

FIGURE 31. Advertisement, *Journal of Scientology*, ca. 1953.

through his life, and by the perceptics of sonic, vision, and so forth, enables him to recover information which has been occluded from him."[23]

In 1960, Burroughs wrote to Allen Ginsberg about the lacunae of con-

23. L. Ron Hubbard, *Science of Survival: Simplified, Faster Dianetic Techniques* (Phoenix: Hubbard Dianetic Research Foundation, 1951), 65–66.

temporary writers: "Have they picked up on the encephalographic research of Grey Walter? He can remove so called [*sic*] hallucinations by direct brain area intervention. Have they picked up on scientology?" Given Burroughs's fascination with occult schemes and provocative readings of the cybernetic, it was perhaps inevitable that he would soon become quite invested in Scientology. As part of his writerly discipline, Burroughs would spend the next decade studying Hubbard's notion of consciousness as a "communication and control system."[24] Hubbard's ideas became yet another source of Burroughs's explication of the mechanics of what he would soon call "the soft machine." Although Burroughs would eventually break with Hubbard's penchant for organizational control, his attraction to Hubbard's blend of neurophysiology and bent mythos distills not simply the cybernetic qualities of Hubbard's metaphysics but also metaphysical directives contained within these early neuromatic models of the brain.

Traversing between the midcentury demimonde of artists and writers interested in questions of occult knowledge and enthusiastic researchers in conversation with sciences of communication and control, Hubbard offered therapeutic riffs on the culture of the neuromatic.[25] In *Self Analysis*, a book advertised in the pages of Garrett's *Tomorrow* and one on which Burroughs took copious notes, Hubbard urges his readers to become practiced in incredible acts of self-awareness.[26] Hubbard then offered a series of exercises for readers to experience themselves as an elaborate commu-

24. Burroughs, *Rub Out the Words*, 58; L. Ron Hubbard, *Scientology: The Fundamentals of Thoughts* (Silver Spring, MD: Hubbard Association of Scientologists International, 1956), 33.

25. On the cold war paranoia of Hubbard and his organizations, see Hugh B. Urban, *The Church of Scientology: A History of a New Religion* (Princeton, NJ: Princeton University Press, 2013).

26. Garrett's journal, *Tomorrow*, often contained notices of other perspectives that she believed complemented her own, including a full-page advertisement for "Self-Analysis in Dianetics" from the Esoteric Book Club in *Tomorrow* 2, no. 1 (Autumn 1953): 2. Garrett and Hubbard anticipated the cybernetic gloss on occult claims of telepathy and ESP within the therapeutic culture of the 1960s, new age venues, and advice to engineer your "success mechanism." Maxwell Maltz, *Psycho-Cybernetics: A New Way to Get More Living Out of Life* (New York: Prentice-Hall, 1960); Sidney Petrie and Robert B. Stone, *Hypno-Cybernetics: Helping Yourself to a Rich New Life* (West Nyack, NY: Parker Publishing, 1973); Claude N. Rosenberg Jr., *Psycho-Cybernetics and the Stock Market: The Key to Maximum Investment Profits and Peace of Mind* (Chicago: Playboy Press, 1971); Benito F. Reyes, *Cybernetics of Consciousness: Theory and Practice of Meditation* (Ojai, CA: World University, 1978); Katherine Cover Sabin, *The Cybernetic ESP Breakthrough* (New York: Award Books, 1966).

nication network. As Hubbard writes, "There are many perceptions, which is to say channels, through which one can contact the physical universe. You are aware of the physical universe because of sight, sound, mouth and other message systems." Hubbard frames each of the senses as a different cognitive system. In addition to the five senses, Hubbard included others: tone, color, loudness, body position, weight, personal motion, and external motion. Given that each capacity was a subsystem, Hubbard then instructs: "Therefore, each time you are asked to recall an incident of a certain kind you will be asked, after you have recalled it, to pay attention to a certain sense channel which was present during the time when you experienced the incident. The circular disc is provided for this purpose." The circular disc, a modified tachistoscope, is labeled with each of the senses and invites the reader to isolate each of these subsystems as they consider their past. As the reader is asked to recall an incident—when "you were happy" or when "you could tell the difference"—the disc facilitates focus on that incident in terms of each of these subsystems. The disc ensures that the reader moves through each of these subsystems of perception, gaining a more complete memory of past experiences. As Hubbard declares, the disc enables a process of data retrieval and computation "as accurate as an IBM card index system and even faster."[27]

Hubbard's program for becoming a more efficient channel for processing the information of one's life was soon extended in a cosmic key with his announcement of Scientology. Whereas in Dianetics the goal was to achieve a state of "clear" in which one's thoughts and actions were unencumbered by the lingering noise of psychic trauma, in Scientology the self was also susceptible to the traumas experienced in past lives. Within the frame of Scientology, the subject was, by definition, overwhelmed by information—from one's childhood, for example, but also from the lifetimes that preceded this one. Indeed, for Hubbard the brain opened up onto cosmic history, for better or for worse, and was part of an epic drama. The flesh within your cranium held the keys to your mundane flourishing and pace of ascent within a vast space opera.[28]

27. L. Ron Hubbard, *Self Analysis* (1951; Los Angeles: American St. Hill Organization, 1971), 64, 13.

28. Hubbard's was an urgent appeal, in part, because the present media environment served to exacerbate the traumas stored inside the individual. "In a TV screen world," wrote Hubbard, "you are apt to be in trouble, because the TV is only a pattern of lights and shadows which is a restimulative mechanism to shuffle your bank around, and give you again some segment of that which you have already experienced." L. Ron Hubbard, "The Parts

Cyberneticians were more than a bit wary of Hubbard's borrowings. Walter likened "the Dianetics of Hubbard" to an undisciplined and selective reading of the cybernetic archive. Hubbard's aim, writes Walter, "was apparently to unite the principles and practices of Freud, Jung, Adler, Pavlov, Behaviorism, Faith Healing, Christian Science, Autosuggestion, Yoga and Theosophy into a single practical system of analysis and treatment. The resultant is the lowest common multiple of all of these cults, incorporating their crudities and exaggerations, ignoring their subtleties and implications. This is something to beware of." Norbert Wiener, who was heavily invested in the public image of cybernetics and his own reputation, was quick to dismiss Hubbard. In 1951, after being introduced to Hubbard's work by a former student, Wiener discovered to his shock and chagrin that he had been listed as an associate member of the Dianetics Foundation. Wiener immediately asked his lawyers to intervene.[29]

The need to differentiate legitimate cybernetics from its popular usurpers was but the flipside of the initial curiosity expressed by Shannon and McCulloch. As Hubbard was developing Dianetics he exchanged letters with Claude Shannon, inviting him to stay at his home on the Jersey shore and claiming that Dianetics opened up new vistas of engineering and cognitive science. Shannon, in turn, introduced Hubbard to Warren McCulloch as a leading science fiction writer who "has been doing very interesting work lately in using a modified hypnotic technique for therapeutic purposes." In 1952 McCulloch permitted Hubbard's Foundation to reprint his "The Brain as a Computing Machine" as a supplement to Dianetics instruction.[30]

In the remainder of this chapter I delve into the lines of communication and influence between Garrett, Gysin, Burroughs, and Hubbard. The lives and ideas of these figures—in their sustained extensions of the neuromatic paradigm—were an integral, if underappreciated, part of the cybernetic moment. Moreover, their compatibility at the level of metaphysical speculation was born of a shared investment in the new vistas of neuromaticity

---

of Man," *Professional Auditor's Bulletin* no. 125 (December 1, 1957), in *Technical Bulletins of Dianetics and Scientology*, vol. 3 (Los Angeles: Scientology Publications, 1976), 150.

29. W. Grey Walter, *The Living Brain* (New York: W. W. Norton, 1953), 264; Kline, *Cybernetics Moment*, 92–93. Even those within the cybernetic fold who were unapologetic about their interest in paranormal phenomena distanced themselves from Hubbard. The neuropsychiatrist J. R. Smythies, for example, panned Dianetics as a sham science of "mumbo-jumbology" (review of *Dianetics* in *Journal of the Society for Psychical Research* 36, no. 663 [March 1951], 406).

30. Kline, *Cybernetics Moment*, 93.

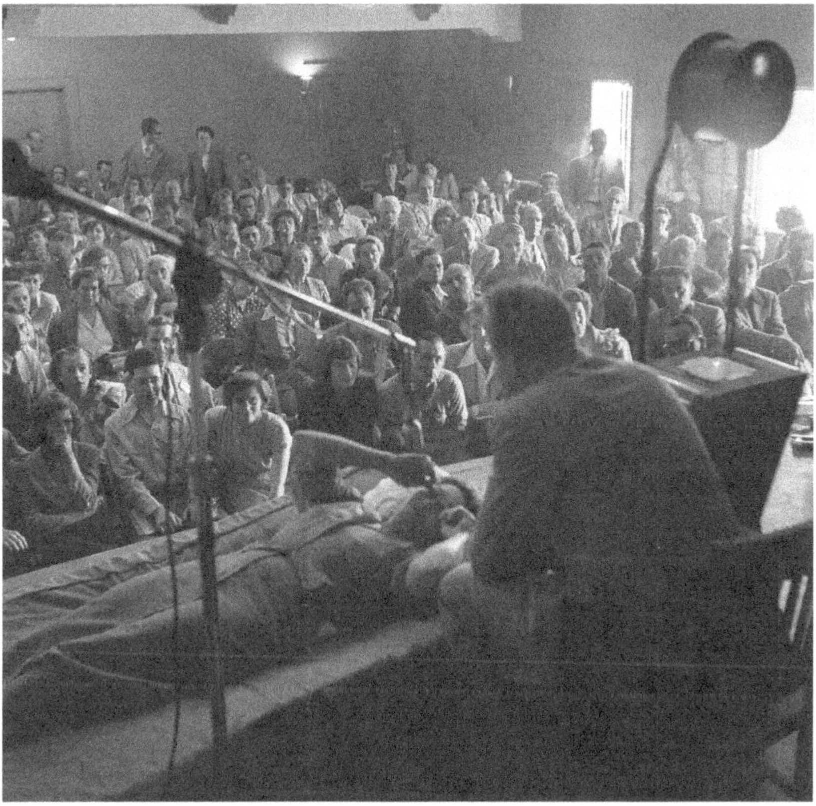

FIGURE 32. L. Ron Hubbard conducting a Dianetics seminar in Los Angeles, California, 1950. *Los Angeles Times* photographic archive.

being forged by midcentury cognitive sciences. Together, they riffed and speculated about information theory, feedback loops, and self-organization before the canonization of these concepts, before the normalization of the cybernetic, before its strangest (and grooviest) edges were smoothed over.[31]

31. See, for example, the general description of those on the margins of the scientific establishment as pursuing a "kind of secularized, quasi-spiritual quest" (David Kaiser and W. Patrick McCray, eds. *Groovy Science: Knowledge, Innovation, and American Counterculture* [Chicago: University of Chicago Press, 2016], 4). The fumbling language is indicative of the difficulty in naming the quality of actions based on science's own secularized vocabulary. The ambiguity of certain activities vis-à-vis the religious-secular continuum is precisely what is so significant about said activities, for they reflect what exceeds categorical incorporation.

3. ELECTIVE AFFINITIES

Eileen Garrett moved to New York City in 1941 where she founded Creative Age Press. As owner and editor, Garrett was not to be crossed. She "looked you over [and] saw your aura" and "fired people ruthlessly."[32] In an effort to Americanize the cosmopolitan aesthetics of British occultism, Garrett's agenda was to introduce parapsychology as humanistic art, that is, as a high culture and high-minded endeavor. In addition to publishing Garrett's *Telepathy* and a book on Nostradamus, Creative Age Press also published literary fiction, including a series of Garrett's own novels written under the name Jean Lyttle.[33] Under the auspices of Creative Age, Garrett founded *Tomorrow: World Digest of Psychical Research and Occult Studies.* This first iteration of *Tomorrow* (1941–51) published works by Thomas Mann, Eudora Welty, Pearl S. Buck, and Ralph Ellison.

In 1951, Garrett sold Creative Age Press to Farrar, Strauss, and Young. By the end of the year she had cofounded the Parapsychology Foundation, Inc. (hereafter PFI) with her friend and patron, Frances Bolton. Upon the founding of PFI, Garrett immediately rebooted *Tomorrow* as the *World Digest of Psychical and Occult Studies* and began to gather and eventually sponsor parapsychological research from around the world. Bolton's financial and political status added a new sense of possibility to Garrett's investment in parapsychology. At the time Bolton had already been serving in the US House of Representatives for over ten years. As a member of the House Foreign Affairs Committee, Bolton oversaw efforts to modernize mining operations in Africa. In her work to advance the flow of capital, she called for mild reforms of British colonial policy and more economic independence for the Federation of Rhodesia.[34]

32. Angoff, *Eileen Garrett*, 1.

33. *Today the Sun Rises* (1942), *You Are France, Lisette* (1943), *Sheila Lacey* (1944), and *Threads of Destiny* (1961). Lyttle had been the last name of Garrett's adoptive aunt and uncle who discouraged talk of her psychic ability. Eileen J. Garrett, *Adventures in the Supernormal* (New York: Paperback Library, 1968), 19.

34. Eileen J. Garrett, "From the Publisher's Desk," *Tomorrow* 1, no. 1 (Autumn 1952): n.p.; Andrew DeRoche, "Frances Bolton and Africa, 1955–58," *Passport: Newsletter of the Society for Historians of American Foreign Relations* (January 2010), 34–37; Andrew DeRoche, "Frances Bolton, Margaret Tibbetts and the US Relations with the Rhodesian Federation, 1950–1960," in *Living the End of Empire: Politics and Society in Late Colonial Zambia*, ed. Jan-Bart Gewald (Leiden: Brill, 2011), 299–325.

Garrett, with Bolton's unwavering support of the PFI, was intent on bringing institutional legitimacy to parapsychology research.[35] Garrett aligned PFI with the "breath-taking strides" made in the physical sciences in the first half of the twentieth century. Her message was that humans could "communicate without physical media" and that a new science of telepathy was evolving to account for that very fact. While "patterns" of psychic phenomena had previously "eluded measurement," wrote Garrett, there was "no reason to believe that, given serious thought by scientifically trained observers, this would continue to be true." In her 1957 anthology of parapsychological research, for example, Garrett wrote of the "passing of the iron reign of late-Victorian mechanistic materialism in the physical sciences" and the "liberating effect in every department of man's thought." As Garrett insisted, something like telepathy was neither pathological nor magical but a universal human capacity that could be cultivated or not.[36]

Garrett was representative of the parapsychological pose insomuch as she imagined herself to be practicing a cutting-edge science of communication.[37] Like many parapsychological researchers at the time, PFI was

35. Bolton, whose uncle was a partner in the Standard Oil Company, was also major seed money for the science of the new age. Previously she had helped establish the McDougall Research Fund at Duke University which supported much of the parapsychological work of J. B. Rhine, one of Garrett's early scientific interlocutors. Bolton found common cause in Rhine, who had coined the term "parapsychology" and championed the use of experimental data to confirm metaphysical principles. R. Laurence Moore, *In Search of White Crows: Spiritualism, Parapsychology, and American Culture* (New York: Oxford University Press, 1977), 202–3. The overriding assumption of this new science was that the brain partakes in processing beyond the merely physical. Consequently, Rhine, Garrett, and others looked to the brain for that "immaterial part of the human personality, a spiritual self that might conceivably be considered capable of survival" (F. H. Cleobury, "Metaphysics and Parapsychology," *Journal of the Society for Psychical Research* 36, no. 664 [May–June 1951]: 444).

36. Eileen J. Garrett, ed., *Beyond the Five Senses* (Philadelphia: J. B. Lippincott, 1957), 9–10; Eileen J. Garrett, *Telepathy: In Search of a Lost Faculty* (New York: Creative Age Press, 1941), 179; Eileen J. Garrett, "Life Is Not Opposed to Death," *Tomorrow* 4, no. 4 (Summer 1956): n.p.; Angoff, *Eileen Garrett*, 182.

37. Moore, *In Search of White Crows*, 212. Garrett was less open to spiritualistic explanations than the scientific researchers that her foundation sponsored. Indeed, Garrett was often quite cagey in her disbelief in spiritualistic explanations, acknowledging them yet confessing her own skepticism while leaving the investigatory door wide open. Writing of the ontology of her controls—Ouvani and Abdul Latif—"For myself, I have never been able to wholly accept them as the spiritual dwellers on the threshold, which they seem to believe they are. I rather leaned away from accepting them as such, a fact which is known to them

highly self-conscious about experimental design.[38] The sobriety propagated by PFI, its publications, and sponsored researchers gave license to the critique of those who would close off inquiry to psi phenomena.[39] Moreover, the "strenuous and painstaking scholarliness" of mainstream science was not simply imperious and power-crazed but also a form of religious enthusiasm. "Science-worship," lamented *Tomorrow*, "was never more passionate than it is today."[40] Consequently, Garrett and other parapsychologists were intent on framing the new science as a secular intervention into both bad religion and bad science.

Over the course of the 1950s and 1960s, PFI was incredibly active, sponsoring over twenty international conferences on topics ranging from psi factors in creativity, spontaneous phenomena, medical implications, psy-

---

and troubles them not at all. They appear to be different in nature. . . . whether these entities do contain a consciousness of their own is a question that will, I am sure, find an answer, although not in my lifetime" (Garrett, *Many Voices*, 94).

38. PFI published the *International Journal of Parapsychology* from 1959 to 1968 as well as a monograph series beginning in 1958 that focused on scientific research. PFI established its own division of research in 1957 with neurophysiology at the center of its mission statement. Citing the research of neurobiologists, PFI promoted inquiry into the physiological correlates of ESP, trance states, and instances of precognition. PFI also advocated "statistical methods [that] reflect current scientific techniques" (*Ten Years of Activities* [New York: Parapsychology Foundation, Inc., 1965], 16–19). For other advertisements of scientific authority, see Whatley Carington, *Thought Transference: An Outline of Facts, Theory and Implications of Telepathy* (New York: Creative Age Press, 1946); Gardner Murphy with the collaboration of Laura A. Dale, *Challenge of Psychical Research: A Primer of Parapsychology* (New York: Harper and Row, 1961); J. R. Smythies, ed., *Science and ESP* (London: Routledge and Kegan Paul, 1967); J. B. Rhine, ed., *Progress in Parapsychology* (Durham, NC: Parapsychology Press, 1971).

39. Gabriel Marcel, "The Limits of Science," *Tomorrow* 2, no. 4 (Summer 1954): 40. As a way to confirm the capaciousness of the parapsychological gaze (that is, its open-minded secularity), Garrett published numerous articles on comparative religion and anthropology, including pieces by Melville Herskovits, Maya Daren, and Robert Lowie. See also Ernesto de Martino, "Parapsychology, Ethnology and History of Religion," International Symposium on Psychology and Parapsychology, April 30 to May 4, 1956, *Proceedings of Four Conferences of Parapsychological Studies* (New York: Parapsychology Foundation, 1957), 160.

40. Roberto Cavanna, ed., *PSI Favorable States of Consciousness: Proceedings of an International Conference on Methodology in PSI Research* (New York: Parapsychology Foundation, 1970), xvii; Allan Angoff, "The Arrogance of Science," *Tomorrow* 6, no. 3 (Summer 1958): 68–69.

chedelics, and pharmacology.[41] With the mingling of psychic investigators, humanists (such as Kenneth Burke), scientists (such as Grey Walter), and LSD pioneers (Timothy Leary and Richard Alpert), there was a shared sense that mind and matter were interdependent and that the study of that interdependence, wherever it might lead, was crucial to the survival of the species. There was an urgency to this staging of the hard problem of consciousness as well as opportunity for resolution. As Garrett wrote, the conferences demonstrated how "men in various fields of scientific research, in philosophy and medicine, are engaged in thoughts and efforts outside the realm provided by orthodox concepts and standards. All of them, earnestly and with devotion, are searching for answers to the mystery of man's very existence." According to Garrett, parapsychology was fast becoming an applied science and one that had "enormous potential force" and could, in the future, generate "whole series of unsuspected mechanical and chemical combinations."[42]

With their mix of natural and social science, the PFI conferences operated as a shadowy extension of the Macy Conference circuit that ended in 1953.[43] Walter, for example, traveled in both circles, attending the final Macy Conference and a number of Garrett's conferences in the 1960s. Walter was also a member of the Society for Psychical Research, delivering its Frederic W. H. Myers Memorial Lecture in 1960. Garrett had been able to recruit Walter into the fold of the PFI given his interest in extreme neurological states. "I had read Dr. Grey Walter's book on the living brain," recalled Garrett, "and I concluded that he would be the one person to work with in a study of the brain, its functioning, and its possible connection with the ego. I

41. *Proceedings of the First International Conference of Parapsychological Studies* (New York: Parapsychology Foundation, 1955); *Proceedings of Four Conferences of Parapsychological Studies* (New York: Parapsychology Foundation, 1957); *Proceedings of Two Conferences on Parapsychology and Pharmacology* (New York: Parapsychology Foundation, 1961).

42. Garrett, "Philosophy, Healing, Parapsychology," *Tomorrow* 2, no. 4 (Summer 1954): n.p.; Garrett, "Life Not Opposed to Death," *Tomorrow* 4, no. 4 (1956): n.p.

43. Like PFI Conferences, the Macy Conferences had been staged as models of interdisciplinarity and open-mindedness. Attendees at both conferences approached psychic phenomena as "related to the very essence of life itself" (Garrett, "Life is Not Opposed to Death," n.p.) and subject to the latest technological advances in recording and measurement. On postwar academic conferences as ritual spaces for curiosity, creativity, and flexibility that "regularly and habitually used academic culture as a model of other parts of society," see Jamie Cohen-Cole, *The Open Mind: Cold War Politics and the Sciences of Human Nature* (Chicago: University of Chicago Press, 2014), 137, 114.

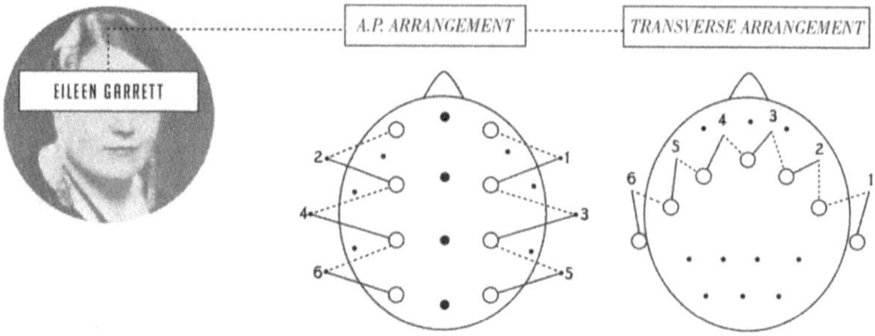

FIGURE 33. Electrode placement on the scalp of Eileen J. Garrett, approximated by Libby Modern from C. C. Evans and Edward Osborn, "An Experiment in the Electro-Encephalography of Mediumistic Trance," *Journal of the Society for Psychical Research* 36, no. 669 (March–April 1952): 591.

wrote to him and told him of my interest, whereupon he replied and invited me to come to Bristol in England, which I did." A short while later, Garrett became Walter's experimental subject. "With the electroencephalographic setup," wrote Garrett, "he was able to make tests during hypnosis, clairvoyance, trance, and finally under the influence of LSD-25."[44]

Walter bonded with Garrett over their mutual insistence that there was, at the end of the day, a neurophysiological explanation for instances of telepathy and clairvoyance. Like Garrett, Walter was extremely cautious in attributing supernatural explanations for paranormal activity. And like the parapsychologists inspired by his work, Walter kept an open mind and suggested a confluence of occult and cybernetic inquiry into the "mechanisms" of psychic phenomena. As Walter confessed, "I assume that all mental—or, if you like psychic—phenomena can be rigorously described in terms of physiological mechanism."[45]

Garrett's recruitment of Walter as well as her own interest in electroencephalography were part of a burgeoning cybernetic-parapsychology nexus at midcentury.[46] Parapsychology, like many fields of scientific inquiry at the time, had taken a hard cybernetic turn. "Extrasensory knowledge" was

44. Garrett, *Many Voices*, 168.

45. W. Grey Walter, *The Neurophysiological Aspects of Hallucinations and Illusory Experiences* (London: Society for Psychical Research, 1960), 3.

46. S. C. Wallwork, "ESP Experiments with Simultaneous Electro-Encephalographic Recordings," *Journal of the Society for Psychical Research* 36, no. 671 (July–October 1952): 697–70; C. C. Evans and Edward Osborn, "An Experiment in the Electro-Encephalography of Mediumistic Trance," *Journal of the Society for Psychical Research* 36, no. 669 (March–

"to be scrutinized and dealt with sanely and systematically" and made "the subject of serious laboratory study."[47] By the late 1950s, parapsychology and cybernetics were part of the same epistemic groove. This confluence had as much to do with shifts within parapsychology as with the metaphysical leanings of cybernetic speculation. Those seeking leverage on the nature of the human by way of the brain, its environment, and the informational surround were often attracted to paranormal investigations.[48] The attraction between parapsychology and cybernetics, in other words, was mutual.

The erotic charge of this shared fascination is, perhaps, best captured by a moment in the transcript of the 1969 Parapsychology Conference hosted by Garrett in St. Paul de Vence, France. After Walter's lecture the question and answer period begins.[49] There is talk of the transitive adaptation process and a suggestion that poems one day may be written by computer. Walter then intervenes with an unconventional digression.

"Yes, I wrote a poem," he announces, "It's just a jingle, suggested by Mrs. Garrett's book called 'Many Voices.' It is a poem about the brain and it's an example of the way in which one can afford to develop a fantasy which may suggest experiments." Before beginning his recitation of "Odysseus Looks at an EEG," Walter informs the conference audience that "Mrs. Garrett and I have a little private fantasy about classical personages." Walter then begins:

> I hear the murmur of your private gossip,
>   I see the gulfs and headlands of your coastline.

April 1952): 588–96. Although it was noted that Garrett's alpha waves became unstable during trance state, the results were considered inconclusive and preliminary.

47. Eileen Garrett, "From the Publisher's Desk," *Tomorrow* 1, no. 4 (Summer 1953): n.p.

48. Other significant figures who were attracted to occult inquiry include Stafford Beer, a recorded member of the Society for Psychical Research in 1962 and 1971, as well as W. Ross Ashby, author of *Design for a Brain* (1952) and Walter's boss at the Burden Neurological Institute (1959–60). As Andrew Pickering has documented, Ashby was a youthful convert to spiritualism after having read Stanley de Brath, *Psychical Research, Science, and Religion* (London: Methuen, 1925) and Sir Oliver Henry Lodge. In August of 1930, shortly after the R101 disaster, Ashby wrote in his journal that the brain was an "elaborate and exceedingly delicate trigger mechanism . . . the obvious place for spirits . . . to step in and seize control." Andrew Pickering, *The Cybernetic Brain: Sketches of Another Future* (Chicago: University of Chicago Press, 2011), 416n62, 425n19.

49. Grey Walter, "PSI and Creativity: Some Neuropsychological Doubts and Discoveries," in Allan Angoff and Betty Shapin, eds., *PSI Factors in Creativity: Proceedings of an International Conference Held at Le Piol. St. Paul de Vence, France June 16–18, 1969* (New York: Parapsychology Foundation, 1970), 125–32.

I hear, too, the surf of my private sea
    That foams and curls around your promontories.
Are your voices, the echoes of my billows?

The descent into a collective state of awkwardness accelerates as Walter specifies Odysseus's sea-level gaze:

on the reefs of your body . . .
I must sacrifice my craft
Run the breakers naked; throw myself breathless
On your beach

Will a princess come with laughter and garments
Welcome me to the circle of maidens?

Upon the conclusion of the reading, the presider cuts Walter off before he can add any further comment: "Thank you, Dr. Walter."[50]

4. THE MECHANICS OF MEDIUMSHIP

The historian R. Laurence Moore has noted that "the flurry of organizational activity, especially during the 1960s, did little to establish inroads for parapsychology in American Universities."[51] Although Moore is not incorrect, he is also not right. For although remaining on the margins of the academy, parapsychology is not absent in a world that dreams of communication across cosmic registers and artificial forms of intelligence that exceed our own. So rather than failure, I am interested in how Garrett's language successfully distilled a metaphysical fundament of the cybernetic imagi-

50. "Open Discussion," in Angoff and Shapin, *PSI Factors in Creativity*, 135–36. Walter's telepathic intimacy with Eileen Garrett resonates with his reduction of sexual desire to a matter of neuromatic processing and, in turn, his recourse to erotic language to describe his relation with machines. William Grey Walter, "Features in the Electro-physiology of Mental Mechanisms," in *Perspectives in Neuropsychiatry*, ed. Derek Richter (London: H. K. Lewis, 1950), 67–78, esp. 69ff.; William Grey Walter, "The Future of Clinical Neurophysiology," *Handbook of Electroencephalography and Clinical Neurophysiology* 1A (1972): 39–50.

51. Moore, *In Search of White Crows*, 210.

nary. Across Garrett's many writings, she frames her psychic powers as part of a corresponding cosmos in which all systems of reality are homologous and interrelated. "There is no reason to believe," wrote Garrett, "that the energy which runs the human system is different from that which runs the universe." Having herself been "exposed to various patterns of existence," Garrett insisted that we lived amid a whirl of matter and spirit. This world was incredibly complex but nonetheless could be decoded. According to Garrett, an appreciation for the ontology of system, for the subtlety of its spirit-matter hybridity, was necessary for a proper understanding of parapsychology.[52]

Even as Garrett cautioned against schemes that would render humans "as robots," her prose spoke of systemic flows between self and environment, humans and other worlds, of congress across categorical divisions. For example, Garrett described telepathy as the translation of the data cloud of another person's consciousness—

> There is a part of mind, or intellect, that appears to pick up, arrange, and coordinate what is taking place in another person's mental atmosphere, and that is what very often happens with me. I am, as it were, thinking with the other fellow's mind. There appears to be a specific place where the spaces between one's own consciousness and the unconscious data of the senses are filled in, to clarify and answer the needs arising in another person's mind.[53]

According to Garrett, telepathy was not so much reading the mind of the other but rather a becoming aware of the brain-environment complex of others in real time—becoming attuned to the flow and processing of the informational surround—not just her own processing but that of others

52. Eileen J. Garrett, "The Bridge of Emotion," *Tomorrow* 4, no. 2 (Winter 1956): n.p.; Eileen J. Garrett, "From the Publisher's Desk," *Tomorrow* 1, no. 3 (Spring 1953): n.p; "Transatlantic Telepathy," *Tomorrow* 1, no. 3 (Spring 1953): 67–74. According to *Tomorrow*, for example, ghosts were thought to be unhappily caught up the binary divide between "our material world of the body, and the next plane of spiritual existence," unable to see the divide as a space of transition, unable to see the fluidity of the situation, unable to grasp the flux and flow of their material habitation. Consequently, they were unhappy and "in need of psychiatric care" precisely because they were ignorant of the spiritual dimension of matter and vice versa. Garrett, "Rockland County Ghost," 10.

53. Garrett, *Many Voices*, 208–9.

as well.[54] As Garrett wrote, "The brain processes the material for our comprehension much in the way a computer deals with the data that is fed to it."[55]

Garrett's capacity to process the input of sensory stimuli was the source of her insight into macrocosmic order. "I had a sense of being volatile within," writes Garrett,

> of various inorganic movements occurring in different parts of my being, but all synchronized with the rhythms of some larger vital activity. I experience an impression of "flowing" within me, and at the same time something moves out from me to an object, yet remains an individual part of myself. By means of an indescribable contact which takes place between the object and me . . . The process is instantaneous and timeless. It takes far longer to describe it than it takes to occur.[56]

Garrett described this capacity to "contact" the environment as a "method of communication"—a "knowing, as apart from thinking" that was intensely and physically felt. According to Garrett, "space has never been empty for me" but rather percolated with a "tremendous 'vitality.' . . . Eventually these impressions of 'thin' sound and infinitesimal particles moving and bursting in all spaces of the outer world shifted to the area of my own self." In traversing the boundaries of her skin, "this *surround*, as I called it for want of a better name" was both utterly abstract yet materially pressing—consisting "of transparent changing colors . . . dense and heavy in character."[57]

Garrett had come to understand her extrasensory perception in terms of information processing. Whatever, exactly, the source of her gifts, Garrett

54. As a resource to explain "communication between minds," other parapsychologists approached this ethereal atmosphere as a mathematical proposition. See, e.g., F. H. Cleobury, "The Theory of Selective Telepathy," *Journal of the Society for Psychical Research* 44, no. 737 (September 1968): 327–28. In attempting to explain the "physical 'stuff' or ether between the bodies," insisted Cleobury, "we have to postulate some kind of physical nexus." For an earlier and more skeptical take on the promise of information theory and parapsychological investigation, see Smythies, "The Extension of Mind," 478. Again, an atmospherics of brain activity was the shared ground of debate.

55. Garrett, *Many Voices*, 208.

56. Garrett, *Adventures*, 26.

57. Garrett, *Adventures*, 24, 26–27; Garrett, *Telepathy*, 178.

FIGURE 34. Eileen J. Garrett, ca. 1930. Courtesy of Chronicle/Alamy Stock Photo.

was convinced that part of her brain could receive "from the entire universe information . . . denied the normal conscious mind."[58] The conceptual and rhetorical turns of information theory had provided Garrett much explanatory leverage on the "mechanics of mediumship." Garrett described this mechanics as a neuromatic proposition, yet one in which different parts of her body worked in concert to maximize their collective input capacities to the brain.[59]

5. IMAGES OF AN ORACLE

In 1964 the Jungian psychotherapist Ira Progoff published *The Image of an Oracle*, a book of transcripts from Garrett's trance sessions. The transcripts are notable for how they both confirm and extend Garrett's own neuromatic musings. According to Progoff, the transcripts were evidence of Garrett's capacious spiritual bandwidth and her capacity to minimize noise. In doing so, she became a significant node in a boundless communication network, assuring that the "whirling . . . levels of experience" came through and that right signals were passed along. According to Progoff, Garrett was able to create lines of communication between temporal and spatial registers. She was sensitive to messages that exceeded whatever categories mere humans

58. Angoff, *Eileen Garrett*, 189. Garrett, here, was in keeping with the cybernetic lean of parapsychological research. In 1952, for example, E. R. R. Holmberg declared that "in a fundamental paper, C. E. Shannon has shown that the physical properties of a channel place an upper limit on the rate at which information can be transmitted through it, and that this limit is independent of the form the signal takes. It is thus of interest to determine whether the same sort of limitations apply to the psi channels"; in "ESP and Information Theory," *Journal of the Society for Psychical Research* 36, no. 668 (January 1952), 574. Over the course of the 1950s and 1960s, parapsychologists continued to adopt information theory to refine their statistical modeling of other dimensions and their understandings of extrasensory perception, telepathy, and telekinesis as varieties of "transmission" and "acquisition of information." Cavanna, *PSI Favorable States of Consciousness*, xviii; Lawrence LeShan, "The Vanished Man: A Psychometry Experiment with Eileen Garret," *Journal of the Society for Psychical Research* 62, no. 1 (January 1968): 115; see also J. Zeman, "Information as Psychic Activity" *Progress in Brain Research* 17 (February 1965): 139–50.

59. Garrett, *Telepathy*, 182. See also PFI's funding of work on information theory in the key of perennial philosophy: C. T. K. Chari, "Different Languages and Philosophies," *Dialectica* 22, no. 3/4 (1968): 300–312; Garrett, *Awareness*, 156.

had constructed. She channeled, precisely, occult patterns, picking out the right signals from all those swirling around her.[60]

Progoff framed Garrett as a detector of agencies, a channel between worlds. With theoretical nods to systems, networks, and information theory, such framing was part of Progoff's larger argument that Garrett's spirit ecology of flux and flow was but an amped-up version of normal psychological processing. Given such hyperactivity, the primary questions animating Progoff's investigation were those of subjectivity—"a question of who is speaking"—during these sessions when supernatural agents were often in possession of Garrett's body. According to Progoff, Garrett was an example of distributed agency. During these moments of trance, Garrett became a node of the informational surround. Something supernatural in her own right. "Eileen Garrett," writes Progoff, is "something much larger than the individual whose name it bears."[61]

As recorded dialogues between Progoff and Garrett's host of spirit controls, the transcripts resembled a mix of séance spectacle and awkward therapeutic encounter—with "OUVANI . . . young Arab soldier of the thirteenth century who had died in battle; and ABDUL LATIF, a Persian physician of the seventeenth century," as well as Ramah and Tahoteh who claimed no previous earthly existence. The onset of communication between Garrett and her environmental controls was disorienting, marked by "*deep breathing, yawns, moans as Mrs. Garrett enters the trance state.*" As filters born of Garrett's own psychological sensitivities, such spirit controls were real and served to regulate a system that was Eileen Garrett. For as Progoff learned from his interviews with them, they were "here to guard against the influx of knowledge, all the forms of knowledge, in a mind that is too open."[62] Consequently, Garrett "could not rid herself of the feeling that there might be another person within her."

Of all of her controls, Ouvani, the so-called "regulator," was the figure most responsible for securing the conditions of self-organization and maintaining the coherence of Garrett's personality from afar. "So when Ouvani refers to himself as the doorkeeper," explained Garrett, "he is not thinking of you, and you, and you, he is thinking of all that is necessary to keep the

60. Progoff, *Image of an Oracle*, 77.

61. Progoff, *Image of an Oracle*, 72, 73, 76, 161, 84, 360–61.

62. Progoff, *Image of an Oracle*, 5, 23, 227, 13, 317; Angoff, *Eileen Garrett*, 78.

equilibrium."[63] During the sessions in which Garrett channeled Ouvani, the question of subjectivity was given a neuromatic spin. On one hand, there was Ouvani's blithe insistence on determinism—"if you will look deeply within her life," Ouvani notes, "it is not her life that she lives, that becomes very obvious to you. If you watch this life that it is not the life of a woman but that it is a life governed, ordinated, and in a sense, projected." Similarly, when questioned by Progoff, Ouvani defers accepting responsibility for his decision to speak through Garrett. "The choice was not made by me," asserts Ouvani. "I was appointed, if you like, to be the doorkeeper, shall we say, certain evidences of factors which were then related to the peoples concerned. The choice was not mine. The choice was made for me by one of those who for want for a better word you will call the masters of her evolution or her destiny." Ouvani, here, conjures an authority structure of system (masters) upon system (controls) upon system (Garrett). And to submit to this scheme of control was positively liberating. "I hope you agree with me," Ouvani continues, that we are the "engineers" of our own existence.[64] The potential to be a steersman—for both Garrett and Ouvani—was predicated not just on acknowledging a system that enveloped them but also accepting its logic of saturation. In other words, the apparent contradictions that were so central to Garrett's self-presentation—agency as submission, negation as amplification—were those engrained in the language of the cybernetic contract. As Garrett announced, during trance sessions, "I become almost simultaneously more and yet less of myself."[65]

Garrett's narrative of self-realization is inconsistent with the notion of freedom as an all-or-nothing proposition. For Garrett conceived of herself as porous and vulnerable—buffered only to the degree that she might interact, as one pattern among other patterns, within a larger environment

63. Progoff, *Image of an Oracle*, 318, 19–20. Or, as Progoff anthropomorphized, Ouvani was "the personification of the regulative function of the psyche . . . that limits and channels the flow" (32–33).

64. Progoff, *Image of an Oracle*, 17, 15, 18. Also, despite his appointment Ouvani claims that he hails from a "place where free-will predominates. . . . All experience is growth. . . . Here is freedom from pain" (110–11).

65. Progoff, *Image of an Oracle*, 240; Angoff, *Eileen Garrett*, 192. As one her controls explained to Progoff: "we speak of the non-identity which man must take on in order to execute this area, this voyage of which we spoke. During this voyage, he is going through the dust of experience, covering and coating himself with it, in order to lose identity with that, in order to gain identity for this" (225).

in an ongoing process of mutual and perpetual transformation. As Tahoteh instructed Progoff, such capacity to be perpetually impressed upon was that which guaranteed Garrett's processual integrity: "For the growth of the whole, it is important that man or whatever pathway of life you like, shall split up, to lose himself, to lose—never quite lose but to almost lose—sight of himself, gathering to himself that dust of experience, which, again, when he has consumed the wholeness of it, he must then begin to shed it again."[66] According to Tahoteh, Garrett was neither an open nor closed subject, neither porous nor buffered but both. Garrett, herself, described changing as she moved through the world, an emergent pattern rather than a static thing—ever interacting with her own celebrity and the expectation of others. For Garrett's "strong dramatic flair" and erotics of self-presentation must be seen as a strategic act of autopoiesis,[67] strategically inviting the gaze of others—scientists, socialites, and readers as integral to her effective performance of an authentic self.[68] Constant traffic between Garrett's brain and environment was foundational to parapsychology, in general, and to the kind of neuromatic subject that Garrett imagined herself to be.[69]

66. Progoff, *Image of an Oracle*, 226.

67. Progoff, *Image of an Oracle*, 3. As Francisco Varela and Humberto Maturana write in their classic work on the metaphysical implications of second-order cybernetics—the book that introduced the term "autopoiesis" to the study of living systems—"we become self-conscious through self-observation; by making descriptions of ourselves (representations), and by interacting with our descriptions we can describe ourselves describing ourselves, in an endless recursive process." In doing so, we remain "in a domain of interactions always larger than that of the representations" (*Autopoiesis and Cognition: The Realization of the Living* [1973; Boston: D. Reidel, 1980], 14). As Hayles notes, an autopoietic unity capable of becoming an observer does so by generating representations of its own interactions (*How We Became Posthuman*, 140ff.).

68. On first meeting Garrett, a colleague recalls that "in the late afternoon, there was a knock on the door and in came an ample woman of fifty or so, richly dressed, wearing a variety of bracelets and rings, and exuding the most delicate scents. She was attractive in a commanding and lusty way." Angoff, *Eileen Garrett*, 5, 31.

69. Garrett was expert in managing her image as an oracle—moving in and out of the present, into the minds of others, as if hovering above their encounters with her in order to reflect on what was and what could be generated in those encounters. Such performativity was part and parcel of her subtle critique of gendered hierarchies and sexual difference. Garrett often cited her friendship with Edward Carpenter, a devotee and acquaintance of Walt Whitman from whom she learned about human sexuality as a fluid proposition. "He wanted me to become aware," recalled Garrett, "that all who possessed these special sensi-

6. THOUGHT DICTATED IN THE ABSENCE OF ALL CONTROL

Garrett did not often dwell on the more disturbing implications of her metaphysics, specifically the prospect that representations could be manipulated for agendas that had little to do with heightened awareness or self-improvement. The more ominous issue of "control" broached by Garrett's ontology was brought to the fore, however, in a 1974 biography by Allan Angoff. According to Angoff, a parapsychology researcher, an associate of PFI, and a librarian trained in information science, Garrett's gifts of "clairvoyance, telepathy, psychometry, trance mediumship" could "be dangerous when controlled by people without compassion."[70] Angoff, here, was distilling a theme within the pages of *Tomorrow* that framed parapsychology as a counter to "Hitler's Black Magicians" and other forms of political totalitarianism precisely because it offered insight into how such manipulation proceeded apace. Rumored by some to be on the payroll of the CIA, Garrett often made the case that telepathy could serve projects of diplomacy and geopolitical security.[71]

Garrett's most overt political practice was her editorial interest in African American literature and culture. In addition to parapsychological writings and fiction, Creative Age Press also published studies on race, including a

---

tivities blended in their natures the qualities of both the masculine and the feminine; in this way he led me to understand that the variations in sexual types, such as the bisexual and the homosexual, were not to be despised." Rather, Garrett learned, queerness comprised "nature's steps to an ultimate and higher form of mankind." Angoff, *Eileen Garrett*, 69–70.

70. Allan Angoff, *Public Relations for Libraries: Essays in Communication Techniques* (Westport, CT: Greenwood Press, 1973), vii. Attuned to how public relations could be used in spreading the political insights of parapsychology, Angoff had previously offered reprints of Edward Bernays's work on the uses of propaganda, and he wrote that "awakening and stimulating an audience that does not fully comprehend its own interests requires skilled practitioners." Even more so than Garrett, Angoff insisted on a strictly secular framing of her powers, noting that Garrett "made no claims to having supernatural powers" and that "there was no magic to it." Her abilities were strictly empirical, her message a therapeutic and political one rather than metaphysical. Angoff, *Eileen Garrett*, ix.

71. Gerda Walther, "Hitler's Black Magicians," *Tomorrow* 4, no. 2 (1956): 7–23. See also Garrett's editorial note for Henry K. Puharich, "Can Telepathy Penetrate the Iron Curtain?," *Tomorrow* 5, no. 2 (1957): 7–9; Brion Gysin, "Goodnight Eileen," in *Here to Go: Planet R-101* (London: Quartet Books, 1985), 3.

biography of Duke Ellington and monographs on slave songs of the Georgia Sea Islands and colonial policies in Africa.[72] In 1946, Garrett published *To Master, A Long Goodnight*, a strange book of African American history by an aspiring Canadian artist named Brion Gysin.

Garrett had met Gysin at a party in New York City attended by the astrologer Evangeline Adams as well as the actor Yul Brynner. Gysin had just completed six weeks of training with the Canadian Infantry Corps at Chatham, Ontario, where he had met a paratrooper named Cleland "Tex" Henson. Henson, as it turned out, was the great-grandson of the Reverend Josiah Henson. Rev. Henson's autobiography, *The Life of Josiah Henson, Formerly a Slave, Now an Inhabitant of Canada, as Narrated by Himself* (1849), was close at hand as Harriet Beecher Stowe wrote *Uncle Tom's Cabin* (1852).[73] Upon hearing Gysin's story of an abolitionist Methodist minister who escaped slavery, was part of the Underground Railroad, and led a black militia unit during the Canadian Rebellion of 1837, Garrett offered him an advance of $1000 for a book on the real Uncle Tom. According to Gysin, "Uncle Tom in fiction was the perfect type of the Negro whom the white folks were willing to free as long as he retained all the characteristics of his former servitude." Gysin looked to the archive and to Henson's own writings in order to challenge what he saw as Stowe's literary, economic, and political exploitation of Henson's story. As Gysin argued, "it was this picture of a subservient Negro, his master's friend, the conciliator and turn-the-other-cheek brother, which Mrs. Stowe presented for the admiration of the nineteenth century." For Gysin, this was an early insight into the power of language—in this case, the text of *Uncle Tom's Cabin*—"to condemn to obscurity, the human counterpart from which it had sprung." The fictions of Josiah Henson, in other words, had been used against him. And they had assumed viable and

72. Lydia Parrish, *Songs of the Georgia Sea Islanders* (New York: Creative Age Press, 1942); Barry Ulanov, *Duke Ellington* (New York: Creative Age Press, 1946); A. A. Nwafor Orizu, *Without Bitterness: Western Nations in Post-War Africa* (New York: Creative Age Press, 1944).

73. John Geiger, *Nothing Is True, Everything Is Permitted: The Life of Brion Gysin* (New York: Disinformation Books, 2005), 65f. The elder Henson had been the overseer of a plantation before he escaped with his family to Canada. In 1841 he bought two hundred acres of land for a freeman settlement and laborer's school that was considered one of the final stops on the Underground Railroad and would, at its height, claim a population of five hundred people. Josiah Henson, *An Autobiography of the Rev. Josiah Henson (Mrs. Harriet Beecher Stowe's "Uncle Tom") from 1789–1881*, ed. John Lobb (London: Christian Age Office, 1878).

criminal force over time. "The man was robbed of his personality and almost of his name by a character in a novel."[74]

Gysin addressed the "common problem" in America (as opposed to the "Negro problem"—that corporate liberal term of polite condescension). And that problem was the "pattern of segregation" and "its attendant evils." Gysin suggested that racialized discourse was a form of mental slavery and madness.[75]

He also framed the violence visited on Henson and other African Americans as a religiopolitical system, multivalent in its reach. "Force is used every day *in one form or another*," wrote Gysin, "by those who wish to ensure that the lines continue to be drawn according to race." With hints of clericalism and totalitarianism, Gysin pointed out a racialized regime of control that was utilizing the "age-old technique of divide and conquer." Words, according to Gysin, were the ends and means of this systematic violence. Words were divisive through and through. Without words to aid and abet, racial difference failed to be persuasive. Gysin was at pains to explain this to his reader, given that the "old methods of subjugation will continue to work if they are not understood."[76] Patterns, not essences. Systematic, structural, organizational.

Gysin's book was a blend of creative nonfiction and analysis of the aesthetic conditions of racial subjugation. What one reviewer called a "history with the narrative pace of fiction," *To Master, A Long Goodnight* was a mashup that borrowed much from Henson's own writing, interspersed with historical documents, social and psychological analyses, and literary asides.[77] Gysin's switching of analytic registers was intended as a disruption of the reader's racialized pattern of recognition. Having run in surrealist circles in Paris and been taken with André Breton's directive to pursue "thought dictated in the absence of all control," Gysin's focus was on the "system" that determined the actions of individuals—both white and black and in between—according to a racial calculus. In calling attention to the characterizations and subplots that animated systematic racism, Gysin argued that the categorical formations of slavery persisted in a "control" system that did not dissipate after the Civil War but had taken hold in the

74. Brion Gysin, *To Master, A Long Goodnight: A Historical Narrative* (New York: Creative Age Press, 1946), 198, 8, 3.

75. Gysin, *To Master*, 199, 5.

76. Gysin, *To Master*, 9.

77. Gysin, cited in Geiger, *Nothing Is True*, 71.

aesthetic structures of the world—language, first and foremost. The codes of slavery, in turn, had been internalized by the industrial North in the twentieth century and taken up residence in the thoughts and habits of the entire nation.[78]

In 1949 Gysin won a Fulbright Fellowship to travel to France and Spain to investigate the formation of laws of the Anglo-American slave trade.[79] Gysin soon abandoned the project, however, and moved to Tangier, Morocco, where he wrote, painted, and became the owner of a restaurant, The 1001 Nights. It was in Tangier that Gysin first met William S. Burroughs, a then relatively unknown writer living off a stipend from his St. Louis family. For years Gysin and Burroughs remained friendly members of the expat community of artists and writers. In 1958 they left Tangier and took up residence at a flophouse in Paris (the infamous "Beat Hotel" at 9 rue Gît-le-Coeur). As Burroughs collated the pages of *Naked Lunch* (1959) for publication, he and Gysin began collaborating, in earnest, on projects that "cut up" established patterns of consciousness and language.

Together, with the aid of Gysin's lover, the mathematician Ian Sommerville, Gysin and Burroughs developed the Dream Machine. The rudimentary stroboscope consisted of a record player spinning at 78 rpm, a cardboard cylinder with perforated patterns, and a 100-watt light bulb hanging down into the cylinder. The light that flickered outward served to disrupt consciousness by affecting the brain's processing of information. As Sommerville surmised, "You look at it with your eyes shut and the flicker plays over your eyelids. . . . Afterwards I found that my perception of the world around me had increased very notably. All conceptions of being dragged

78. André Breton, "Le Manifeste du Surréalisme," in Patrick Waldberg, *Surrealism* (New York: McGraw-Hill, 1971), 66–75; Gysin, *To Master*, 12, 232–33. Gysin's account of the vectors of power emanating from word and image was not unprecedented. See, for example, W. E. B. Dubois's struggle to dissect the microphysics and metaphysics of white hegemony and his appreciation for unsettling experiences of black subjectivization that could be experienced, for all intents and purposes, as otherworldly. Such degradation—at once blatantly systematic and subtle—was integral to what DuBois called the "new religion of whiteness" (*Darkwater: Voices from within the Veil* [New York: Harcourt, Brace, 1920], 31).

79. Brion Gysin, "Games, Drama, and Murder," *Tomorrow* (April 1947): 39–41; Geiger, *Nothing Is True*, 78–79.

FIGURE 35. Scene from the "Beat Hotel" at 9 rue Gît-le-Coeur, ca. 1963. Approximated by Jenny Schulder.

or tired had dropped away."[80] According to Burroughs, the Dream Machine was a "non-chemical method of expanding consciousness."[81]

The Dream Machine owed much to Walter's *Living Brain* (1953) and was designed to affect the brain waves of those who sat for prolonged periods in its presence. In September of 1960, Burroughs and Sommerville traveled to London in order to attend Walter's lecture to the Society for Psychical Research.[82] As Burroughs wrote a few weeks later, "Dear Doctor Walter, I heard your very interesting talk in London recently. I [too] am concerned with the phenomena of flicker. . . . As a writer I am interested in the effect of flicker on the creative process."[83] Burroughs, Gysin, and Sommerville were very much taken with Walter's research and his demonstration that brain waves responded to repetitive stimulation of stroboscopic light.[84] Whereas Walter and other electroencephalographers imagined flicker as a way to learn more about how to adjust and normalize the cerebral rhythms of epileptics, Gysin and Burroughs imagined the Dream Machine as a tool of deconditioning and discovery of heretofore hidden processes of cognition. Quoting from *The Living Brain*, Burroughs noted that "The rhythmic series of flashes appeared to be breaking down some of the physiological barriers between different regions of the brain."[85]

The Dream Machine disrupted the brain at the level of statistical infer-

80. Ian Sommerville, cited in Brion Gysin, "The Dream Machine," *Olympia* 2 (1962): 31.

81. Daniel Odier, *The Job: Interview with William S. Burroughs* (New York: Penguin Books, 1989), 134.

82. Ian Sommerville, "Flicker," *Olympia* 2 (1962): 37. Barry Miles, *The Beat Hotel: Ginsberg, Burroughs, and Corso in Paris, 1958–1963* (New York: Grove Press, 2000) 219; Walter, "Neurophysiological Aspects of Hallucinations"; John Geiger, *Chapel of Extreme Experience: A Short History of Stroboscopic Light and the Dream Machine* (Brooklyn, NY: Soft Skull Press, 2003) 48.

83. Burroughs, *Rub Out the Words*, 52. Soon after his initial contact with Burroughs, Walter used the stroboscope in discovering what he called the Contingent Negative Variation (discussed in chap. 1) in research supported, in part, by Garrett's Parapsychology Foundation. W. Grey Walter et al., "Contingent Negative Variation: An Electric Sign of Sensorimotor Association and Expectancy in the Human Brain," *Nature* 203 (July 25, 1964): 380–84.

84. W. Grey Walter et al., "Analysis of the Electrical Response of the Human Cortex to Photic Stimulation," *Nature* 158 (October 19, 1946): 540–41. See also Walter's article "Patterns in Your Head," *Discovery* (February 1952): 56–62.

85. William S. Burroughs, "Points of Distinction between Sedative and Consciousness-Expanding Drugs," in *LSD: The Consciousness-Expanding Drug*, ed. David Solomon (New York: G. P. Putnam's Sons, 1964), 171.

ence. In its effect on the frequency of the alpha wave, the neural algorithm of expectancy was momentarily scrambled. In the presence of the Dream Machine, one could, quite literally, begin to see differently. With eyes closed, new patterns of light and color emerged that were not immediately part of what Burroughs called the Control Machine. With changes at the level of neural processing came changes in thinking and being. With face up against the swirling lightscape shooting out of the cardboard cylinder, one could begin to appreciate how the senses came to recognize patterns and distill meaning from them. For Burroughs and Gysin, the brain had been habituated by the media that surrounded it. Breaking such habits of sensory attention was essential in reversing the "fast death of a culture overwhelmed."[86]

7. THE CUT-UP EXPERIMENTS

The Dream Machine was part of ongoing collaborations between Gysin and Burroughs, all of which used insights from neurophysiology to rethink the practice of writing as a sensory technology. Referring to their cut-up collaborations, Burroughs declared, "what I want to do is to learn to see more of what's out there, to look outside, to achieve as far as possible a complete awareness of surroundings." As Gysin explained, the cut-up method was revelatory—"a sort of peeling inside my head, like an ether experience."[87] Calling on their readers to move beyond scripted modes of attention, Burroughs and Gysin understood the written word neuromatically. And they were committed to sharing the good news:

> we have come to let you out
>
> here and now we will show you   what you can do
> with and to
> the word
> the words

86. William S. Burroughs Papers, 1951–72. New York Public Library, Henry W. and Albert A. Berg Collection of English and American Literature, box 68, folder 11, "Evaluation of Information, August 11, 1964."

87. William S. Burroughs and Brion Gysin, *The Third Mind* (New York: Viking, 1978), 2, 45.

any word
all the words

Pick a book any book cut it up
cut up
prose
poems
newspapers
magazines
the bible
the koran
the book of moroni
lao-tzu
confucius
the bhagavad gita
anything
letters
business correspondences
ads
all the words

slice down the middle dice into sections
according to taste
chop in some bible pour on some Madison Avenue
prose
shuffle like cards toss like confetti
taste it like piping hot alphabet soup. . . .

you will soon see just what they really are
saying this is the terminal method for
finding the truth
. . . .

the writing machine is for everybody
do it yourself until the machine comes
here is the system according to us.[88]

88. Brion Gysin, "Minutes to Go," in Sinclair Beiles, *William Burroughs, Gregory Corso, and Brion Gysin: Minutes to Go* (Paris: Two Cities, 1960), 3–5.

Like the Dream Machine, the cut-up experiments assumed readers to be processors of information, for humans experienced word blocks and surface patterns. They noted deviations in pattern. They could become sensitive to the deeper layers of self and world in the moment of reading. They could attend to the potential for the most commonsense articulations to mean differently. "Whatever you do in your head bears the prerecorded pattern of your head," insisted Gysin. "Cut through that pattern and all patterns if you want something new."[89]

As conceived by Gysin and Burroughs, the cut-up method was a critical performance that owed much to the burgeoning currency of cybernetics and the neuromatic brain.[90] The cut-ups took Shannon's premise that the "semantic aspects of communication are irrelevant to the engineering problem" to its logical conclusion. Burroughs, for example, insisted that the cut-up method revealed the author to be embedded in an elaborate communication network, dealing not in meaning-making but rather in the selection of messages to code and transmit.

> People say to me, "Oh, this is all very good, but you got it by cutting up." I say that has nothing to do with it, how I got it. What is any writing but a cut-up? Somebody has to program the machine; somebody has to *do* the cutting up. Remember that I first made selections. Out of hundreds of possible sentences that I might have used, I chose one.[91]

As per Shannon's version of information theory, Burroughs's artistic freedom was manifest in both his choice of messages and his coding of them at their source through splicing techniques.[92] For Burroughs, the cut-up

89. Brion Gysin, "Cut-ups: A Project for Disastrous Success," in Burroughs and Gysin, *Third Mind*, 44.

90. "The Human Being," wrote Burroughs, "are strung lines of word associates that control 'thoughts feelings and apparent sensory impressions.' Quote from Encephalographic Research, Chicago, Written in TIME." William Burroughs, "The Exterminator," in Burroughs and Gysin, *Third Mind*, 71.

91. "Interview with William Burroughs," in Burroughs and Gysin, *Third Mind*, 8.

92. In Burroughs's book *The Soft Machine* (1961; New York: Grove Press, 1967), Shannon is mentioned to have been a part of a tragic expedition that ended as it began, that is, in ontic hybridity. "Just who died is uncertain since one member of the party has not been found alive or dead and identity of the missing person is dubious. The bodies were decomposed when found and identification of bodies was based on documents. But it seems that the party was given to exchange of identifications, and even to writing in each others' dia-

expressed less his own creative genius than his "soft machine" scanning the horizon for heretofore hidden patterns.[93] As an anthropologist of the virtual, Burroughs's mode of participating in and observing media culture sought to lessen interference from the ego. "I prefer not to use *my own words*, I don't like *my own words* because *my own words* are prerecorded. . . . my words are prerecorded for me as yours are prerecorded for you."[94]

Gysin was even more explicit in his dismissal of the authorial voice. Manifesting as a "writing machine," Gysin would systematically break down such habitual prerecordings—"I Am That I Am," for example—into their component parts, algorithmically exhausting all potential combinations in order to "explore . . . all the potential sounds and meanings of the sentence." Such a full account of a sentence's surface permutations offered a fleeting glimpse of an underlying structure that determined much more than the

---

ries" (175). In addition to Shannon's notion that information makes sense only in relation to a horizon of possibilities, John R. Pierce, Shannon's colleague at Bell Labs, also sheds light on Burroughs's cybernetic imaginary. Pierce, a founding figure in the development of satellite communications, defended the narrow application of Shannon and was highly critical of Wiener and the universalist claims he made for cybernetics (Kline, *Cybernetics Moment*, 127–28). Despite this critique, Pierce used information theory as the basis for his own brand of metaphysical speculation about "authorless" texts. Pierce was interested in the algorithm that defined the conditions of human being and broached the possibility that communication was not about pure intentionality and choice. Moving past a psychoanalytic rendering in which words and stories were indexes of the messy play of primitive passions within the self, Pierce theorized that the human unconscious was a statistically driven process in which words were generated according to fundamental mathematical principles. "English and its statistics," he wrote under the pseudonym J. J. Coupling, "reside in the human brain and they can be tapped at the source" ("Chance Remarks," *Astounding Science Fiction* 44, no. 2 [October 1949]: 109). This article appeared alongside "The Automagic Horse" by L. Ron Hubbard. For a perceptive discussion of Pierce in relation to information theory, see Lydia H. Liu, *The Freudian Robot: Digital Media and the Future of the Unconscious* (Chicago: University of Chicago Press, 2011), 127f.

93. Burroughs's grandfather, it should be noted, had founded the Burroughs Adding Machine Company, an important precursor to the digital computer and "magnetic-core devices in information processing." James W. Cortada, *Before the Computer: IBM, NCR, Burroughs, and Remington Rand and the Industry They Created, 1865–1956* (Princeton, NJ: Princeton University Press, 1993). The development of digital devices was revolutionary in overcoming the inevitable "noise" generated by the imprecision and moving parts of analogue machines. Albert J. Meyerhoff, ed., *Digital Applications of Magnetic Devices* (New York: John Wiley and Sons, 1960), vii.

94. William S. Burroughs, *APO-33 Bulletin: A Metabolic Regulator* (San Francisco: Beach Books, 1966), 16–17.

FIGURE 36. Brion Gysin and William S. Burroughs beneath the International Monument to the Reformation in Geneva; from Burroughs and Gysin, *The Third Mind* (New York: Viking, 1978).

accepted meaning of that sentence: a *logos*, in this case, operating within a widely recognized phrasing for the *logos*—"I AM THAT I AM / AM I THAT I AM / I THAT AM I AM / THAT I AM I AM"—printed in the anonymity of computer type. For according to Gysin, the cut-up method "upset semantic order" and broke through the material surface of texts. "The truth" of language lay below the surface patterns, in the occult infrastructure of their design—a heretofore hidden logic that shaped how words were received by readers and, in turn, how words acted on readers. "Henceforth, every form of writing will consist of an operation of decoding," declared Gysin. "All this because all language is essentially mystification."[95]

According to Burroughs and Gysin, the power of language resided in its capacity to channel implicit directives that had nothing, essentially, to do with the words that comprised them. The cut-up experiments, in laying bare that words carry with them all manner of affordances and occult demands, aspired to disrupt the processing capacities of the reader in order to reveal that which exceeded the words on the page. "No one can conceal what is saying cut up," wrote Burroughs and Gysin. "You can cut The Truth out of any written or spoken words."[96] Such insight could then be used to "rub out the word" in an ongoing battle against the deterministic pressures of language. As Burroughs instructed:

> By this time you will have gained some insight into the Control Machine and how it operates. You will hear the disembodied voice which speaks through any newspaper on lines of association and juxtaposition. The mechanism has no voice of its own and can talk indirectly only through the words of others. . . . speaking through comic strips . . . news items . . . advertisements . . . talking, above all, through names and numbers.[97]

95. Gerard-Georges Lemaire, "23 Stitches Taken," in Burroughs and Gysin, *Third Mind*, 21, 14–15. Shannon, in similar terms, would advocate the application of the "terms of engineering" to the "answers" that life demanded, in general. Claude E. Shannon, lecture, "Creative Thinking" (1952), 7, https://archive.org/details/ShannonMiscellaneousWritings/page/n829/mode/2up. Thanks to Bernard Dionysius Geoghegan for providing me a copy of this manuscript. See also Claude Elwood Shannon, *Miscellaneous Writings*, ed. N. J. A. Sloane and Aaron D. Wyner (Murray Hill, NJ: AT&T Bell Laboratories, 2013).

96. William Burroughs and Brion Gysin, *The Exterminator* (San Francisco: Auerhahn Press, 1960), 5.

97. William Burroughs, "Inside the Control Machine," in Burroughs and Gysin, *Third Mind*, 178.

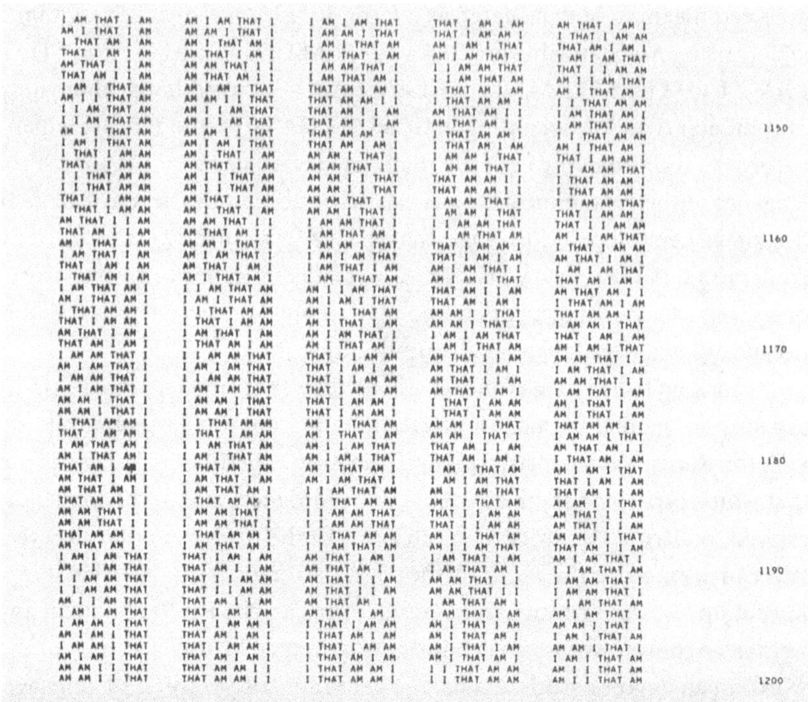

FIGURE 37. Excerpt from Brion Gysin's "I Am That I Am" (1960).

By bringing this voice to the surface, cut-ups staged the act of reading as a process of neuromatic deconditioning—of exposing the reader to the "real significance" of the sentences they encountered as shot through with pressing epistemological and political demands. For Gysin and Burroughs, the desired effect of cut-ups on the reader occurred at the level of "psychosensory process[ing] that is going on all the time."[98] As artists "interested in all the structural problems of sensation," their practice was, in Gysin's words,

98. Lemaire, "23 Stitches," 14–15. On the cut-up method as a form of mystical deconditioning, see my *The Bop Apocalypse: The Religious Visions of Kerouac, Ginsberg, and Burroughs* (Urbana: University of Illinois Press, 2001), 237–38. See also Robin Lydenberg, *Word Cultures: Radical Theory and Practice in William S. Burroughs' Fiction* (Urbana: University of Illinois Press, 1987); "Interview with William Burroughs," 4. Cut-ups were designed to explore "what words actually are, and exactly what is the relationship to the human nervous system" and to "destroy in your nervous system the effect of a verbal formulation" (cited in Odier, *The Job*, 207, and Eric Mottram, *William Burroughs: The Algebra of Need* (Buffalo, NY: Intrepid Press, 1971), 24. See also Eric Mottram, "Recontre avec William Burroughs," *Les Langues Modernes* 58 (1965): 80.

"parascientific." For to become aware of "word dust" clinging to the texts around us—that infinitesimal entity that conditioned the reception and meaning of language—was to become aware of oneself as an information-processing machine. Indeed, "word-lock[s] of static intention" were material manifestations of a power that was not strictly human. For what was really "written on the door [was] pure information."[99]

The phenomenology of cut-ups, then, was an increased awareness of how language registered its biological effects. As such, the cut-ups were a necessary defense against unseen forces of control that operated through words. In serving to break "synaptic attachments to language," the cut-ups were a form of execution, or "extermination," to use Burroughs' preferred term. "*Communication must become total and conscious before we can stop it.*" The cut-ups were language laid bare, deactivated, reduced to pure pattern, divorced at some level from both its author and potential host, an inert specimen on an autopsy table, awaiting reanimation. For within every word was the "code for the organization of life," waiting to find its cellular host, to become a living system at your expense.[100]

8. FROM VOODOO DEATH TO VIROLOGY

Burroughs had arrived at Harvard in 1931 just as the physiologist Walter Cannon was finishing up his masterwork, *Wisdom of the Body* (1932).[101] Cannon was a founding member of the Institute for the Unity of Science, a benefactor of Rockefeller Foundation largesse, and an inspiration to self-described "brain-wavers" at midcentury.[102] Before the official nomination of informa-

99. Brion Gysin, "Brion Gysin: Eight Units of a Permutative Picture," *Art International* 6 (Summer 1962): 87; Burroughs, *Soft Machine*, 90, 121; Gysin, cited in José Férez Kuri, ed., *Brion Gysin: Tuning into the Multimedia Age* (London: Thames and Hudson, 2003), 96; Burroughs, *The Ticket That Exploded* (New York: Grove Press, 1967), 65.

100. Gysin, *Here to Go*, 51; Burroughs, *Ticket That Exploded*, 51; "Cracking Life's Code," *Science News-Letter* 69, no. 18 (1956): 275.

101. Walter B. Cannon, *The Wisdom of the Body* (New York: W. W. Norton, 1932).

102. In his 1950 presidential address to the American Electroencephalographic Society, for example, Hallowell Davis positioned Cannon's work within the "language of cybernetics." Davis breathlessly noted Cannon's influence on his continued thinking "about alpha rhythms and delta rhythms and kappa rhythms and beta waves and more recently about digital computers and analogue calculators and negative feed-back circuits and scanning

tion theory, Cannon's neurological work at Harvard Medical School had been of long-standing interest to Norbert Wiener. Cannon's experiments in neural stimulation (with Wiener's frequent collaborator Arturo Rosenblueth) were driven by his interest in the body's capacity for self-regulation and the interdependence of the internal organs, including the brain, with the environment.[103] According to Cannon, the body itself was a self-organizing mechanism par excellence. For Cannon, bodily systems, like all systems, were "relatively constant." They were "homeostatic" in that they shared a condition of vibrant stasis, a drive toward constancy, every movement a consolidation of their integrity.[104] While "made of extraordinarily unstable material," the body was able to achieve stability in its constant interaction with environmental "agencies."[105] As Cannon wrote, "the high degree of instability of the matter of which we are composed explains why drowning, gas poisoning, or electric shock promptly brings on death." Yet, as Cannon gushes, we are capable of incredible feats of perseverance. As he would later declare, "our bodies, which are the culmination of ages of experience, have learned secrets of management that are worthy of our study."[106]

Shortly after the publication of *Wisdom of the Body*, Cannon adapted his physiological approach to political economy in a reform-minded response to the Great Depression. The concept of homeostasis now covered—in addition to the molecular, the nervous, the respiratory, and the digestive—the

---

mechanisms and what the relations among all those things might be." Hallowell Davis, presidential address at the third annual meeting of the American Electroencephalographic Society, "Homeostasis and Cerebral Excitability," *Electroencephalography and Clinical Neurophysiology* 2 (1950): 243–44.

103. Arturo Rosenblueth and W. B. Cannon, "Cortical Responses to Electrical Stimulation," *American Journal of Physiology* 135 (1941): 690–741. Cannon's framing of physiology in engineering terms inspired the work of Rosenblueth and Norbert Wiener and shaped their foundational statement on goal-seeking mechanisms: Arturo Rosenblueth, Norbert Wiener, and Julian Bigelow, "Behavior, Purpose, and Teleology," *Philosophy of Science* 10, no. 1 (1943): 18–24.

104. Cannon's notion of homeostasis was given its proper cognitive spin in Roger Walsh's term of "neurostasis" in Roger Walsh, *Towards an Ecology of the Brain* (Jamaica, NY: Spectrum Publications, 1981), 121.

105. As Jessica Riskin has noted, Cannon strategically refrained from naming these agencies as such but assumed them to be active given their homeostatic effects (*The Restless Clock: A History of the Centuries-Long Argument over What Makes Living Things Tick* [Chicago: University of Chicago Press, 2016], 308).

106. Cannon, *Wisdom*, 24, 20, 19; Walter B. Cannon, "The Body Physiologic and the Body Politic," *Science* 93, no. 2401 (January 3, 1941): 4.

market, ecosystems, machines, God, and any system that displayed purposive organization and control as opposed to mere order.[107] Cannon began his 1933 article, "Biocracy: Does the Human Body Contain the Secret of Economic Stabilization?," by quoting the Congregationalist minister Henry Ward Beecher: "Civilization is in a hell of a mess." Cannon then goes on to describe all the efforts to impose some order on the human form—from financiers, to labor leaders, to politicians, to technocrats—all those efforts intent on replacing "the present hell of hunger, nakedness, demoralizing idleness, and fear, with the plenty and security of the kingdom of heaven." Cannon then retorts—"I wish to suggest that the kingdom of heaven is within us!" In setting the stage for his own social gospel, Cannon looks forward to a moment when "social stabilization" would "release the highest activities of the brain for adventure and achievement." For Cannon, then, to get inside the ways and means of molecular organization and its internal communications was to also gain insight into other forms of natural systematicity—industrial, economic, political, and linguistic.[108]

Cannon was turning his attention to the implications of homeostasis on the study of language and culture just as Burroughs was attending graduate courses in anthropology at Harvard University. The research that Cannon was conducting at that time resulted in "'Voodoo' Death," published in *American Anthropologist* in 1942. In it Cannon sought to distill a physiological mechanism from ethnographic reports of sudden death caused by curses. Having spent his career demonstrating the protective function of the autonomic nervous system, Cannon was now broaching instances when language appeared to disrupt the system or else turn the "wisdom of the body" against itself. Cannon strongly suggested that "voodoo death . . . may be real." According to Cannon, words could actualize the fear of death in an individual at the molecular level, activating the central nervous system

107. James Beniger, *The Control Revolution: Technological and Economic Origins of the Information Society* (Cambridge, MA: Harvard University Press, 1986), 35.

108. Walter Cannon, "Biocracy: Does the Human Body Contain the Secret of Economic Stabilization?," *Technology Review* 35, no. 6 (March 1933): 203–6, 227. The body, in other words, was potentially scalable. For like our biological systems, society, the economy, and politics are capable of self-organization ("Body Physiologic," 4). In Cannon's analogical vision, social systems might one day be brought under control because they inevitably replicated the bodily organism in their pace and pattern of development. So while the individual body may fail, a well-organized society might last indefinitely if its "nature" were properly identified and channeled (*Wisdom of the Body*, 24–25, 305, 315, 310, 319).

to either compensate for imminent threat or else prepare itself for traumas that it deemed inevitable.[109]

Although there is no evidence that Burroughs encountered Cannon at Harvard, the latter's explanation of the biological processing of messages was eerily resonant with the underlying premise of the cut-up experiments: Burroughs's interest in the neurophysiological effects of language resonated with the cybernetic drift of the human sciences. But it was his encounter with the science of virology that would make a direct and lasting impact on his experimental approach to the physiology of language. As Burroughs would later write in *The Wild Boys* (1971), curses were contagious: "Words that cut like buzz saws. Words that vibrate the entrails to jelly. Cold strange words that fall like icy nets on the mind. Virus words that eat the brain to muttering shreds."[110] While viruses had first become visible to scientists in the 1930s, in the 1950s new technologies allowed for cellular dissection and theorizations of viral control and communication. The reproduction of a virus was now conceived of as a transfer of information. A virus was not alive per se. But in communicating with a host organism, a virus integrated itself within and assumed control over that organism's capacities for self-

109. Walter B. Cannon, "'Voodoo' Death," *American Anthropologist* 44, no. 2 (1942): 169–81. In dwelling on the physiological registers of language, Cannon's musings resonated with the structuralist revolution then brewing within anthropology. Claude Lévi-Strauss, for example, would soon address "The Effectiveness of Symbols" and "The Sorcerer and His Magic" as a way to investigate the links between language and the body, specifically, how the body was a product of linguistic sculpting and therefore susceptible to influence, for better or for worse. In writing about Mauss's earlier attempt to secure the link between the biology of humans and their social nature, Lévi-Strauss drew from information theory a way to explain and revise Mauss's insight into how shamans and other magical operators manipulated the social in such a way as to do bodily harm. See, e.g., Claude Lévi-Strauss, *Introduction to the Work of Marcel Mauss* (1950), trans. Felicity Baker (London: Routledge, 1987) and Claude Lévi-Strauss, *Structural Anthropology* (New York: Anchor Books, 1967), 161–201. Cannon's insights, as filtered through structuralism, were picked up by parapsychologists decades later. See, e.g., Joan Halifax-Grof, "Hex Death," a lecture delivered at the Parapsychology and Anthropology International Conference in 1973 in London, with Grey Walter in attendance. For a more recent CSR study on the matter, see Nora Parren, "The (Possible) Cognitive Naturalness of Witchcraft Beliefs: An Exploration of the Existing Literature," *Journal of Cognition and Culture* 17, no. 5 (2017): 396–418.

110. William S. Burroughs, *The Wild Boys: A Book of the Dead* (1971; New York: Grove Press, 2007), 165–66.

organization.[111] A virus, like Burroughs's notion of language, made its way in through a combination of patience, pluck, and pattern.

9. ENGRAMS, AUDITING, AND THE APPEAL OF SCIENTOLOGY

In 1959, Gysin introduced Burroughs to the linguistic theories of L. Ron Hubbard. Burroughs's interest intensified beginning in 1963. So began Burroughs's fervent period of study of Dianetics and Scientology at St. Hill Manor in East Grinstead, England. This season of revival would culminate in Burroughs becoming Clear #1163 in 1968.[112] Burroughs shared with Hubbard an interest in the biology of language, control, and information theory. In addition to Hubbard's directive to think in sensations and not in words, Burroughs was attracted to Hubbard's focus on how representations—linguistic, visual, and mental—were sources of unhappiness, illness, madness, and death. From the beginning, Hubbard wrote in a self-conscious cybernetic script, addressing with therapeutic intent the "computational ability of the individual" and the problems of being compromised by "incorrect data."[113]

111. See, e.g., A. T. Bharucha-Reid, "On the Stochastic Theory of Epidemics," *Proceedings of the Third Berkeley Symposium on Mathematical Statistics and Probability, 1954–1955*, vol. 4 (Berkeley: University of California Press, 1956): 111–19. In the 1950s and 1960s virologists increasingly looked to information theory for explanatory leverage. See also Priscilla Wald, *Contagious: Cultures, Carriers, and the Outbreak Narrative* (Durham, NC: Duke University Press, 2008), 181, 305n65, 158.

112. Harris, "Can You See a Virus?," 247; "William Burroughs: Clear 1163," *Advance!* 2 (1968): 3. While living in Tangier, Gysin had met John and Mary Cooke—wealthy American bohemians who became Gysin's patrons. Both Cookes were dabblers in Scientology and had experimented with auditing visitors on the coast of Algiers. As Geiger notes, "John Cooke eventually returned to California where he developed a 'New Tarot Deck for the Aquarian Age' and directed a small band of acid evangelists known as the Psychedelic Rangers, who selected high-profile people for LSD initiation" (*Nothing Is True*, 109). Peter Manseau and others also report on rumors that John Cooke was the catalyst for Hubbard's transforming Dianetics into the religion of Scientology (Peter Manseau, *One Nation under Gods: A New American History* [New York: Little Brown, 2015]).

113. L. Ron Hubbard, "Terra Incognita: The Mind" (1950), *Technical Bulletins of Dianetics and Scientology*, vol. 1 (Los Angeles: Church of Scientology of California, 1976), 6–10; David S. Wills, *Scientologist! William S. Burroughs and the "Weird Cult"* (UK: Beatdom Books, 2013), 65.

Like Wiener, writing in the same year that *Dianetics* was published, Hubbard viewed "the individuality of mind" as lying "in the retention of its earlier tapings and memories."[114]

In *Dianetics* Hubbard offered details of what he called the "Mind Schematic" that demonstrated how humans processed information.[115] In foregrounding "circuit pathways to show the flow of signals and messages," Hubbard claimed that the brain was divided not in terms of consciousness and the unconscious but rather in terms of its analytic and reactive dimensions. At one level there was the Analytical Mind, which "behaves like a computing machine, yet [was] more fantastically capable than any computing machine ever constructed and infinitely more elaborate."[116] The Reactive Mind, by contrast, was imbued with a flawed operating system. It corrupted the Analytical Mind with unprocessed information stored at the cellular level. The Reactive Mind was thinking in terms of habitual response and circular confirmation.[117] According to Hubbard, "the villainous reactive mind is the human evil which Dianetics is assailing, and has dedicated itself to eradicate from mankind: it is the leech upon our rational behavior, the hypothetical cyst which occludes the proper functioning of the Analytical Mind, the root of all of our psychosomatic ills, the barrier which prevents us from attaining the optimum of our abilities and our aspirations."[118]

According to Hubbard's Mind Schematic, the Reactive Mind was the

114. Wiener, *Human Use*, 101–2. Despite his constant talk of computational capacity and data ("now data," "then data," and "value data"), Hubbard assured his readers that they were more capable than any existing computer. The brain Hubbard claimed he was interested in was not something crass like a computer but rather the platonic ideal of a computer. L. Ron Hubbard, "Arithmetic," a lecture delivered on September 10, 1951, in *Research and Discovery Series*, vol. 7 (Los Angeles: Bridge Publications, 1984), 178–84.

115. See also L. Ron Hubbard, *Dianetics: The Modern Science of Mental Health* (New York: Hermitage House, 1950), 44; L. Ron Hubbard, "Motion and the Tone Scale," a lecture given on August 17, 1951, in *Research and Discovery Series*, vol. 7 (Los Angeles: Bridge Publications, 1984), 103–13; and L. Ron Hubbard, "Clearing the Human Mind as an Electronic Computer," a lecture given on October 22, 1951, in *Research and Discovery Series*, vol. 8 (Los Angeles: Bridge Publications, 1984), 189–211.

116. Hubbard, *Dianetics*, 421, 43. This part of the mind was "arranged to analyze each situation in the light of available data and to determine and direct the next acts of the Organism so as best to enable the individual, his progeny, associates, and environment to survive" (422).

117. Hubbard, *Dianetics*, 62.

118. L. Ron Hubbard, *Dianetics: A Brief Discussion* (Elizabeth, NJ: Hubbard Dianetic Research Foundations, 1950): 2.

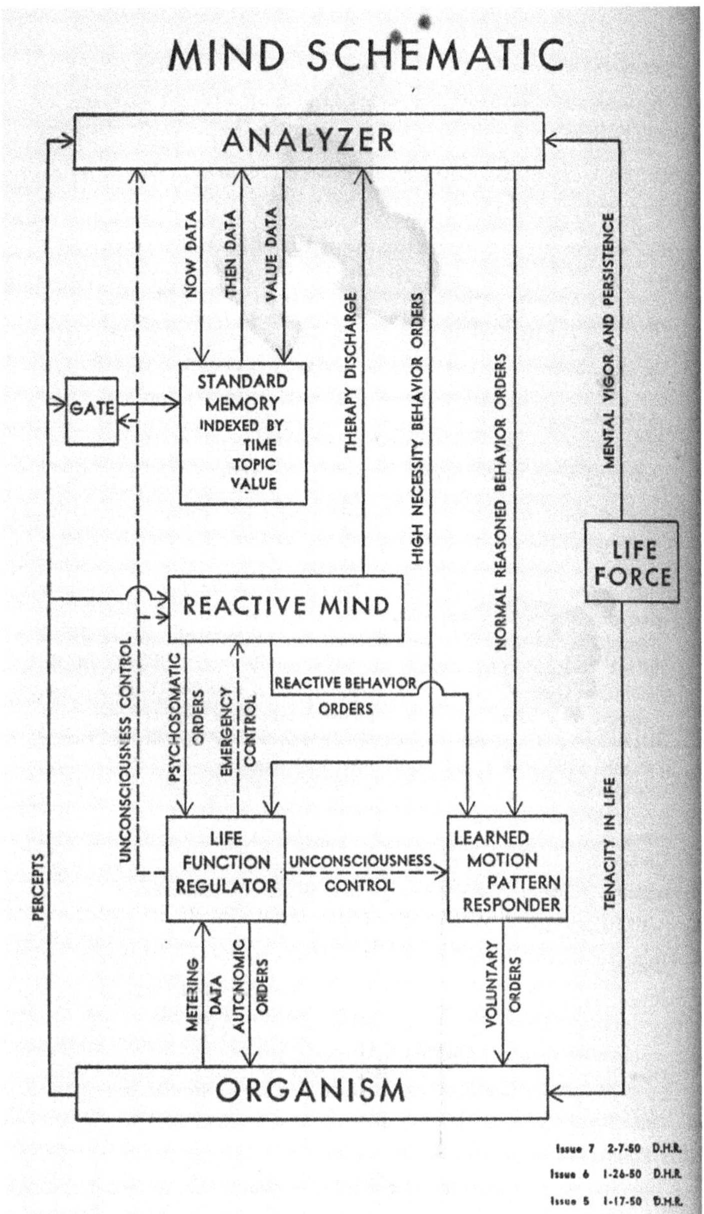

FIGURE 38. The Mind Schematic, from L. Ron Hubbard, *Dianetics: The Modern Science of Mental Health* (New York: Hermitage House, 1950). "The mind schematic is a block diagram illustrating analogically the observed operation of the mind in the organism. It may be likened to the schematic diagram of an electrical circuit, in which the position and relationship of elements derive from the connections to them and not from their placing on the page. The connecting lines represent paths for the transmission of messages which control activity."

site where "engrams" were recorded and stored.[119] Having entered "from the exterior world into the hidden recesses below rational thinking," an engram was "information . . . unappraised by [the] conscious mind" that "takes over the motor controls of the body and causes behavior and action to which the conscious mind, the individual himself, would never consent." In Hubbard's vision of the engram as a "parasitic" entity "which protects itself in various ways," Burroughs found yet another, more refined layer of explanation for his working theory of language as a virus.[120] Or, as Burroughs would insist, Hubbard had defined *the* "problem," namely, that "the brain artifact has a built in mechanism that prevents it from solving problems and that mechanism is The Word."[121]

In his 1967 essay, "The Engram Theory," Burroughs begins by describing engrams as a word sequence, a pattern whose meaning is not primarily semantic but, in the words of Hubbard, "traced on a cell." Engrams were the hinge of control and the source of the individual's failure to recognize the truth of the situation. As Burroughs wrote, "*Engram words are always spoken in a voice not your own. . . .* Engram words are the voice of dogmatic authority." All engrams, he continues, are "parasitic organisms similar in many respects to the virus." According to Burroughs's reading of Hubbard, engrams are empty, presemantic in their ontology. "Like the engram," wrote

119. "The reactive bank," wrote Hubbard, "does not store memories as we think of them. It stores engrams. These engrams are a complete recording, down to the last accurate detail, of every perception present in a moment of partial or full 'unconsciousness'" (*Dianetics*, 60). Hubbard, here, was borrowing from the German word *Engramm*, coined by the zoologist and evolutionary biologist Richard Wolfgang Semon in 1904 who suggested that memories left a physiological imprint on the structure of the brain. In the 1960s, neurologists still employed the term "engram" to discuss the "permanent impression left behind by psychical experience in the brain's cellular network." Wilder Penfield, "Engrams in the Human Brain: Mechanisms of Memory," *Proceedings of the Royal Society of Medicine* 61 (August 1968): 831–40.

120. Hubbard, *Dianetics*, xiii, 138, 208; L. Ron Hubbard, "The Anatomy of the Engram," a lecture delivered on August 15, 1950, in *Research and Discovery Series*, vol. 3 (Los Angeles: Bridge Publications, 1982), 28. Hubbard's dissertations on engrams were useful to Burroughs in refining his own theories on how language registered its viral effects on the individual and society. For according to Hubbard, "Words contained in engrams" were, in essence, "contagious. Like germs they respect none and carry forward from individual to individual, from parents to child, respecting none until they are stopped by dianetics." For "with the invention of language," wrote Hubbard, "man brought upon himself an unexpected source of aberration" (*Self Analysis*, 63; Hubbard, *Dianetics*, 86, 134f.).

121. Odier, *The Job*, 96.

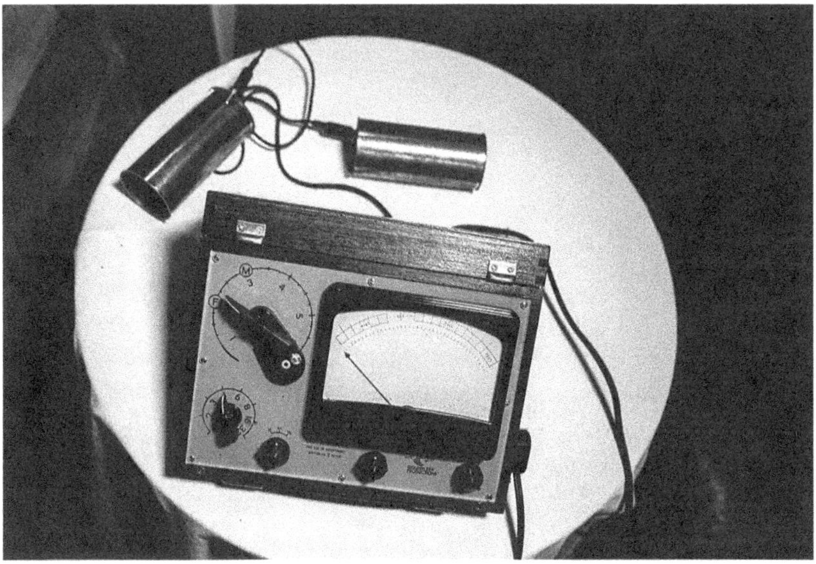

FIGURE 39. Mark V E-meter on table at the "Beat Hotel" at 9 rue Gît-le-Coeur, ca. 1963. Approximated by Jenny Schulder.

Burroughs, "the virus has no image. The image of the virus is the effect it produces. The image of the engram is that it can be associated with any image." Like a virus, engrams manifest in their "precision of communication" and their capacity to "permanently fuse . . . into any and all body circuits."[122]

In 1951, a year before the founding of Scientology, Hubbard chose the E-meter as the privileged device for facilitating the program outlined in *Dianetics*. The "Hubbard Electrometer" was used to facilitate ideal states of computation and cognitive wherewithal by eradicating "engrams." In regis-

122. Engrams, Burroughs reported, were currently being used by the secret service against enemies of the state. The state, in other words, no longer needed police because engrams had become the police: "With engrams [one] can achieve computerized control of every citizen while preserving the forms of democratic government" (William Burroughs, "The Engram Theory," *Mayfair* 2, no. 11 (1967): 28–31). On such linguistic paranoia in Scientology, see the anonymously penned essay, "Language as a Symbol," in which language becomes a form of magic by which "men seek to order the lives of men for their own advantage" (*Journal of Scientology* 14-G [Philadelphia, 1953]: 11). William Burroughs, "By Far the Most Efficient and Precise Language We Possess Is the Common Cold," *Mayfair* 3, no. 3 (1968): 54–56. See also William Burroughs, "Scientology Revisited," *Mayfair* 3, no. 1 (1968): 30.

FIGURE 40. Scene from the "Beat Hotel" at 9 rue Gît-le-Coeur, ca. 1963. Approximated by Jenny Schulder.

tering shifts in thought, "it sees all, knows all. It is never wrong." A Mark-V E-meter, the one used by Burroughs during his course of study, consisted of a low-voltage battery, adjustment knobs, a display that measures electric current, and two attached metal cans ("electrodes") that are held in each hand so that the body itself completes an electric circuit. The E-meter was designed to measure electric resistance within the human form. Honing in on engrams within the ecology of the Reactive Mind, charting their affective flows within the individual, using them to discover deep-laden trauma, the E-meter announced itself to be a technology of liberation—psychic, political, and biological: "This instrument," wrote Hubbard, "gives Man his first keen look into the heads and hearts of his fellows. The nimble needle of the electropsychometer can detect with accuracy things which would have been otherwise hidden from Man forever."[123]

The science of reading the electrical peaks and valleys of engram resistance within the Reactive Mind was called auditing. During an auditing session, a lower-level initiate (the "preclear") interfaced with the E-meter as a trained guide (auditor) asked questions. The auditor then monitored

123. L. Ron Hubbard, *Electropsychometric Auditing: Operator's Manual* (Phoenix: Office of L. Ron Hubbard, 1952), 25, 51, 16–17.

readings of electrical resistance elicited by these questions. Sitting across the table from one another with the E-meter in between, the auditor mapped the contours and contents of the subject's mind. A skilled auditor could identify nodes of resistance "before the preclear becomes conscious of the datum." Auditing, first and foremost, was an intimate ritual with the machine. During training Hubbard instructed auditors to "get familiar with the meter by holding it, watching it, turning it on and off. Touch it. Reach and withdraw from it. Play catch with it. Don't just read books about it." Such intimacy was necessary because the auditor was crucial in establishing the circuit between the E-meter and the subject being audited. Through detailed questioning and rundowns the auditor was able to pinpoint past traumas in the preclear in order to facilitate a direct confrontation with them. By working through the repressed emotions and "attention unit patterns" that originated in the moment of trauma, an engram could be identified, contained, and eventually drained of its charge.[124]

Instructing the preclear to "scan the relationship between you and your body," a traumatic incident is located and turned "up like a volume on a radio." The auditor can then help the preclear work through the incident. "Recall any part of incident," writes Burroughs, summarizing Hubbard. "Find exact date. Go to beginning of incident. Run through from beginning to end and say what is happening to you as you go along. There is a moment when the incident is quite clear and then suddenly fades like old film."[125]

124. L. Ron Hubbard, *E Meter Essentials* (1962; Edinburgh: Publications Organization Worldwide, 1967), 7–8; Evans Farber, *How to Develop Theta Perceptics* (Los Angeles: Emergency Dianetics, 1952), 5–7. Each attention unit referred to a "quantity of awareness existing in the mind" in "varying quantity from person to person" (*Hubbard Communication Office Bulletin* [May 11, 1965]). These units were in the brain, wrote Hubbard, "energy flows of small wavelengths and definite frequency" that were "measurable on specifically designed oscilloscopes and meters." L. Ron Hubbard, *Scientology* 8–80 (Phoenix: Hubbard Association of Scientologists, 1952), 45; Burroughs, "Engram Theory," 30.

125. Farber, *How to Develop Theta Perceptics*, 14, 12; Burroughs, "Scientology Revisited," 80. As Burroughs wrote in his study notes on the word, "clear" was "a standard term which originally came out of the field of adding machines. There is a button that says clear and that is what clear means. . . . Why is that there. In the world of electronic computers they sometimes have . . . trouble [with], nos of circuits, someone with a soldering iron, drops on drops of solder that adds some figure to all the computations; all figures have this additive figure because short circuited with a drop of solder, multiplies or mutates the soldered figure. You have added yesterday . . . if you forgot to push the clear button you still have a total unseen in the machine. . . . Old data modifying the present time answer." The processing of information, in other words, was susceptible to glitches in the system. Like an errant drop

From his first encounter with Hubbard's ideas, Burroughs was taken with the practice of auditing and its confirmation of the neurophysiological properties of language.[126] In Burroughs's estimation, auditing complemented the logic of his cut-up experiments. Both were concerned with addressing "exterior determinism" and the viral qualities of words. And both sought to free the reader from established and predictable responses to words. During auditing, the preclear was tasked with a particular kind of interior inspection—to break down the affective registers of language in terms of information flow—to understand emotions as embodied, as visceral effects of a particular cognitive "training pattern" that had become "hooked into the circuits." As the auditor honed in on the preclear's responses in relationship to electrical changes within, the traumatic source of the engram revealed itself. The auditor would then identify the pattern—a memory, word, or image that bowled up the neurophysiological works. In turn, the subject dwelled on it until its charge was neutralized. Once the problematic pattern had been decoded, the auditor and preclear would move on to the next engram to be discovered. As Burroughs noted to Gysin, the critical edge of auditing was immanent. The engram "is *inside* the precise neural area you are trying to confront it with," wrote Burroughs, "like trying to measure a [*sic*] object of unknown length inside a ruler with the ruler itself."[127]

Through auditing and the readings generated by the E-meter, one learned how one's mind worked and how to process its multimedia processes. When one's mental state was in perfect working order, the subject achieved the designation of "clear." In 1952, at a Dianetics auditing training course in Wichita, Kansas, Hubbard declared the new revelation of Scientology that would address human consciousness beyond the state of "clear."

---

of solder on the circuit board, such glitches were utterly material. William S. Burroughs Papers, 1951–72, box 68, folder 38, "Clearing"; dated Sept 3, 1964.

126. In 1959, Burroughs declared that "I have a new method of writing and do not want to publish anything that has not been inspected and processed. I cannot explain this method to you until you have necessary training. So once again and most urgently . . . I tell you: 'Find a Scientology auditor and have yourself run.'" Oliver Harris, ed., *The Letters of William S. Burroughs, 1945–1959* (New York: Viking, 1993), 429.

127. William S. Burroughs Papers, 1951–72, box 68, folder 5, "E-meter Errors"; L. Ron Hubbard, "Dianetics: The Evolution of a Science," *Astounding Science Fiction* 45, no. 3 (May 1950): 76; Wills, *Scientologist!*, 65; Hubbard, *Dianetics*, 78; Burroughs, *Rub Out the Words*, 273.

> "Scientology" is a new word which names a new science. It is formed from the Latin word "Scio" which means KNOW or DISTINGUISH. . . . It is formed from the Greek word "logos," which means THE WORD, or OUTWARD FORM BY WHICH THE INWARD THOUGHT IS EXPRESSED AND MADE KNOWN. Thus, SCIENTOLOGY means KNOWING ABOUT KNOWING or SCIENCE OF KNOWLEDGE.[128]

In its commitment to metareflexivity, Hubbard situated Scientology as the culmination of two centuries of neurophysiological advance. In his ongoing investigations of the electrical basis of cognition and the relationship between the brain and its environments, Hubbard presented Scientology as nothing less than a general theory of communication in which individuals (preclears) would understand themselves vis-à-vis "theta being" and a cosmological backdrop.[129] Consequently, the new revelation of Scientology transcended the "mumbo jumbo of mysticism or spiritualism or religion." These forms of "bad religion" were "a sort of theta trap to keep men in their bodies, in apathy, ill and tied to superstition."[130]

10. PAST LIVES OF THE NEUROMATIC BRAIN

Shortly after the publication of *Dianetics*, Hubbard had discovered that "preclears were recalling . . . incidents further back in time than their present lifetime span."[131] Whereas Dianetics had instructed individuals to intervene

128. "What Is Scientology?," *Journal of Scientology* 1G (Phoenix, AZ, August 1952), 2.

129. L. Ron Hubbard, *Scientology: Control and the Mechanics of S.C.S.*, ed. Johann Templehoff (Washington, DC: Hubbard Association of Scientologists, International, 1957), 2–3. Hubbard's discussion of the "formula" of communication bore the mark of his engineering surround. See, for example, "Two-Way Communication in Action" (January, 21 1955) in *Technical Bulletins of Dianetics and Scientology*, vol. 2 (Los Angeles: Scientology Publications, 1976), 136. "It does not matter," Hubbard wrote elsewhere, "whether the communication particle is a bullet, a word, a thought or a light particle." "The Theory of Communication" (October 1953) in *Technical Bulletins of Dianetics and Scientology*, vol. 1 (Los Angeles: Scientology Publications, 1976), 464–66.

130. L. Ron Hubbard, *History of Man* (1952; East Grinstead: Department of Publications World Wide, 1961), 56.

131. L. Ron Hubbard, *Science of Survival: Simplified, Faster Dianetic Techniques* (Phoenix: Hubbard Dianetic Research Foundation, 1951); L. Ron Hubbard, *The Book of E-Meter*

in their own lives at the level of communication between brain and environment, Scientology expanded the field of intervention to include past lives. With the unveiling of Scientology, Hubbard sanctioned the E-meter to facilitate levels of human communication beyond the merely human standard of clear. In beginning to talk openly about past lives in *Science of Survival* (1951), Hubbard doubled down on the cybernetic wager of *Dianetics* and its emphasis on the human as a processor of information. For with the advent of Scientology the traumas that affected the Analytical Mind were not only psychic but cosmic in character. The path to becoming an "Operating Thetan" ("transcendence over death and mortal existence") was a matter of cultivating the ability to process information vis-à-vis the "whole track"—a cosmic narrative that began "about seventy trillion years ago."[132] Information from all past lives were stored in the thetan. The thetan, according to Hubbard, was the mainframe of self, witness to all of one's past lives and others' besides, storing and organizing the facsimiles that these minds had produced.

Hubbard attributed his discovery and mapping of the "whole track" to the technological advance of the E-meter and its superiority to the electroencephalograph. In *Have You Lived Before This Life?* (1958), Hubbard provided a "scientific survey" of forty-two auditing case reports detailing individual improvement after an engram from a past life was run. The stories of each engram—"engram reports"—take up the bulk of the book. Dramas of having lived as a Roman soldier, a British Naval Officer in battle or a French aristocrat, a citizen of Vesuvius on the eve of the eruption, a manta ray, a Tibetan diplomat, a young girl born in London in 1666, a squire in Shropshire born in 1792, of having been "born of space parents . . . nine galaxies ago," of being in a robot body 78 trillion years ago or of being a swashbuckling space fleet hero. It took approximately fifteen to twenty hours to audit a past life and to generate these reports. Among other things, the stories of past lives guaranteed an endless supply of story lines about how the self does or does not manage itself in the present.[133] One preclear, for example, relates

Drills, compiled by Mary Sue Hubbard (Los Angeles: American St. Hill Organization, 1965), 50.

132. Hubbard, *History of Man*, 6; L. Ron Hubbard, *The Book of Case Remedies* (1964; Los Angeles: Church of Scientology of California, 1977), 11–12. See also L. Ron Hubbard, "Auditing the 'Whole Track,'" *Professional Auditor's Bulletin* no. 52 (May 13, 1955).

133. Hubbard, *History of Man*, 7; L. Ron Hubbard, *Have You Lived Before This Life? A Scientific Survey* (1958; Los Angeles: Department of Publications World Wide, 1968), 48, 72, 39, 41, 58.

a story from 17,543 years ago when Martian automatons indentured him to be a "solitary observer on a space post." The preclear is "given a metal body with every conceivable electronic gadget" yet the "monotony of robot life" took its toll. When his replacement arrived, the Martians ripped out his innards and "threw my empty shell of a body on a scrap heap. I remained in the right lobe of my head, while my body rusted and disintegrated."[134]

Stories of past lives were therapeutic in their allegorical register. The trajectory of the auditing process in Scientology came to involve an increasingly detailed mapping of past lives in order to disentangle them and to reveal their common referent. Each of these lives had been mediated by one's theta being—that layer-on-layer of neuromatic subjectivity building up over time.[135] One's thetan was that which carried across the lifetimes, burdened with the responsibility of processing, organizing, and storing the sensory records of each one of them. As an applied religious philosophy, Scientology viewed the entwined traumas of past lives—physical and mental—as present obstacles to rational behavior, spiritual development, freedom, happiness, and success. These traumas (still called "engrams") were carried forward in time by way of one's theta being and prevented a person from taking an active position in relation to matter, energy, space, and time (MEST).

Within Scientology, one's neuroses assumed both an otherworldly *and* erotic charge. Indeed, psychic traumas could themselves originate in one's encounter with "body thetans" who wandered in and out of MEST, unattached to any human host yet affecting them unbeknownst. Body thetans were, like the Diakka before them, devious (and anthropomorphized) glitches in your system. According to Hubbard, body thetans were themselves vulnerable, subject to political manipulation and distraction. They could interfere with communication patterns of the human body and disrupt normal modes of processing emotion. By "emitting a considerable electronic flow," a thetan could "give somebody a very bad shock, [or] put out his eyes or cut him in half."[136] As early as 1952, Hubbard had lectured on the "Sex Practices of Thetans." According to Hubbard, "the sexual activity of the unbodied thetan absorbs our attention in processing." A wandering thetan

134. Hubbard, *Have You Lived Before*, 162–63.

135. "I was able to demonstrate in 1952," wrote Hubbard, "the actual existence of this thing called a spirit." Hubbard coined the word thetan "to avoid 'spiritualism' or even 'soul' because both had such bulky histories as words" (*Book of Case Remedies*, 11).

136. Hubbard, *History of Man*, 14–15, 42; L. Ron Hubbard, "Routine 3: The Nature of Formation of the GPM," *Hubbard Communication Office Bulletin* (July 14, 1963).

may become impressed by "the frenzied wave-lengths of mating" and "he will approach to investigate." The curiosity of the thetan, however, may soon become a clear and present danger to humans. "The thetan can become enthusiastic," warned Hubbard, "and in his ignorance of the frailty of MEST bodies [those within the continuum of matter, energy, space, and time] he may embrace or blanket the two in order fully to savor their emotions. Possibly he may cause their death by the over-enthusiasm of his embrace." Such sadism was pleasurable and habit-forming for the thetan. For "much as narcotic users become addicted, he may find a similar pleasure simply in 'nipping' MEST bodies in order to see them react. It is not his purpose to kill them, but, unfortunately, they often just curl up and expire after one of his nips."[137]

11. EXTERIORIZATION, OR THE RITUAL OF BEING THREE FEET BACK OF YOUR HEAD

In addition to steeling themselves against the incursions of such body thetans, preclears revisited each of their past lives, isolated the traumas associated with them, and cleared them of their debilitating charge. Auditing sessions continued until there was nothing more to clear and one's theta being became the perfect processor of information. The goal was to become free

137. L. Ron Hubbard, *Sex Practices of Thetans: Technique 88 (A Summarization of Tape Lecture 88L4C; Professional Course Booklet 43)* (Phoenix: Hubbard Association of Scientologists, 1952); Compiled in Written Form by D. Folgere aka Richard De Mille (reissued by Ron's Org Grenchen, Switzerland), 12–13. Body thetans were given the full cosmological treatment in 1967 when Hubbard penned their myth, which began about 75,000,000 to 95,000,000 years ago with the genocidal actions of Xenu, the head of the Galactic Confederation. Soon after Hubbard revised the content and parameters of OT-III and OT-V to reflect the Xenu myth, Burroughs underwent a two-month training session at St. Hill Manor, the headquarters of Scientology in the United Kingdom. Although there is no firm evidence of Burroughs attaining any state beyond that of "clear" that year, circumstantial evidence suggests that he may have achieved the level of at least OT III, or The Wall of Fire, the level at which members learn of the Xenu myth and the origin of body thetans. Mikael Rothstein, "'His name was Xenu. He used renegades . . .': Aspects of Scientology's Founding Myth in Scientology," in *Scientology*, ed. James R. Lewis (New York: Oxford University Press, 2009), 365–87; James R. Lewis, "Technological Exorcism, Body Thetans, and Scientology's Secret Mythology," *Numen* 63, no. 1 (2016): 33–53.

from MEST for all time. MEST itself was a kind of conspiracy against theta being, a kind of ontic noise that corrupted the processing of signals within. "The intention of the physical universe," wrote Burroughs in his study notes, "is to make a thetan solid, immobile, decisionless. . . . The fight of the Thetan is to remain unsolid, mobile or immobile at will and capable of decision. This is principal [*sic*] unsolved problem."[138]

The problem Hubbard set before each of his readers was to correspond, clearly and without noise or interruption, with their theta being. Such correspondence could be achieved during the auditing process. As Burroughs wrote: auditing offered "complete freedom from past conditioning and a level of functioning that has not been seen before on this planet. Mr. Hubbard insists that we are not our animal bodies, that we can exteriorize from the body through advanced processing." According to Hubbard, the discovery of exteriorization "proved beyond any question the existence of a thetan, that the individual *was* a thetan, not a body, and disproved that man was an animal, and that he was a spiritual being, timeless and deathless."[139]

Exteriorization signaled "separation from the body" and was indicated by "a peculiar rapid action of an E-Meter needle" called a Theta-Bop. Exteriorization was the experience of oneself as a clear channel and as seeing, however fleetingly, from the theta perspective. As Burroughs noted during his study, exteriorization was initiated during auditing with the following instructions:

> Ask the preclear to locate the inside of his forehead. Ask him to put a pressor beam against it and push himself out the back of his head. Supplement this by asking him to reach out through the back of his head and grab the wall with a pulling beam and pull himself out. Ask him to steady himself outside and then by means to raise and lower himself and to move to various parts of the room.[140]

138. William S. Burroughs Papers, 1951–72, New York Public Library, box 68, folder 26.

139. Burroughs, "Scientology Revisited," 80; L. Ron Hubbard, "Exteriorization and High TA" (January 4, 1971) in *Technical Bulletins of Dianetics and Scientology*, vol. 7 (Los Angeles: Bridge Publications, 1991), 168–71.

140. William S. Burroughs Papers, 1951–72, New York Public Library, box 68, folder 23. Writing to Gysin about his auditing courses at St. Hill, "There are now eight grades above clear where you learn to leave the body at will and be at cause [ease?] over your environs. We shall see. I plan to take the clearing course and one of the levels above that then see how the abilities gained can be applied to writing" (Burroughs, *Rub Out the Words*, 275).

According to Hubbard, exteriorization was a ritual effect, achieved during an auditing session, in which the individual could, momentarily, encounter their theta being.[141] Exteriorization was, perhaps, an inevitable discovery given Hubbard's cybernetic focus on the human as a perfectible channel of information flow. For in exteriorization, the self becomes transparent to the information being processed, free from the inevitable noise of the MEST universe. Indeed, according to Hubbard, a prerequisite for "perfect communication" with one's thetan was to *understand* the "problem of communication." For in order "to exteriorize from and interiorize into any and all objects and spaces at will," preclears must first recognize that their cognitive processing is distributed by definition, their brains ever processing their encounters with MEST and the experiences stored by their thetans. To gain knowledge of how and why information was being processed by one's thetan was to be free, to have become, at the level of neuromation, part and parcel of the universe. In order to exteriorize, wrote Hubbard, "one must stop considering the brain to be comprised only of the 'content of the skull.'"[142]

In the moment of exteriorization, one's cognition became a node in a living network. For to experience oneself as a thetan was to become an abstract biological being. As Hubbard wrote, the thetan "does not need a body to think, make facsimiles, experience emotions, remember or perceive." The thetan, in other words, perfectly embodied the formal principles of neuromatic processing. For at its best, the thetan was a matter of pure attention, in Hubbard's words, "the awareness unit which has all potentialities but no mass, no wave-length and no location." In its optimal state—when it was not

141. L. Ron Hubbard, "Parts of Man," 5–8. As Hubbard gloated, "when you exteriorize somebody—this is a known operation to a spiritualist—and he says at once, 'Oh, you mean you made the person step out of his corporeal body in his astral body?' . . . Well, we go further than that. We even know what an 'astral body' is." L. Ron Hubbard, "Universes" (April 7, 1959), in *The Skills of a Theta Being: Transcripts and Glossary*, pt. 1 (Los Angeles: Golden Era Productions, 1994), 17. As the historian Hugh B. Urban notes, Hubbard distinguished exteriorization from astral projection as it was outlined in Sylvan Muldoon's *The Projection of the Astral Body* (1929); see Hugh B. Urban, "The Occult Roots of Scientology: L. Ron Hubbard, Aleister Crowley, and the Origins of a Controversial New Religion," *Nova Religio* 15, no. 3 (2012): 91–116.

142. L. Ron Hubbard, "Certainty of Exteriorization" (April 16, 1954), in *Technical Bulletins of Dianetics and Scientology*, 3:47–48; Hubbard, *Scientology: Control and the Mechanics of S.C.S.*, 6ff.

bogged down by MEST and its MEST body—the thetan was disembodied precisely because it was "aware of being aware."[143]

Exteriorization, then, was a cybernetic ritual par excellence: leaving your brain while still remaining within it, becoming observant of yourself—in the moment and in real time—to the point of omniscience. Observing yourself as part of a communication system. Reflecting on your experience as a channel. Seeing the world, as in the cut-up, as the effect of your communication with it. The preclear could achieve such perspective in an act akin to remote viewing—becoming the moving camera eye and master manipulator of the scene. As Hubbard wrote, "*Exteriorization is the phenomenon of being in a position of space dependent on only one's consideration, able to view from that space, bodies and the room, as it is.* That is exteriorization."[144] Here was a moment when past and present, self and environment coexisted in a single space. Information everywhere. A moment of being you and watching yourself as if you were on TV—seeing yourself seeing the world becoming the world that you see.

12. BREAK THROUGH IN GREY ROOM

In framing Garrett, Gysin, Burroughs, and Hubbard against a cybernetic backdrop, I have sought to better appreciate the kinds of cosmologies that can flow from the chance instantiations of the neuromatic purview. There is something both jarring and, by now, oddly familiar to each of their struggles as they come to grips with their informational surround. In articulating each of their fugitive theories of a neuromatic cosmos, each broached the possibility of there being a register of reality that impinged and very much determined the biological experience of consciousness but whose existence was *almost* completely abstract. You could feel it but not see it. You could respond to it but not touch it. This invisible layer of reality activated the

143. Hubbard, *History of Man*, 14–15; L. Ron Hubbard, *Dianetics and Scientology Technical Dictionary* (1975; Los Angeles: Bridge Publications, 1987), 431–32; Hubbard, *Science of Survival*, 53.

144. Hubbard, "Parts of Man," 149. As Mark Seltzer has astutely pointed out, exteriorization is not about the externalization of inner experience but rather the experience of one's immediate outside on the inside (*The Official World* [Durham, NC: Duke University Press, 2016], 15).

senses even as it evaded them. What was in the body, within the neural skein of the brain, opened out onto the cosmos. As a matter of course, the physical body was vulnerable to a system beyond its purview—construed, variously, as the deep psyche, the spirit world, the forces of magical control, or the ecology of thetans.

Garrett, Gysin, Burroughs, and Hubbard each viewed language as the hinge and/or access point of an *almost* immaterial surround. Words were the link between the phenomenal world of flesh and synapse and a world elsewhere. In each of their respective riffs on cybernetics, words were infused with metaphysical demand, their operations and living effects being neurophysiological matters, for better or for worse. Each of these individuals wove together neuromatic insights to explain the otherworldly vitality of language, to imagine systems beyond the pale yet speaking to you and through you. From this angle, the cybernetic framing of Garrett's mediumship was of a piece with Gysin and Burroughs's study of "word-locks" and Hubbard's promise to reveal "the role language plays in human life."[145] To be sure, there is something both ominous and exhilarating about each of their worldviews in which humans are subject to systems beyond their control. In the moment of exteriorization or mediumistic utterance, or in the reading of a cut-up in which the future "leaked" out, a vast system made itself known by way of the frail human form.

Garrett, Gysin, Burroughs, and Hubbard posited themselves as embodied extensions of a an informatic system, with patterns of its code ever coursing through them. So even as they borrowed insights from the neuromatic brain, each tapped the visceral excesses of cybernetic abstraction and pretenses to disembodiment. In Garrett's trance sessions, for example, the body is front and center as the conduit between worlds and across time. In the investigations of Gysin and Burroughs into the biology of language and its patterned demands, the body became the battleground for control. And in Hubbard's mythology, the human body was inscribed with all matter of cosmic codes unbeknownst to the uninitiated.

Each of these members of the cybernetic demimonde imagined the human brain as hardwired to communicate—with the past, with one's memories, with one's culture and its micropolitics, with spirit entities from another era, with one's past lives, with the drama of the cosmos itself. To experience oneself as a channel was to experience not disembodiment but rather the body, differently. Even in Hubbard's devotion to cybernetic desire

145. Hubbard, *Self Analysis*, 11.

for perfect and immaterial communication, he could still describe exteriorization as a radical form of cognitive embodiment—a heightening of the senses to the point of revealing the external source code of one's physical existence. As Hubbard quipped in 1956, "The way out is the way through. You've got to be in a body before you can get out of it. And to be in a body, you've got to accept that you have it." So, even as Hubbard relied on computational metaphors, he sought to conjure a human whose body transcended the strict concern with "neurons, automaticities, etc." For such a preoccupation with neurophysiology, mass, and energy—at the expense of his cosmic view—would only serve as a "foundation for a slave society."[146] According to Hubbard, training your body to be a machine of the highest order, beyond the analog, demanded that one integrate levels of micro and macro processing.

Here, at the margins of cybernetic propriety, we witness the vivid dream of the neuromatic brain becoming effervescent, operationalized into an affective program for living. But while Garrett, Gysin, Burroughs, and Hubbard each pursued the cybernetic promise of "perfect communication," their mathematically inspired reduction of the human brain to a matter of information processing was never quite complete.[147] For in their desire to free rationality from the flesh and tap its communicative potential, they drew on the language of neuromation. Yet neither numbers nor the machines that ran the numbers could do justice to the chance encounters that they experienced.

146. Hubbard, "Exteriorization"; Hubbard, *Scientology: Control and the Mechanics of S.C.S.*, 6.

147. Donna Haraway, *Simians, Cyborgs, and Women: The Reinvention of Nature* (New York: Routledge, 1990), 176. On the visceral and disciplinary remainder of bodies imagined as virtual, see Emily Martin, "The End of the Body?," *American Ethnologist* 19, no. 1 (1992): 121–40.

SYNAPTIC GAP: WHITE MACHINERY

> "He looked back to where the white beam filtered from the projection room above the balcony. It started small and grew large, specks of dust dancing in its whiteness as it reached the screen. It was strange how the beam always landed right on the screen and didn't mess up and fall somewhere else. But they had it all fixed. Everything was fixed."
>
> RALPH ELLISON, "King of the Bingo Game" (1944)

AT THE SAME MOMENT THE WORLD was being introduced to new visions of the human-machine interface, Ralph Ellison published one of his first short stories that addressed, among other things, the link between those visions and their racializing effects. "King of the Bingo Game," published in 1944 in Eileen Garrett's *Tomorrow*, staged the drama of control and communication as an existential endgame. "He felt vaguely that his whole life was determined by the bingo wheel," wrote Ellison of the main character whose grasp on the difference between clichéd movie scenes and his own life serves to illuminate the power of machines and the enabling fictions of racial difference that machines generate. The Bingo King, a southern transplant, is desperate to be recognized as a full participant in a gathering of his fellow African Americans in this northern movie hall. But rather than experience the effervescent bonding with people, the Bingo King encounters a different and more disturbing kind of solidarity. When it is his turn to activate the spinning wheel all eyes are on the Bingo King. But rather than the sympathy of the crowd he feels instead the biological effects of the machine. "Didn't they know that although he controlled the wheel, it also controlled him? . . . There was only one chance; he had to do whatever the wheel demanded. And gripping the button in despair, he discovered with surprise that it imparted a nervous energy." And while "his spine tingled" and "he felt a certain power," the sense of agency instilled in this moment is as fleeting as it is ambiguous. For the Bingo King remains subject to the whims of the wheel rather than achieving lasting camaraderie with those who had also been marginalized by the techniques of white society.[1]

Ellison's blistering and beautiful rendition of immanent critique also shone forth in *Invisible Man* (1952) and, for a brief moment, took explicit aim at the neuromatic imaginary that was emerging in his midst. Briefly, midway through his account of the ecstatic dimensions of the "calm, scientific gaze," when the narrator of *Invisible Man* finds a job at Liberty Paints, he is charged with mixing their most valuable commodity—"Optic White"—with black drops that, counterintuitively, give it its purity and have made it the government's paint of choice for maintaining the white sheen of national monuments. "Optic White," among other things, speaks to the black experience that has been covered up and over in order to deny African Americans full participation in democratic life. Like Ishmael, the narrator of *Moby-Dick*, Ellison's narrator plays on the ambiguity of whiteness, its seductive and horrific quality of contrastlessness and seeming unendingness. He soon concludes that "Optic White" is really about the power both to imagine and to manufacture a world without difference. As such, "Optic White" and the paint factory where it is produced serve to cover up the technological means and the men who have created it. The medium had become the message. "Perhaps," suggests the narrator, "the real quality of the paint is *always* determined by the man who ships it rather than by those who mix it."[2]

After an explosion at the paint factory for which he becomes the chief suspect, the narrator finds himself strapped to "a cold, white rigid chair" wearing "new overalls, strange white ones" and later wakes to find "two indefinite women in white attending him." He wakes to find himself encased, enclosed, imprisoned—"a whirring began that snapped and cracked with static, and suddenly I seemed to be crushed between the floor and ceiling." An electric shock therapy machine, as the narrator soon learns, has already punished him and, with the right adjustments, will deliver him from sin. "My little gadget will solve everything!" exclaims the attending physician. "From now on do your praying to my little machine. I'll deliver the cure." In contrast to the "negative effects of the knife," ECT will make Invisible

1. Ralph Ellison, "King of the Bingo Game," *Tomorrow* 4 (November 1944): 29–33.

2. Ralph Ellison, *Invisible Man* (1952; New York: Vintage International, 1995), 245, 202.

"physically and neutrally whole" and able to live "with absolute integrity."[3] In being subjected to electric shock therapy, Invisible loses his memory and becomes privy to the deployment of the paranoid view—a world bent on fungibility, producing difference in order to extinguish it, to incorporate it into a seamless vision of whiteness. Invisible describes the scene as a first-hand encounter with something akin to a biblical plague—"the air seemed to grow thick with fine white gnats, filling my eyes, boiling so quickly . . . a live white cloud mixing with the tones upon the torpid air."[4]

In this brief foray in the middle of his novel, Ellison captures the paranoia induced by the cybernetic gaze. When asked about his name and identity, his next of kin and blood relations, it becomes clear that technology has all but replaced genealogy. "A machine my mother?," he considers. "Clearly, I was out of my head." This was the kind of subject idealized within the neuromatic imaginary in which any "short-circuiting [of] the machine" was equivalent to suicide. In narrating his incorporation into whiteness, Invisible becomes a machine of a very particular kind, normalized according to a neuromatic script, absorbed into a "vast whiteness in which I myself was lost."[5]

The irony here is that such an abstraction of identity, one that defies the existence of one's own navel, removes one from one's body and, moreover, occludes those structures that have conditioned that removal. This is not the metaperspective promised by exteriorization and Scientology (founded in the same year that *Invisible Man* was published) but, rather, a paranoid revelation in which one becomes fungible with the technological surround.[6]

3. See, for example, the more recent argument that "successful ECT treatment is effective because it facilitates and restores the function of specific memory systems that are deficient in the course of a severe depressive episode. It is these memory systems that link a person to the very essence of their existence and personal identity." Anthony T. Frais, "Electroconvulsive Therapy: A Theory for the Mechanism of Action," *Journal of ECT* 26, no. 1 (March 2010): 60–61.

4. Ellison, *Invisible Man*, 232–36.

5. Ellison, *Invisible Man*, 240, 243, 238. For an inspired reading of this scene, see Johnnie Wilcox, "Black Power: Minstrelsy and Electricity in Ralph Ellison's *Invisible Man*," *Callaloo* 30, no. 4 (2007): 987–1009.

6. And it this first principle of the cybernetic sciences, more so than any critique of representation, that best captures their racializing directives. For when the human can be abstracted to the degree that it is deemed fungible with the technological surround, the political, material, and historical contexts that cling to, intervene in, infuse and/or violate that human life become, at best, matters of peripheral concern. Unless, of course, they can be quantified and measured through an algorithm. See, e.g., calls to "create an academic

For within this "vast whiteness" there is no you and me, us and them. There are only patterns, ever shifting as information is processed. A transcendence that is pure precisely because it is mathematical. The invisible man, in considering the question of whether a machine was his mother, catches himself as he becomes incorporated into system—here a biomedical regime compatible with the virulent politics of whiteness and, moreover, a medium for its evolutionary refinement.[7] Like the doctors and nurses that hover around him as his head is "pounded between crushing electrical pressures," he, too, has lost contact with his human origins and, in this technologically induced amnesia, become what the factory engineer, Lucius Brockway, had prophesied before the explosion—one of many "machines inside the machine."[8]

---

discipline of AI that takes the complexity of human behavior into account"—all in order to combat increasingly recognized racial biases built into existing algorithms. Gary Shiffman, "We Need a New Field of AI to Combat Racial Bias," *TechCrunch* (July 3, 2020), https://techcrunch.com. On the practical application and business necessity of this new field, see Tomas Chamorro-Premuzic, "Technology Can Help Organizations Reduce Racism, But Will It?," *Forbes* (June 6, 2020), https://www.forbes.com.

7. On the "full erasure of people of colour from the White utopian imaginary" advanced by contemporary algorithms, see Stephen Cave and Kanta Dihal, "The Whiteness of AI," *Philosophy & Technology* 33 (2020): 685–703.

8. Ellison, *Invisible Man*, 232, 217.

04

HISTORIES OF ELECTRIC SHOCK THERAPY CIRCA 1978

"It's like a big thing, and uh it—takes both sides of your head, it goes *boom* like that, and all of a sudden you feel something. And after that you—don't feel anything."

"Are you conscious after that?"

"No, you're out completely."

"For about how long—have you any idea?"

"For about a half-hour."

"Then what?"

"Then you wake up—then you find you've been under shock."

UNIDENTIFIED FEMALE PATIENT AT NAPA STATE MENTAL HOSPITAL, ca. 1960[1]

1. OF SYSTEMS, SEX, AND SECULAR CONVERSION

A significant branch of cybernetics was operational research (OR), a kind of totalizing approach to "the fundamental problem of management" born of Cold War strategies and the Rand Corporation. Within corporate struc-

1. This patient was interviewed at Napa State Mental Hospital, circa 1960, as part of a study of women diagnosed with schizophrenia. Patients were "all white, were from lower to lower middle class backgrounds, had at least one child, were currently married, and ranged in age from 26 to 40." Carol A. B. Warren, "Electroconvulsive Therapy, the Self, and Family Relations," *Research in the Sociology of Health Care* 7 (1988): 284.

tures in the 1960s and 1970s, OR was a self-consciously gendered enterprise as "systems men" succeeded in redefining the computer as an information management system.[2] No longer was the computer a mere data processing instrument, left largely in the hands of female punch-card operators. The computer, instead, had become a machine of corporate power, of men mostly, and an essential tool for upper management grappling with the intricate problems of organization. During these postwar years, as the risk management of data became a general good, OR was thought to be commonsensical and nonideological—an approach that could "be used by any political system or private enterprise." As such, OR was also a "revelation" in which mathematical formalization could be extended into previously uncharted territories—beyond military planning and antiaircraft ballistics into business, society, mental health, and biology.[3]

Smitten with the vast promise of algorithms, OR consultants aggressively, and with military meticulousness, began thinking of the world around them in terms of self-organizing systems. In doing so, they took up Walter Cannon's question of what "agency in civilized society . . . correspond[ed] to that feature of our bodily arrangements?"[4] In the postwar years giddy plans for reproducing the power of agency leaned on the cybernetic compulsion to elucidate "such things as men, materials, machinery and money—*and their intersection*" (my emphasis).[5] The intersections added up to the network that added up to explanatory purchase on just about anything. For the brain, in transmitting information from one neuron to another, served as the ideal of systematicity. It became the resource for explaining the microphysics of control in a variety of domains. It was the model for how things always added up. Drawing much of its inspiration from cutting-edge attempts to think through the staggering density of neural network connectivity, OR honed in on the problem of complexity in general: how to visualize it, how to measure it, how to generate it, how to manipulate it in any circumstance that might arise.

In 1958, for example, Stafford Beer defined OR as "the organization of

2. Thomas Haigh, "Inventing Information Systems: The Systems Men and the Computer, 1950–1968," *Business History Review* 75, no. 1 (2001): 15–61.

3. Harvey M. Wagner, "Practical Slant on Operations Research," *Harvard Business Review* 41 (May–June 1963): 61–71; Stafford Beer, "Operational Research as Revelation," *Operational Research Quarterly* 21, no. 1 (March 1970), 13.

4. Walter B. Cannon, *The Wisdom of the Body* (New York: W. W. Norton, 1932), 313.

5. Stafford Beer, *The Brain of the Firm: The Managerial Cybernetics of Organization* (London: Allen Lane, 1972), 15.

unthinkable systems." In framing all of nature—organic and inorganic—in terms of system, cyberneticians like Beer saw the neural network of the brain as the most definitively complex and unthinkable yet readily accessible object on which to build their models. So while the system of the brain may have been unknowable in its detail, it was for Beer an opportunity to think and to know anew. For he was quite confident, despite definitional stances on complexity, that he could model just about any system, including the brain, in terms of inputs and outputs. While perfect prediction could never be achieved, one could approach the degree zero of probability when dealing with the future trajectories of any system. Given the driving question of how to help systems self-regulate and survive, OR consultants like Beer strategically conflated a model of the brain's reality as a model for all of reality.[6] The desire to do so was clear enough, wrote the psychologist and computer scientist J. C. R. Licklider in 1960 in his aptly titled "Man-Computer Symbiosis." "The hope," Licklider announced, "is that in not too many years, human brains and computing machines will be coupled together very tightly, and that the resulting partnership will think as no human brain has ever thought and process data in a way not approached by the information processing machines we know today."[7]

OR and its promise of technical solution achieved its greatest currency in these early years of Johnson's Great Society. But by the late 1960s, OR had begun to generate much negative publicity with the hippies and the protests and the debacle of Vietnam military strategies. And indeed, the failures of OR in the face of looming crises and social fissures were widely reported.[8]

6. Beer's electroencephalogram of Templeborough Rolling Mills steelworks, a subsidiary of his employer, United Steel (see fig. 26) analogized the neuromolecular all the while using a computer to imagine the brain in order to imagine the corporation as a kind of brain that could be modeled by that very same computer. Stafford Beer, "Towards the Automatic Factory," in H. von Foerster and G. Zopf, eds., *Principles of Self-Organization: Transactions of the University of Illinois Symposium on Self-Organization, Robert Allerton Park, 8 and 9 June, 1961* [*sic*: actually 1960) (New York: Pergamon, 1962): 25–89. For a discussion of Beer's article in terms of its open-ended epistemology, see Andrew Pickering, "The Science of the Unknowable: Stafford Beer's Cybernetic Informatics," *Kybernetes* 33, no. 3–4 (2004): 499–521.

7. J. C. R. Licklider, "Man-Computer Symbiosis," *IRE Transactions on Human Factors in Electronics* (March 1960): 4.

8. Agatha Hughes and Thomas Hughes, eds. *Systems, Experts, and Computers: The Systems Approach in Management and Engineering, World War II and After* (Cambridge, MA: MIT Press, 2011). One is not hard pressed to read many accounts of OR as jeremiads about the unfulfilled promise of the Great Society. See, e.g., Robert H. Hayes and William J. Aber-

But rather than mark a retreat of cybernetic reasoning, the so-called dissipation of OR's authority marked its diffusion into other forms and other registers—epistemic, economic, medical, biological, sensual, affective, philosophical, carceral. Outside the laboratory and computer room, in other words, the neuromatic paradigm was bending, looping, and mutating in strange ways.[9]

As a practical matter OR made the most significant headway in mental health services, initially in its engagement with the nuts and bolts of running hospitals and the distribution of resources within them. Throughout the 1960s, operational researchers imagined the hospital as a set of overlapping systems—"a physical system of buildings," "a system of many interacting staff; a complex logistics system; a system for treating patients; [and finally,] an information system." During this time OR-inspired reformers also looked to "computer data collection systems" to adjust staffing levels and evaluate patient care data as well as the implementation of closed-circuit television in therapy sessions.[10]

The analysis of neural systems processing information had reached a

---

nathy, "Managing Our Way to Economic Decline," *Harvard Business Review* (July–August 1980): 67–77, as well as the jeremiadic diptych of Russell Ackoff in his back-to-back articles in 1979 issues of the *Journal of Operational Research Society* 30, no. 2–3—"The Future of Operational Research Is Past" and "Resurrecting the Future of Operational Research."

9. See, e.g., the increasingly expansive arc of Jay W. Forrester's trilogy: *Industrial Dynamics* (Cambridge, MA: MIT Press, 1961), *Urban Dynamics* (Cambridge, MA: MIT Press, 1969), *World Dynamics* (Cambridge, MA: Wright-Allen Press, 1971). See also the summary appraisal in Robert Trappl and Franz R. Pilcher, eds., *Progress in Cybernetics and Systems Research*, vol. 1: *General Systems, Engineering Systems, Biocybernetics and Neural Systems* (New York: John Wiley and Sons, 1975).

10. William Horvath, "Planning for Delivery of Mental Health Services," *Health Services Research* 6 (1971): 96–100; Donald A. B. Lindberg, *The Computer and Medical Care* (Springfield, IL: Charles C. Thomas, 1968), vii, 189; Arthur Spindler, "Social and Rehabilitation Services: A Challenge to Operations Research," *Operations Research* 18, no. 6 (1970): 1112–24; F. Blewett et al., "Computer Simulation Models for a Multi-Specialty Ward," *Operational Research Quarterly* 23, no. 2 (June 1972): 139–49; G. M. Luck et al., *Patients, Hospitals and Operational Research* (London: Tavistock, 1971); David H. Stimson and Ruth H. Stimson, *Operations Research in Hospitals: Diagnosis and Prognosis* (Chicago: Hospital Research and Educational Trust, 1972); Janic M. Messick, "A Systems Analysis Approach to Planned Change in a Clinical Psychiatric Program," *Journal of Psychiatric Nursing and Mental Health Services* 13 (1975): 7–11. See also Allen Newell and Herbert A. Simon, "Computers in Psychology," in *Handbook of Mathematical Psychology*, vol. 1, ed. R. Duncan Luce, Robert R. Bush, and Eugene Galanter (New York: John Wiley and Sons, 1963), 361–428.

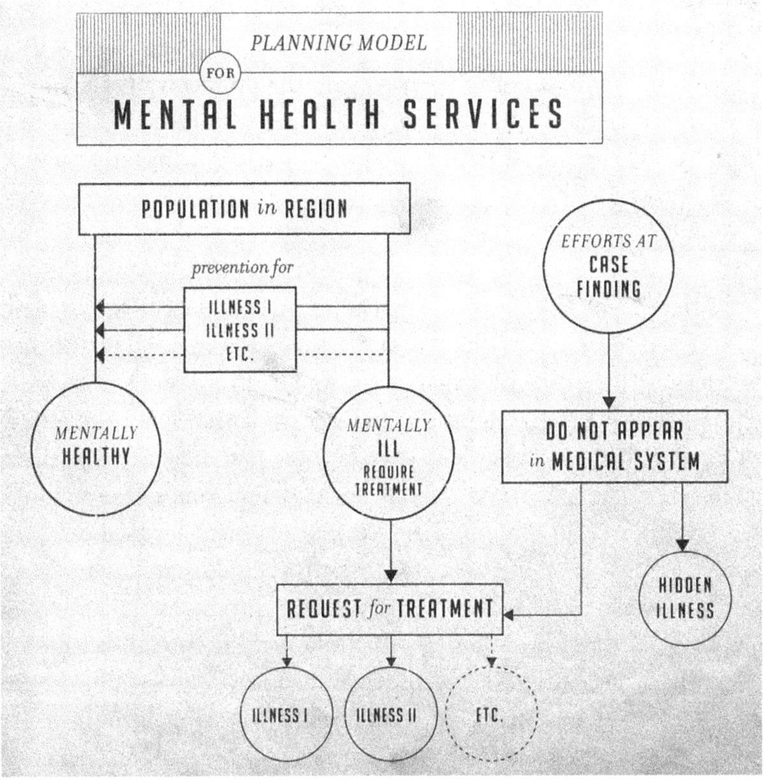

FIGURE 41. Planning Model for Mental Health Services. Model by Libby Modern. "The chart allows for the possibility that systematic efforts will be made to induce persons to enter the mental health treatment system, whether by use of screening procedures or by such a simple measure as requesting [doctors] to suggest psychiatric consultation for patients they believe to be mentally ill." William Horvath, "Planning for Delivery of Mental Health Services," *Health Services Research* 6 (1971): 98.

threshold moment. As an approach modeled on the brain made its way into ever deeper levels of hospital administration, that very same approach would, and perhaps inevitably, return to the flesh, finding its target in the medical modulation of neural systems of information processing.[11] Aided by increasingly precise machines for measuring cognitive responses to environmental stimuli, the treatment of mental illness came to be premised on the manipulation of the neural network of the brain. For as the brain

11. See, e.g., the conference proceedings of *The Neural Control of Behavior*, ed. Richard E. Whalen et al. (New York: Academic Press, 1970).

was increasingly considered the engine of social life, in general, the brain became *part of* the cultural system to be managed effectively.[12] And as it turns out, the diffusion of cybernetics into the protocols of mental health institutions and then into the brain itself only amplified the disciplinary dimensions of psychiatric power that had originated in an earlier moment.

"Moral treatment" was an agenda that emerged over the course of the nineteenth century. It, too, was a kind of systems analysis. According to a Quaker reformer in 1825, moral treatment was a totalizing approach to the patient environment, encompassing "the habitation, classification, cleanliness, diet, coercion, punishment, treatment of the feelings, treatment of the intellectual faculties, and occupation of the patients."[13] Originating in the writings of the French physician Philippe Pinel, moral treatment focused on a population of suspect psyches and offered a holistic understanding of the relationship between mental disturbance and the environment. This move away from physical restraint toward psychic intervention spread rapidly in the United States and evolved alongside and often in tandem with the ends of liberal Protestant practice.[14] Curing madness was often code in these circles for taming the excesses of revivalism, drawing as they did on the fact that revivals in the early nineteenth century were often cited by early reformers in their successful quest to build asylums around the United States.[15] "In a camp meeting you will observe unmistakable mad-house exhibitions!" declared Andrew Jackson Davis, ever quick to needle his Protestant rivals in their skirmish to define the terms of benevolence and true religious piety.

12. See, e.g., Karl U. Smith and Margaret Foltz Smith, *Cybernetic Principles of Learning and Educational Design* (New York: Holt, Rinehart, and Winston, 1966), 237f.

13. Robert Waln Jr., *An Account of the Asylum for the Insane, Established by the Society of Friends, Near Frankford, in the Vicinity of Philadelphia* (Philadelphia: Benjamin and Thomas Kite, 1825), 13.

14. On the Quaker roots of moral treatment in England, see Anne Digby, "Moral Treatment at the Retreat, 1796–1846," in *The Anatomy of Madness: Essays in the History of Psychiatry*, vol. 2, ed. W. F. Bynum, Roy Porter, and Michael Shepard (London: Tavistock Publications, 1985), 52–73. Gerald Grob, *The Mad among Us: A History of the Care of America's Mentally Ill* (New York: Free Press, 1994), 30.

15. Ronald L. Numbers and Janet S. Numbers, "Millerism and Madness: A Study of 'Religious Insanity' in Nineteenth-Century America," in *The Disappointed: Millerism Millenarianism*, ed. Ronald L. Numbers (Knoxville: University of Tennessee Press, 1993), 95.

Mental health reformers like Davis were confident in their approach, given that the brain organs of the most positive minds "send" out "a kind of nerve-aura, an atmosphere of psychic potency—which touches, penetrates, kindles, and inflames similar organs in persons more passive and sensitive." Within these strong versions of moral treatment, the atmosphere, itself, had become a lever for a different kind of conversion. "The very architecture of an asylum," wrote Davis, "should be . . . not 'merely four square walls with pictures hung and gilded;' but a grand place of helpfulness and affection, filled with that holy magnetism which attracts and heals the desolate, the homeless, and the broken-hearted."[16]

Moral treatment was, at once, a ritual of sympathetic magic and biopolitical performance. Reformers assumed an impressible brain of the lunatic, one that could be managed in the name of civilized progress. Under regimes of moral treatment, the nervous organization of an individual patient became an object to sculp and refine by manipulating the culture that contained it. Because mental illness was born, in part, as a response to the environment, the best treatment was to strategically change the environment so as to allow the patient to learn to respond differently. Combining scrutinizing attention to aesthetics with encompassing material instruction, moral treatment sought to interrupt patterns of attention and memory. In its efforts to reprogram the mind using the levers of culture, the hallmark of moral treatment was its formal faith in systematicity and routine. "By our whole moral treatment, as well as by our religious services," wrote the trustees of Worcester State Lunatic Hospital in 1841, "we inculcate all the habits and obligations of rational society."[17]

More often than not, moral treatment was aimed at women in the service of taming their irrationalities and enthusiasms. When it set its sights on the psychic excess of men, their madness or simply their race was framed as an effeminate condition. In addition to leveraging sexual differences, Michel Foucault writes that "moral treatment" in the nineteenth century sought to "validate the power exercised within the asylum as being quite simply the

16. Andrew Jackson Davis, *The Temple: Concerning Diseases of the Brain and Nerves; with Full Directions for Their Treatment and Cure* (Boston: Colby and Rich, 1871), 212–13; Andrew Jackson Davis, *Mental Disorders; or, Diseases of the Brain and Nerves, Developing the Origin and Philosophy of Mania, Insanity, and Crime, with Full Direction for Their Treatment and Cure* (New York: American News Company, 1871), 88.

17. *Ninth Annual Report of the State Lunatic Hospital at Worchester* (Boston: Dutton and Wentworth, 1842), 88; see also Amariah Brigham, "The Moral Treatment of Insanity," *American Journal of Insanity* 4 (July 1847): 1–15.

power of reality itself." Notions of what constituted "reality" have, indeed, changed over time, yet the desire to reduplicate the system of reality within the asylum has not. What Foucault calls the "asylum tautology"—"giving power to reality and founding power on reality"—remained largely intact over the course of the twentieth century.[18] For as the management of "culture" came to encompass not just the surfaces of the feminized body but the feminized brain as well, the "reality" within the asylum, in turn, assumed a different character.

Which is to say that while sexual norms, not to mention the theorizations of correspondence between the inside and outside of the asylum, may have changed dramatically in terms of substance over the course of the twentieth century, the will to organize the relationship between brain and environment remained remarkably consistent. The cybernetic incorporation of mental health, in other words, still entailed "something like a residue" of moral treatment.[19] Indeed, such discipline continued to generate an outside for purposes of retaining and consolidating its own authority. For in its will to "classify, hierarchize, supervise, and so on," psychiatric power "come[s] up against those who cannot be classified, those who escape supervision, those who cannot enter the system of distribution, in short, the residual, the irreducible, the unclassifiable, the inassimilable." Such failures of discipline, according to Foucault, are built in, necessary even, as they generate "supplementary disciplinary systems" and "new recovery systems" in the "never-ending work" of conserving the rule of discipline.[20]

18. Michel Foucault, *Psychiatric Power: Lectures at the Collège de France, 1973–1974*, ed. Jacques Lagrange, trans. Graham Burchell (New York: Picador, 2003), 175. For a devastating account of the conflation of blackness and insanity in contrast to white idealizations of religious freedom in the early twentieth century, see Judith Weisenfeld, "Normal Minds: Psychiatry and African American Religions," delivered at the Religion and the American Normal conference, Princeton University, February 8–9, 2018.

19. So, for example, the troubling of sexual difference (itself coded as feminine) can now be explained neurologically and by way of the MRI. Julie Bakker, "Brain Structure and Function in Gender Dysphoria," *Endocrine Abstracts* 56 (2018); Judy van Hemmen, et al., "Sex Differences in White Matter Microstructure in the Human Brain Predominantly Reflect Differences in Sex Hormone Exposure," *Cerebral Cortex* 27.5 (2017): 2994–3001.

20. Foucault, *Psychiatric Power*, 53–54; E. Mansell Pattison, "Systems Pastoral Care," *Journal of Pastoral Care* 26, no. 1 (1972): 2–14. As medical anthropologist Simon Cohn has noted, this persistent neurological "faith" in madness (or, say, something like religion) that is present "beneath and prior to the symptoms" ensures that "the new biology of mental

FIGURE 42. Ediswan Electric Convulsion Apparatus built to the specifications of Dr. William Grey Walter, scientific director at the Burden Neurological Institute. On display at Glenside Museum, Bristol, England.

Electric convulsive treatment (ECT) was a nineteenth-century fever dream of reform come to life.[21] With precursors in galvanic treatments and crude applications of electrical wires applied to the ears and forehead, ECT emerged in the 1930s alongside insulin shock therapy and psychosurgery as technologies of normalization and reform.[22] ECT also developed in tan-

disease will inherit much of the old social nature of mental illness" ("Disrupting Images: Neuroscientific Representations in the Lives of Psychiatric Patients," in *Critical Neuroscience: A Handbook of the Social and Cultural Contexts of Neuroscience*, ed. Suparna Choudhury and Jan Slaby [Oxford: Blackwell Publishing, 2016], 187).

21. See, e.g., M. B. Ray, *Doctors of the Mind: The Story of Psychiatry* (Boston: Little Brown, 1942) in which he writes that in the case of "chronic schizophrenics, as with committed criminals . . . we must use more drastic measures to silence the dysfunctioning cells and so liberate the activity of the normal cells. This time we must kill the too vocal dysfunctioning cells. But can we do this without killing normal cells also? Can we select the cells we wish to destroy? I think we can" (250).

22. Insulin shock therapy was a labor-intensive treatment in which the patient was injected on a daily basis with high levels of insulin. Such doses were used, over the course of months, to induce daily comas in the patient. Such comas, although exhausting and physically debilitating, were thought to be psychologically calming and restorative for patients.

dem with electroencephalography (EEG) and the institutionalization of the cybernetic imaginary.[23] Grey Walter, in addition to his pioneering work on the alpha wave, studied and patented methods of convulsive therapy.[24] Both EEG and ECT were premised on a neuromatic understanding of cognition. And as Warren McCulloch suggested in his 1944 address before the American Society for Research on Psychosomatic Problems, both held the highest promise in defining the physiological spectrum of neural activity and acting on this knowledge. Indeed, McCulloch and Pitts's "Logical Calculus" was an effort, in part, to provide scientific foundation for initiating effective shock treatment.[25] As part of his long-standing effort to isolate madness at the neuronal level, McCulloch was hopeful that ECT could soon, with unprecedented accuracy, alter the firing thresholds of neurons and disrupt positive feedback loops that were the source of dysfunction.[26] Within

---

In the 1950s ECT rapidly displaced insulin shock therapy as the medical establishment became increasingly skeptical of its therapeutic benefits. Harold Bourne, "The Insulin Myth," *Lancet* 262, no. 6793 (1953): 964–68, and his "Insulin Coma in Decline," *American Journal of Psychiatry* 114, no. 11 (1958): 1015–17.

23. Lothar B. Kalinowsky, "Cerebral Localization and the Contributions of Psychosurgery and Shock Therapy," *Archives of Neurology and Psychiatry* 69, no. 5 (1953): 582–86. The Electroshock Research Association was founded in 1944, followed a year later by the Society for Biological Psychiatry. Enthusiastically welcomed by the medical insurance industry, the majority of American hospitals had incorporated ECT by the end of the 1950s. In large facilities, patients were lined up in large dormitories with their heads at the foot of their beds. The ECT apparatus could be wheeled on a tray from bed to bed and administered to each patient, one after the other. Timothy W. Kneeland and Carol A. B. Warren, *Pushbutton Psychiatry: A Cultural History of Electroshock in America* (Walnut Creek, CA: Left Coast Press, 2008), 50, 58. See also R. W. Russell et al., "A New Apparatus for the Controlled Administration of Electroconvulsive Shock," *Journal of Psychology* 26 (1948): 71–82.

24. Gerald Fleming, Frederick Golla, and Grey Walter, "Electric-Convulsion Therapy of Schizophrenia," *Lancet* 234 (1939): 1353–55; and Frederick Golla, Grey Walter, and Gerald Fleming, "Electrically Induced Convulsions," *Proceedings of the Royal Society for Medicine* 33 (1940): 261–67.

25. E. W. Davis, W. S. McCulloch, and E. Roseman, "Rapid Changes in the O2 Tensions of Cerebral Cortex During Induced Convulsions," *American Journal of Psychiatry* 100, no. 7 (1944): 825–29. See also W. S. McCulloch's contribution to "The Modern Concept of Schizophrenia," *Medical Clinics of North America* 29, no. 1 (1945): 147–64. The article was funded by the Macy Foundation and written with L. J. Meduna, an ECT pioneer from Budapest, Hungary.

26. McCulloch also describes a neurotic brain in terms of the failure of negative feedback. "The Brain Computing Machine," *Electrical Engineering* 68, no. 6 (1949): 496. For a more recent and compatible claim, see Jinkun Zeng et al., "Reorganization of Anatomical

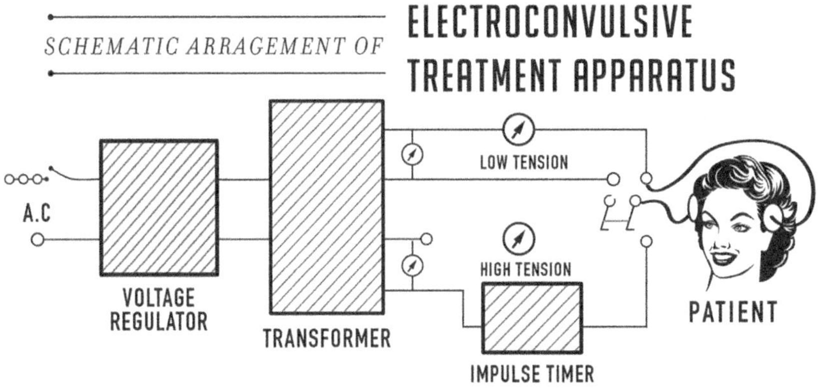

FIGURE 43. Schematic arrangement of electroconvulsive treatment apparatus. Created by Libby Modern.

the cybernetic fold, computers would be able to quantify madness like never before, isolating its neural substrate and suggesting a target point for maximum leverage.[27] With brains equated with machines and mental illness with circuit glitches, ECT was the equivalent of switching parts of the machine off and starting again.[28]

From the very beginning ECT has been enlisted to induce whatever version of reason or normal personhood was thought to lie deep within the circuits of the brain.[29] Indeed, ECT was a technological means to manage

---

Connectome following Electroconvulsive Therapy in Major Depressive Disorder," *Neural Plasticity* (2015): 1–8.

27. The most visible defender of ECT in the 1960s and 1970s, Max Fink, wrote the definitive article on computers and pathology in 1967: "Computers and Psychiatry," in *Comprehensive Textbook of Psychiatry*, ed. Alfred M. Freedman and Harold I. Kaplan (Baltimore: Williams and Wilkins, 1967), 232–38.

28. W. Ross Ashby, "Cybernetics," in G. W. T. H. Fleming, ed., *Recent Progress in Psychiatry* (London: J. and A. Churchill, 1950), 107. Wiener likens ECT to the erasure of a computer's memory; see *Cybernetics: or, Control and Communication in the Animal and Machine* (1948; Cambridge, MA: MIT Press, 1965), 149–59. See also Claude E. Shannon's concept of an "antineurotic circuit" that might aid in the development of machine intelligence: "Presentation of a Maze-Solving Machine," in *Cybernetics: Circular Causal and Feedback Mechanisms in Biological and Social Systems*, Transactions of the Seventh Conference, March 23–24, 1950, ed. Heinz von Foerster (New York: Josiah Macy Jr. Foundation, 1951), 173–80.

29. F. M. Lorimer, M. M. Segal, and S. N. Stein, "Path of Current Distribution in Brain during Electro-Convulsive Therapy," *EEG and Clinical Neurophysiology* 1 (1949): 343–48.

the conversion to sanity. If you were depressed, then ECT would make you happy. If you were manic, it would calm you down. If your personality was split, it would make you whole. If you were gay, it would make you straight. If you woke up one day and believed yourself to be the Virgin Mary, then ECT would cure you of this delusion.[30] If you had unreasonably converted to Orthodox Judaism, then it would alleviate your desire to grow a bushy beard and obsess over dietary restrictions. ECT, in other words, was a machine bent on purification, blasting out the bad, the blurry, and the loose feelings that stubbornly refused to fit into prescribed categories. In its advertised function, ECT distilled something about the aggressive and gravitational push to maintain clear boundaries—between man and woman, straight and gay, madness and civilization, and so on.

The technology of ECT inaugurated a season of contagious normality. As one journalist wistfully wrote in 1947,

> Take the case of the middle-aged woman who had been a civic leader in her community until she became a victim of mental depressions that virtually confined her to her home. She gave up her outside interests and became careless in her personal appearance, spent a great deal of time in weeping, and required constant supervision. Following six electric shock treatments, this woman was able to resume her normal way of life. She was not only able to manage her household as before and return to her clubs and her civic organizations, but she took on the added task of hunting out other persons she thought might be benefited by electric shock treatment.[31]

ECT was, from its origins, a technology of sexual differentiation.[32] As sexual deviance became a visible matter of public concern in the 1950s, ECT was enlisted in "the replacement of homosexuality with normal sexual desire."[33] But in the majority of cases, ECT was administered to women who did not

30. Oliver Swigert, "Myth and Psychosis," *Napa Quarterly* (Winter 1972–73), n.p.

31. Henry Shryock, "Is Shock Treatment Successful?," *Science Digest* 22 (October 1947): 60–65.

32. Jonathan Sadowsky, *Electroconvulsive Therapy in America: The Anatomy of a Medical Controversy* (New York: Routledge, 2016), 75–86.

33. George N. Thompson, "Electroshock and Other Therapeutic Considerations in Sexual Psychopathy," *Journal of Nervous and Mental Disease* 109 (1949): 531–39. ECT was enlisted alongside other somatic treatments to cure sexual perversion of one kind or another. See also M. Owensby Newdigate, "Homosexuality and Lesbianism Treated with Metrazol," *Journal of Nervous and Mental Disorders* 92, no. 1 (1940): 65–66. J. N. Marquis,

conform to the accepted standards of Cold War domestic bliss.[34] A patient in 1948, for example, was considered sane only if she could "manage her home efficiently" or become "active in society."[35]

ECT, in its normalizing capacities, did much to rationalize sexual difference with the authority of cybernetic science.[36] As Joan Baker responded when asked about her incarceration at Napa State Mental Hospital in the 1950s, the promise of ECT was often irresistible: "I don't care what they do, as long as it helps me—helps me not to be depressed—helps me to be a different person, to like people. I want to forget—I don't know if I can or if I know what I mean when I say it—but my father never liking me as a child made me feel I was a monster, I was different, making me hide in my bedroom." ECT, here, served to generate the conditions for transcending the past and achieving a new identity. The differences between past and future, old self and new, were each coordinated, with the attendant sublimation and erotic charge, in relation to sexual difference. In Baker's framing of ECT as that which allowed her to become the best version of herself, the admittedly hazy discourse of secularism assumes contour and neurological character at the level of deployment.[37]

"Orgasmic Reconditioning: Changing Sexual Object Choice through Controlling Masturbation Fantasies," *Journal of Behavior Therapy and Experimental Psychiatry* 1 (1970): 263–71.

34. Paul Popenoe, *Marriage Is What You Make It* (New York: MacMillan Company, 1950), 124.

35. N. K. Rickles and Charles G. Polan, "Causes of Failure in Treatment with Electric Shock," *Archives of Neurology and Psychiatry* 59, no. 3 (1948): 337–46.

36. Londa L. Schiebinger, "The Gendered Brain: Some Historical Perspectives," in *So Human a Brain: Knowledge and Value in the Neurosciences*, ed. Anne Harrington (Boston: Birkhauser, 1992), 111–20.

37. Carol A. B. Warren, *Madwives: Schizophrenic Women in the 1950s* (New Brunswick, NJ: Rutgers University Press, 1987), 132. According to historian Joan Scott, the rationalization of sexual difference has been the central feature of secularism since the nineteenth century, which is to say that it has become the central feature of so-called secular orders as well as so-called religious orders. Secularism, one might go so far as to say, makes sense only in relation to the sexual difference that it continually conjures. Sexual difference, suggests Scott, is the fetish behind which the spirit of secularism has circulated since the nineteenth century. Sexual difference, then, is the key to thinking about the buzzing compatibility between good government (of self or society) and good religion. To reframe this as a claim about change over time, one could say that the categorical tension between true and false religion—a formal proposition ever filled with evolving content—has been integral in determining the propriety of sexual difference at any given time. Joan Wallach Scott, *Sex and Secularism* (Princeton, NJ: Princeton University Press, 2018).

While the adoption of ECT was checked momentarily in the late 1950s with the development of new antidepressants such as Deprol and Tofranil, by the 1970s the practice of ECT had become widespread. With the rise of ECT came questions about its politics given the fact that mainstream physicians who practiced ECT were, for the most part, male while the majority of their patients were female.[38] Even as proponents in the 1960s and 1970s hailed ECT as a salvific technology, a "miracle" that achieved spiritual renewal through secular means, the gender imbalance of ECT did not go unchallenged. Feminist critics framed their own resistance to ECT as part of the unfolding of modern progress, the latest iteration of a Protestant Reformation bent on liberating the individual from the clutches of the "state religion" of psychiatric practice. Yet despite the fact of having very different understandings of therapeutic ethics and the viability of somatic treatments such as psychosurgery and ECT, both practitioners and critics subscribed to a strikingly similar vision of what constituted human freedom. Both were invested in the promise of voluntary conversion. And both were concerned with securing the viability of a liberal rights–bearing subject capable of voluntary reason.

Over the course of this chapter I chart moments in the morphing of moral treatment into an even kinder, gentler, and more precise mode of individuation. It is a selective story of how the brain became a primary target by which gender, and much else besides, became mobilized in the reproduction (ever flexible) of secularism's common sense.

38. George Gulevich et al., "The Decreasing Use of Electroconvulsive Therapy," *American Journal of Psychiatry* 118, no. 6 (1961): 555–57; Fred H. Frankel et al., *Task Force Report 14: Electroconvulsive Therapy* (Washington DC, American Psychiatric Association, 1978), 5. In 1975, for example, women made up 70.1 percent of ECT patients treated for affective disorder nationwide. By the 1970s ECT was deemed "significantly more effective for women than for men." The biological difference between male and female anatomy was, here, "clinically significant." For this difference served, simultaneously, as principle, premise, and promised outcome. Studies have shown that women often respond more readily to ECT because men have, on average, a 50 percent higher seizure threshold than women due to thicker skin, thicker skulls, and a longer distance between electrodes when placed on the male cranium. Michael E. Thase and Susan G. Kornstein, "Gender Differences in Responses to Treatments of Depression," in *Gender and Its Effects on Psychopathology*, ed. Ellen Frank (Washington, DC: American Psychopathological Association, 2008), 120, 119.

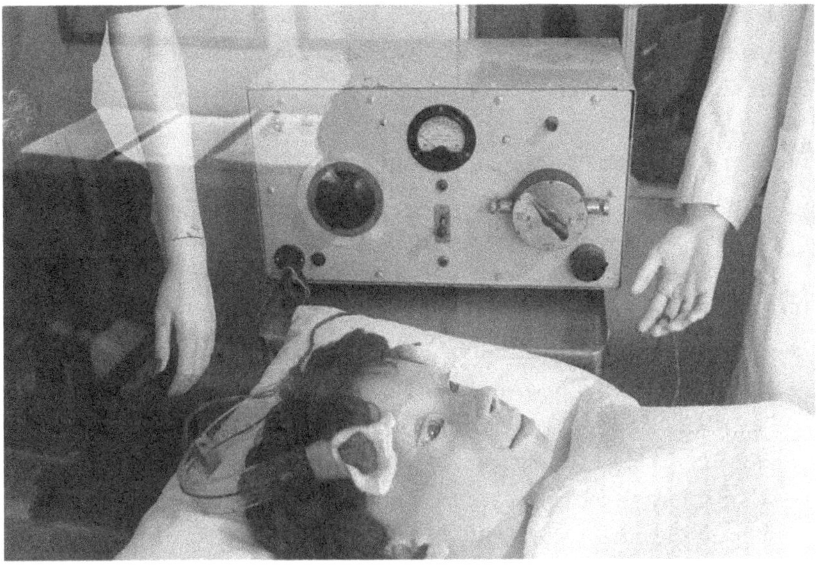

FIGURE 44. Electroconvulsive therapy machine on display at Glenside Museum, Bristol, England.

This chapter begins by revisiting the early career of Eliza Farnham and her stints as the matron of Sing Sing Penitentiary in the 1840s and, later, as the matron of the Female Department at Stockton Insane Asylum. In her pursuit of "true religion" Farnham did much to establish and articulate the metaphysical case for nineteenth-century American reform of prisons and mental asylums. From the very beginning, the "moral treatments" of Farnham and others possessed a physiological cast, leveraging her ward's environment as a means to manage not only bodily surfaces but, in Farnham's case, the interior spaces of cognition.

Farnham's mix of systematicity, spiritualist sentiment, and gendered investment in the brain helps make sense of the discursive diffusion of OR after World War II into mental health industries, in general, and into the brains of particular patients.[39] In Farnham, as in the cybernetic expanse

39. For more background on Farnham, see "The Touch of Secularism," in my *Secularism in Antebellum America* (Chicago: University of Chicago Press, 2011), 243–46. By insisting on the corporeal dimensions of Farnham's practice I am pushing on, and slightly against, Benjamin Reiss's argument that the metaphysics of moral treatment only lingered in "echoes" into the twentieth century. Whereas Reiss and others have largely affirmed the decline of moral treatment after the Civil War, rendering it a ghost haunting twentieth-century practices, I argue that with the so-called demise of the category of "moral treatment," the

I have surveyed throughout this book, there was a direct correspondence between the organization of psyche and the organization of all natural matter. Whereas Farnham leaned on the ethereal paradigms of her metaphysical moment (animal magnetism, mesmeric fluids, etc.) to explain the correspondence between brain and environment, mental health reformers in the cybernetic vein drew on the universalist assumptions of information theory to reframe psychiatric practice as the modulation of neural networks.[40]

In order to track this change over time I then move into a discussion of the triumphs and trials of liberal reform as it played out at Napa State Mental Hospital in the twentieth century. Built in the image of the Stockton Asylum, Napa Asylum for the Insane began receiving patients in 1876 and by 1954 had expanded to over three hundred buildings on over two thousand acres of land.[41] Napa State soon became a front line in the battle between mental health institutions and critics of ECT, including many former patients who publicized their own stories of abuse. By the 1970s reformers had spearheaded the patients' rights movement in California and began projects of grassroots organization in the form of books, zines, and political lobbying. One the movement's most outspoken critics, Leonard Roy Frank, was incarcerated at Napa State in 1962 against his will after his parents had become alarmed at his intense interest in Orthodox Judaism. As a patient, Frank's religion and sexuality became suspect. As a means to address his double-coded deviance, Frank was drugged and subjected to eighty-five insulin and electric shock treatments. After his release Frank cofounded the Network Against Psychiatric Assault (NAPA).

The remainder of the chapter digs deep into ECT as a technology of con-

---

practices associated with it intensified in a neurobiological key. Benjamin Reiss, *Theaters of Madness: Insane Asylums and Nineteenth-Century American Culture* (Chicago: University of Chicago Press, 2008), 191–97.

40. See, e.g., Lawrence Kubie, "Some Implications for Psychoanalysis of Modern Concepts of the Organization of the Brain," *Psychoanalytic Quarterly* 22 (1953): 21–52, and W. Ross Ashby's "The Application of Cybernetics to Psychiatry," *Journal of Mental Science* 100 (1954): 114–24, in which Ashby argued for an inherent compatibility between psychiatry and the new imagination of cognitive life offered by cybernetics and information theory. See also Ashby's *Design for a Brain* (New York: John Wiley and Sons, 1952). My used copy of Ashby's text was discarded from the Juvenile Research Library at the Illinois Neuropsychiatric Institute at the University of Illinois, Chicago.

41. In 1897, Napa Asylum for the Insane officially changed its name to Napa State Hospital. Following this change, Napa State Mental Hospital sometimes appeared in popular media and court documents.

version and, quite explicitly, of sexual differentiation. It does so by revisiting the practice of Dr. H. C. Tien, the independent electroconvulsive therapist who ran the Michigan Institute of Psychosynthesis in Lansing, Michigan. Dr. Tien's practice was a hybrid of cybernetic science and the reform impulses of antipsychiatric critique. According to Tien, ECT should be wholly voluntary, part of a process by which patients would willingly participate in their own psychic rebirth. Tien promised his patients happier and more equitable marriages, more fulfilling sex lives (or just a "normal" sex life for frigid women and men who were "longstanding impotent transvestite[s]"), and parental control over sexually licentious children. Throughout the 1970s, as the industrial economy crumbled around him, Tien advocated a "systems approach" to family psychiatry. Tien's innovation was ELT, which stood for electrolytic therapy or, as Tien was quick to point out, Electric Love Therapy. ELT was a system "infused with warm human communication." It was a system that one entered into voluntarily. And it was a system in which a "person undergoes an actual electro-chemical transformation of a personality."[42]

In Tien's practice one can glimpse not only the endgame of moral treatment but also how the liberatory kernel of religion and the truth of sexual difference continued to regulate one another well into the 1970s.[43] For Tien's practice was part of an emerging paradigm in which neural nets—in their all but transcendent numeration—became not only subject to mathematical formalization but also the hinge for conversion, for inducing the freedoms of one's reason.[44] By subjecting the promises of voluntary conversion made by Tien and others to genealogical scrutiny, I am not simply interested in the gendered means and ends of that promise. I am also considering how

42. H. C. Tien, "The Psychosynthesis of Stan," *World Journal of Psychosynthesis* 2, no. 5 (May 1970): 28; H. C. Tien, "Pattern Recognition and Psychosynthesis," *American Journal of Psychotherapy* 23 (1969): 53–66; H. C. Tien, "Psychosynthesis: Telefusion of Modern Electrical Psychiatry," *World Journal of Psychosynthesis* 2, no. 5 (May 1970): 49–53; H. C. Tien, "Psychosynthesis with Television Technology," *Current Psychiatric Therapies* 15 (1975): 233–36; Tien, "100 Questions and Answers on ELT: The Electrolytic Therapy of Psychosynthesis," *World Journal of Psychosynthesis* 6 (February 1974): 37.

43. As Scott points out, sex may have long been a matter of public concern and governing rationality, but it was with new cries of sexual liberty in the postwar years that biopolitical machinations of secular governance became explicit. Congealing around a rights-based sexual subject, sexual freedom, as much as religious freedom had long been, becomes the horizon of a late secular age (*Sex and Secularism*, 148).

44. On the potential programming of individual neurons, see Dan-Anders Jirenhed et al., "Learned Response Sequences in Cerebellar Purkinje Cells," *Proceedings of the National Academy of Sciences* 114, no. 23 (2017): 6127–32.

this fierce desire for transformation operationalizes sexual difference by way of neural plasticity.

Conversion as a source of a *newfound* autonomy. The right to choose *to negate* one's less than best self—conceived of as inauthentic, unintegrated, deviant, sickly, sinful. This *longue durée* logic of conversion—from Protestant pieties to self-help guides of neural plasticity—consists of a perpetual *promise* of choice and enduring *demand* for decision. An ontological genre of carrots and sticks securing much of what passes as normal at any given moment. Conversion as the sine qua non of true religion, broadly construed. Sane sexuality as a bulwark against all that is coded as false. *You have the right to change yourself, to liberate yourself, to conquer addiction and depression, to be on the winning side, to make meaning and be your best self now.*[45] The life of this all-too-familiar subject is consumed with ever taking steps, with striving for correspondence with truths it has already deemed self-evident. Such striving, first and foremost, is imagined in this secular age to be solitary endeavor, a matter of changing one's brain in accordance with a universal and proprietary notion of cognition. Not surprisingly, perhaps, this subject leaves little room to consider the possibility that whatever is *above* may never be known *below*, as it were, let alone correspond with it.

2. MORAL TREATMENT AND HEADS THAT DIFFER IN SHAPE

As historian Jodie Boyer has demonstrated, the forerunner of the American Psychiatric Association—the Association of Medical Superintendents of American Institutions for the Insane—was filled with members who magnetized their Calvinist heritages and applied those metaphysical insights to

45. Raymond F. Paloutzian, Erica L. Swenson, and Patrick McNamara, "Religious Conversion, Spiritual Transformation, and the Neurocognition of Meaning Making," in *Where God and Science Meet: How Brain and Evolutionary Studies Alter Our Understanding of Religion*, ed. Patrick McNamara, vol. 2 (Westport, CT: Praeger, 2006), 151–204. For earlier research on pathology and religious conversion, see Kenneth Dewhurs and A. W. Beard, "Sudden Religious Conversions in Temporal Lobe Epilepsy," *British Journal of Psychiatry* 117, no. 540 (1970): 497–507; Charles Spellman et al., "Manifest Anxiety as a Contributing Factor in Religious Conversion," *Journal of Consulting and Clinical Psychology* 36, no. 2 (1971): 245–47; Jesse O. Cavenar, "Depressive Disorders and Religious Conversions," *Journal of Nervous and Mental Disease* 165, no. 3 (1977): 209–12.

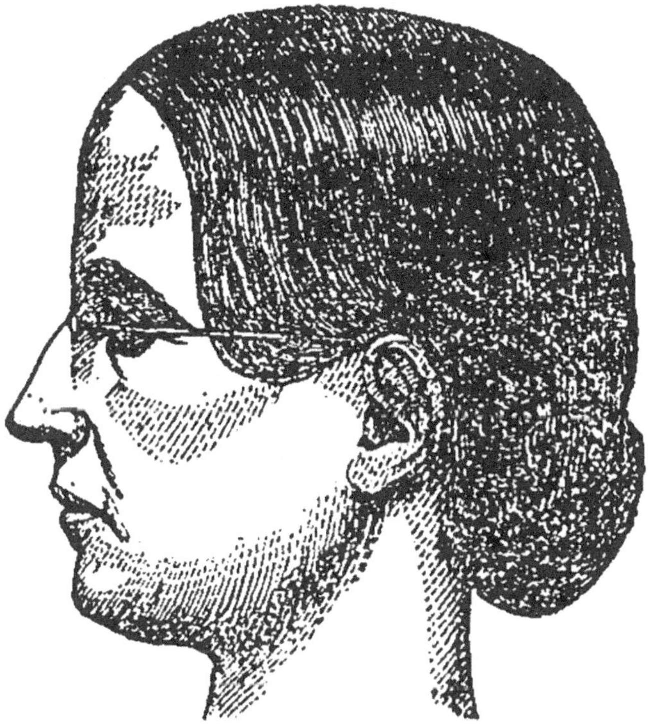

FIGURE 45. Eliza Farnham, portrait in *American Phrenological Journal* (June 1857).

their therapeutic profession.[46] Indeed, the goal of many Unitarian, Quaker, and liberal-leaning Congregationalists, whether they were involved in asylum reform, magnetic cures, or were merely practitioners of progressive politics, was to allow the world, and each individual within it, to correspond to an overarching and energetic system that may have been incomprehensible yet was known to be definitively divine, definitively true, and definitively American. *As above, so below.*[47]

The occult metaphysics undergirding this asylum reform may be glimpsed with striking clarity in the career of Eliza Farnham, matron of the women's wing of Sing Sing State Penitentiary and later at Stockton Insane

46. Jodie Elizabeth Boyer, "Sin and Sanity in Nineteenth-Century America" (PhD diss., University of Toronto, 2013).

47. For the definitive history of this theory of correspondence at the center of metaphysical piety, see Catherine L. Albanese, *A Republic of Mind and Spirit: A Cultural History of American Metaphysical Religion* (New Haven, CT: Yale University Press, 2007).

Asylum. Farnham's stated goal was to produce "a sound state of discipline" through "the systematizing of details." Farnham was an acknowledged "phrenological convert" to the reforms of "moral treatment."[48]

As such, Farnham privileged questions of how to mobilize and harness the energies pulsing to and from the mind. Through the "charming" activation of sight, smell, taste, sound, and the "sense" of democracy, wrote Farnham, "the lash may be removed as gently and effectually as the icy drapery of winter is melted away in the genial atmosphere of Spring." In hopes of cultivating both sanity and citizenship and securing them both within each inmate, Farnham organized a "natural" sensory environment in which each object and each individual was connected to every other within a closed semiotic circuit.[49]

At Sing Sing, Farnham was quite explicit about the panoptic mode of discipline she deliberately orchestrated in the women's ward. The array of women under her care would watch Farnham in patient imitation of the skill she was demonstrating (flower arranging, basket weaving, recitation, a phrenological self-examination). Farnham would watch the women demonstrating that skill. The women would watch each other watching Farnham. And so on and so forth. Drawing inspiration from the spiritualist cosmology of Andrew Jackson Davis[50] and the "spiritual side" of Emanuel Swedenborg, Farnham's was a specific strategy for "channeling" and perpetu-

48. Prison Association of New York, *Second Report of the Prison Association of New York* (New York: Prison Association of New York, 1846), 38–39. The language of conversion was a common trope among antebellum phrenologists competing for the hearts and minds of an expanding evangelical populace. "Remarkable Case of Change of Character and Pursuits, with Corresponding Change in the Form of the Head," *Phrenological Journal* 13, no. 65 (1840): 341–44; Alpha, "Phrenological Conversion," *American Phrenological Journal* 4, no. 7 (July 1842): 176–78. On George Combe's "conversion" to the truth of phrenology, see discussion of his "phrenological conversion" in George Combe and Robert Cox, *Moral and Intellectual Science: Applied to the Elevation of Society* (New York: Fowlers and Wells, 1848), 427. On the phrenological conversion narrative of Harriet Martineau, see Shalyn Claggett, "Harriet Martineau's Material Rebirth," *Victorian Literature and Culture* 38, no. 1 (2010): 53–73.

49. "Influence of Flowers," *Prisoner's Friend* (July 7, 1847), 105; "Visit of the Hutchinson Family to Sing Sing Prison," *Prisoner's Friend* (April 29, 1846), 66. Farnham, here, was seeking to displace the common disciplinary practice of flogging.

50. And he from her. See, for example, Davis's rewriting of the Bible with new chapters, including "The Gospel According to St. Ralph" and "The Gospel According to St. Eliza [Farnham]" in his *Arabula; or, The Divine Guest: Containing a New Collection of Gospels* (Boston: William White, 1867).

ating the visible power of "good resolutions" within the public setting of the ward.[51]

Like many nineteenth-century alienists and reformers in the liberal vein, Farnham sought nothing less than to refine the organization of her patients' psyche. She assumed that such refinement was not an external imposition but rather an invitation to religious freedom. Farnham, for example, believed that her success was grounded in the principle that each inmate under her care possessed the faculty of Spirituality ("discovered" by the phrenologist Orson Fowler in 1842). This moral faculty was the essence of true religion, empowering one to encounter divinity on one's own terms as it were, apart from institutions, doctrine, and external discipline. Spirituality, according to the phrenological script, was a purified mechanism of religious conversion. The moment of "phrenological conversion" was defined as "the reception of the truth by the spiritual faculty," a moment in which one was "awakened to a higher sense than intellect or reason can ever reach." Spirituality was the faculty that enabled one to experience the truth of phrenology and to become expert in shaping one's own architecture of cognition.[52]

Once activated, the faculty of Spirituality would enable a conversion from a subject defined by passion and impulse to a subject who would know how to manage those impulses and passions for the greater good.[53] An active faculty of Spirituality, in turn, activated self-control and self-organization in light of continual intercourse with "spiritual influence." This faculty, located just to either side of the crown of the head, would enable a vigilant assessment of one's interactions with others and the establishment of equilibrium between interior energies and environmental forces. Without the proper attunement of Spirituality, one's descent into madness was all but assured.[54]

51. Eliza W. Farnham, *Woman and Her Era*, vol. 2 (New York: A. J. Davis, 1864), 198–99, 113; Prison Association of New York, *Third Report of the Prison Association of New York*, pt. 2 (New York: Prison Association of New York, 1847), 56.

52. "A Change of Heart," *American Phrenological Journal* 48, no. 2 (August 1868): 74; Orson S. Fowler, *The Octagon House: A Home for All; or, A New, Cheap, Convenient, and Superior Mode of Building* (New York: Fowlers and Wells, 1848).

53. Indeed, phrenology would often compare its pedagogy to morphologies of conversion and offer itself as the superior, scientifically sanctioned, truth to which to convert. See, e.g., Nelson Sizer, "Phrenology and 'Conversion': A Letter to a Clergyman," *Phrenological Journal and Science of Health* (December 1878): 280–82.

54. On the construction of madness over and against the moral faculties, see Jan Verplaste, *Localising the Moral Sense: Neuroscience and the Search for the Cerebral Seat of Morality, 1800–1930* (Springer 2009), 191ff.

Farnham's goal, then, was to enable one to lay claim to one's own volition within a cosmos (and a social environment) of de facto relationality.

Undergirding Farnham's version of moral treatment was a fierce advocacy of woman's rights and the natural difference of sexuality.[55] According to Farnham, "the Superiority of Woman" was premised on the natural proclivity for self-organization: "Life is exalted in proportion to its Organic and Functional Complexity; Women's Organism is more Complex and her totality of Function larger than those of any other being inhabiting our earth; Therefore her position in the scale of Life is the most exalted—the Sovereign one."[56] Like complex cellular structures, women possessed the most internal differentiation and variation within their "nerve-matter" and "complexity" of "nervous structure."[57] Consequently, Farnham sought leverage on the nervous structures of her female wards in order to convert them from disequilibrium to the "true feminine type." The "Truth" of sexual difference for Farnham—as displayed in her detailed anatomical descriptions—was biological. But it was also Biblical, theological, and, most significantly, metaphysical.[58]

3. GENDERED ELECTRICITY IN THE NEUROMATIC GROOVE

Farnham resigned from Sing Sing in 1848. Afterward she made a name for herself as a spokesperson for reforms based on a common sense blend of

55. As Farnham wrote, "The feminine is the spiritual. That is its essential character. Woman is feminine in proportion as she is spiritual and the feminine side of Man is his spiritual side—his best side. An unspiritual Woman is masculine, and is felt to be therefore repulsive" (*Woman and Her Era*, 2:387).

56. Farnham, *Woman and Her Era*, 2:311; *Woman and Her Era*, 1:13, 73–74. Despite her boldness, Farnham aligned her argument about women's superiority with a conception of the sentimental nuclear family as the incubator of labor power (Scott, *Sex and Secularism*, 84).

57. Farnham, *Woman and Her Era*, 1:26–30, 72–73. Given their superior nerve endowment, women possessed more potential to "hold relations with the Objective world," to benefit from the congress between brain and environment, what she called the "me and the not me" (1:94). The superior designation did not necessarily extend to "women of the savage and barbarous races and nations" given "their departure in personal development from the true feminine type" (1:88). According to Farnham, civilized progress was premised on "reverence for Maternity"—the object lessons being American Indians and South Sea savages (89).

58. Farnham, *Woman and Her Era*, 1:88, 11.

phrenological materialism and a refined spiritualist cosmology. She traveled to California in 1849, encouraging women's rights advocates from the East Coast to migrate west. In 1861 Farnham was recruited by the State of California to be the matron of the Stockton Insane Asylum. Farnham's arrival at Stockton was heralded as a much-needed jolt of the management energies of East Coast liberals: as the *San Francisco Times* reported,

> A GOOD APPOINTMENT. Mrs. Eliza W. Farnham [is] a woman of mind, of thought, of human sympathies. . . . She has experience as Matron of Sing Sing, and from her experiences and observation, wrote and published an interesting book on the subject. We would give great odds in a bet, if making one, that she will do more, if allowed to exercise her knowledge and experience, towards curing the insane, than all the doctors and political attachés that have ever belonged to the hospital.[59]

Given the directive at Stockton that "in the management of patients, unvarying kindness must be strictly observed by all," Farnham was an ideal choice. At Stockton, Farnham's so-called Science of Humanity would leverage the systematic relationality between individuals and their surrounding environment.[60] As she declared, "organization as a means, not an end: as the clothing which life puts on that it may have adequate expression in a material world." No detail was too small. Farnham's first project was to build a small library with curative intent. She then obtained a sewing machine in order to establish an atmosphere of precision acoustics as well as a work routine for patients. Their assignments of making dresses, sheets, aprons, and American flags possessed a semiotic agenda on Farnham's part—the coordination of an active domestic charge with patriotic affect. According to Farnham, such systematic "organization" rehearsed "the primitive language of Deity in the world of life." For in the asylum, "the methods of the scientists and spiritualists will be at one."[61]

59. *San Francisco Times*, cited in Jo Ann Levy, *Unsettling the West: Eliza Farnham and Georgiana Bruce Kirby in Frontier California* (Santa Clara, CA: Santa Clara University Press, 2004), 203.

60. As the resident physician at Stockton confidently wrote, "she shall direct the employment and amusements of all the inmates of the female wards; in short, it will be expected of her to look frequently and carefully into every interest connected with her department." *Rules and Regulations of the Insane Asylum of California, Prescribed by the Resident Physician, August, 1861* (Stockton: Armor and Clayes, 1861), 6, 10.

61. Farnham, *Woman and Her Era*, 1:18, 36–37, 22–23; Levy, *Unsettling the West*, 207.

Phrenology, for Farnham, was the perfect blend of spiritualism and science. The phrenological worldview of animal magnetism coursing in and through the individual inspired Farnham to theorize a new form of treatment for insanity and "all mental conditions in which the mind is deprived of the power of true moral perception." At Stockton her focus was on the looping relationship between the mind and the environment of the women she served. "We may calculate with much exactness," surmised Farnham, "what effect certain causes will produce on [a patient's] faculties, and what, in turn, will be the consequences of these acting outwardly upon surrounding objects."[62]

Farnham's emphasis on affective attachments to the world as privileged access points, as the hinges between brain and environment, lived on in different parts of the moral treatment movement. In 1881, for example, a doctor at the Utica Insane Asylum provided a diagram of the "central inward psycho-physical tracts and centers of the brain," how they related to one another, and how their organization was determined by incoming stimuli from the environment.

Here was a tentative blueprint for moral reform of the patient that imagined a nervous ecology in which faculties of mind, neural physiology, and sensory impressions interacted systematically. As the doctor surmised, the diagram was "useful, for our main purpose, to render possible an analytical and synthetical conception of the nervous organism, which is concerned in the manifestations of mental life, and to bring the phenomenon of so-called mental alienation, or aberration of mind, at least for a part, within the reach of physical explanation."[63]

Farnham, too, was fond of electrical and galvanic metaphors when describing affective attachments to the world. Although she herself did not apply electricity to the brains of her patients, Farnham's metaphysical position would assume a more intense charge as electricity was increasingly deployed by alienists in the second half of the nineteenth century.[64]

62. M. B. Sampson, *Rationale of Crime, and Its Appropriate Treatment; Being a Treatise on Criminal Jurisprudence Considered in Relation to Cerebral Organization with Notes and Illustrations by E. W. Farnham* (New York: D. Appleton, 1846), xiv–xvii.

63. Theodore Deecke, "The Condition of the Brain in Insanity," *American Journal of Insanity* 37 (April 1881): 376, 381.

64. Eliza W. Farnham, *My Early Days* (New York: Thatcher and Hutchison, 1859), 45, 89, 110, 138, 262; Eliza W. Farnham, *The Ideal Attained* (New York: C. M. Plumbe, 1865), 161, 223; Edward Stainbrook, "The Use of Electricity in Psychiatric Treatment during the Nineteenth Century," *Bulletin of the History of Medicine* 22, no. 2 (1948): 156–77.

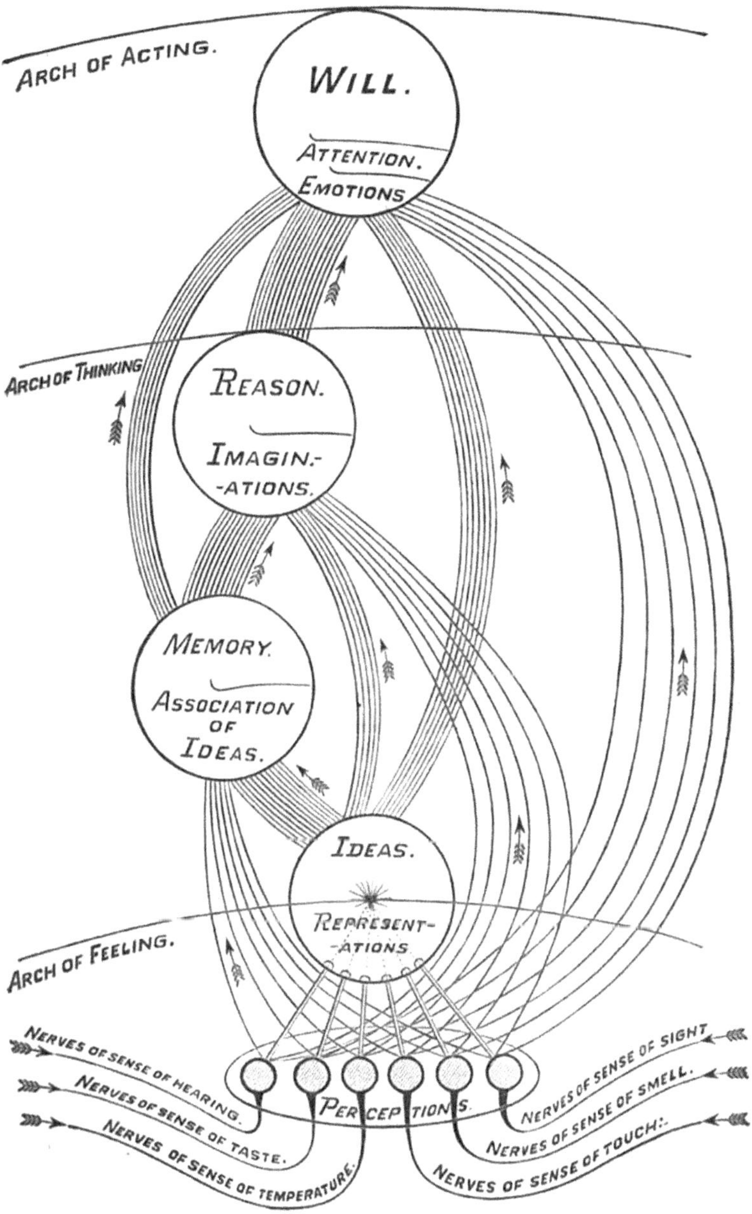

FIGURE 46. Nervous schematic for moral treatment, from Theodore Deecke, "The Condition of the Brain in Insanity," *American Journal of Insanity* 37 (April 1881): 377.

Although moral and medical therapy possessed different lineages and emphases, the distinction between them was often nominal as both forms of treatment were practiced simultaneously in nineteenth-century asylums. Indeed, in situations in which "subtle energies" were thought to pervade the living world, the application of electricity to bodily surfaces became a logical (and visceral) means of manipulating the environment of the patient.[65]

Among advocates of moral treatment in the nineteenth century, Farnham was not exceptional in her consideration of religion and its truth to be intimately bound up with the truth of sexual difference. Nowhere is this more evident than in the diagnosis of religious insanity as a hysterical condition. In the nineteenth century, the gendered edge of mental health—and its decidedly conservative politics, often more so than Farnham's—culminated in galvanic treatments for hysteria,[66] ovary compressors, and the general regulation of womanly disorder—racially coded, of course, as irrational manifestations of bad belief.[67] By the late nineteenth century, hysteria had

65. In discussing "The Electro-Neural Pathology of Insanity" in 1878, for example, alienist A. H. Newth described "infinitesimal particles, acting on the nervous system" as "capable of producing very decided effects." In concluding that the "disarrangement of the nerve molecules" cause "disturbed mental activity," Newth identifies a "neural force" which can be manipulated by "some variation in the surroundings of the nerves" (*Journal of Medical Science* 24, no. 105 (1878): 76, 78. See also George M. Beard, MD, "The Treatment of Insanity by Electricity," *Journal of Medical Science* 19, no. 87 (1873): 355–60; Joseph Wiglesworth, MD, "On the Use of Galvanism in the Treatment of Certain Forms of Insanity," *Journal of Mental Science* 33, no. 143 (1887): 385–95.

66. Take, for example, an early nineteenth-century discussion of medical electricity in which a young woman is given a galvanic bath to overcome her domestic failures (T. Gale, *Electricity, Or Ethereal Fire, Considered: 1st. Naturally, as the Agent of Animal and Vegetable Life: 2d. Astronomically, Or as the Agent of Gravitation and Motion: 3d. Medically, Or Its Artificial Use in Diseases: Comprehending Both the Theory and Practice of Medical Electricity: and Demonstrated to be an Infallible Cure of Fever, Inflammation, and Many Other Diseases: Constituting the Best Family Physician Ever Extant* (Troy, NY: Moffit and Lyon, 1802), 126–27.

67. On early iterations of electrotherapeutics and sexual difference, see the physician William F. Channing's *Notes on the Medical Application of Electricity*, 5th and enl. ed. (Boston: Thomas Hall, 1859), 220. Channing (son of William Ellery Channing, the so-called apostle of Unitarianism) reported good results from using battery currents to treat hysteria. William F. would go on to work with Alexander Graham Bell on the development of the telephone. See also the work of Roberts Bartholow, an American physician inspired by the work of Hitzig and Fritsch, whose electrical experiments on a female patient suffering from a brain ulcer were recorded in 1874. "Experimental Investigations into the Functions of the Human Brain," *American Journal of the Medical Sciences* 67 (1874): 305–13. Bartholow was an army surgeon in the Utah War of 1857–58. In observing practices of Mormon polygamy,

become the quintessential condition of breaking bad both religiously and sexually—ever tinged with anti-Catholic tropes of religious excess and sexual frustration.[68] Indeed, the expansion of the California asylum system coincided with the admission in the West of a disproportional number of cases of insanity "attributed to undue religious interest."[69]

In the late nineteenth century the observation that "undue religious excitement produces or helps produce insanity" was consistently directed toward women and made in the service of explaining the propriety of sexual difference.[70] Bad religion remained bound up with repressed sexual desire given the shared psychic mechanics between proper sexuality and true religious feeling. "Repressed feeling," reported the *Boston Medical and Surgery Journal* in 1877, "tends to unhealthy and exaggerated religious feeling. This is particularly true of the sexual instincts when unsatisfied. There is, therefore, at all times a large number of persons ready to give themselves up to

---

Bartholow aligned sexual difference with religious difference and concluded that the moral depravity of the Mormons manifested in neurological degeneracy. See Roberts Bartholow, "The Physiological Aspects of Mormonism and the Climatology and Diseases of Utah and New Mexico," *Cincinnati Lancet and Observer* 10:4 (April 1867), 193–205.

68. When men were diagnosed with hysteria they exhibited "wild" talk about "religious matters" or an excessive concern "with the state of his neighbors' souls" (J. Michell Clarke, "On Three Cases of Hysteria in Men," *Brain: A Journal of Neurology* 14, no. 4 [1891]: 529). On the grammars of religious possession that were deployed in the making of this diagnostic, see Elizabeth Lunbeck, *The Psychiatric Persuasion: Knowledge, Gender, and Power in Modern America* (Princeton: Princeton University Press, 1994), 209–28. It should not be surprising, then, that by the 1890s paranoia was achieving its gendered distinction alongside hysteria. "Paranoia" was a form of religious debilitation among an otherwise "well, strong man," reported a doctor in 1899. "Mental derangement" occurred, in part, because the faculty of reason had become overactive. Consequently, paranoia happened most often in men whose "fixed ideas of various kinds grow to be real and the individual is possessed of all kinds of insane notions" (F. P. Warner, "Paranoia," *Eclectic Medical Journal* 59, no. 2 [February 1899]: 89).

69. Theodore W. Fisher, MD, "Insanity and the Revival," *Boston Medical and Surgical Journal* 97, no. 3 (July 19, 1877): 62.

70. William Sims Bainbridge, "Religious Insanity in America: The Official Nineteenth-Century Theory," *Sociological Analysis* 45, no. 3 (Autumn 1984): 223–40. Religious insanity, it should be noted, could also be used to throw shade at evangelicals—"religious worship . . . carried to an unreasonable extent." Amariah Brigham, *Observations on the Influence of Religion upon the Health and Physical Welfare of Mankind* (Boston: Marsh, Capon and Lyon, 1835), 331. On the alignment of insanity with bad religion in a decidedly metaphysical key, see T. Romeyn Beck, "Swedenborg on Insanity," *American Journal of Insanity* 4 (1848): 61–64.

FIGURE 47. Napa State Hospital, from *Appendix to the Journals of the Senate and Assembly of the Session of the Legislature of the State of California* (1903).

emotional excess when encouraged by the example of others or upheld by popular sentiment." The religiously insane, in other words, lacked control and a stable identity. They were "childish, weak-minded," "rebellious under family restraint, [and] unmanageable."[71]

4. THE OPERATIONALIZATION OF NAPA STATE HOSPITAL

As a recent mission statement of Napa State Hospital reads, "Our current treatment philosophies, rooted in the moral treatment concept of the late 1800's, have reflected the numerous prevailing practices over the decades. We are moving toward the next century with a bio-psychosocial rehabilitation philosophy for the patientized treatment of patients."[72] This seamless vision of "patientizing patients" is a consummation of sorts, a harkening

71. Bad religion could also cut the other direction as an excess of reason—"undue scrupulosity and excessive conscientiousness" (Fisher, "Insanity and the Revival," 59, 61).

72. Napa State Hospital Mission Statement, https://www.dsh.ca.gov/Napa/. See also https://web.archive.org/web/20130625011022/https://www.dsh.ca.gov/Napa/.

back to the strategies of California reformers who, in the late nineteenth century, sought to model their own asylums on East Coast practices of liberal benevolence.[73]

Napa Asylum for the Insane opened in 1876 and was modeled, in part, on the cosmic cut of Farnham's systematicities and reform blueprint at Stockton. From the beginning, Napa Asylum pursued a systematic science that would overcome not only the excesses of religious piety among patients but also those excesses that hobbled public opinion: The "spirit of progress" was soon to overcome the "almost universal opinion . . . that insanity was a state of demoniacal possession."[74] In the decades that followed, however, California reformers remained frustrated with the state's framing of asylums as primarily custodial rather than therapeutic.

Glitches and outright failures were routine. By 1891, Napa Asylum housed twice the number of patients as its intended capacity. But such irregularities also became routinized as opportunities for the reform of treatment within Napa Asylum. In that same year the state superintendent of asylums reported that "Everywhere there is an upward tendency toward improved comforts for . . . those housed in asylums, and at the same time the distrust with which the public has heretofore viewed asylum management is fast passing away." As a result of this newfound confidence, Napa trustees were able to wage a successful political campaign for more capacious standards regarding asylum admissions.[75]

Over the course of the twentieth century, Napa trustees made much progress in their drive to systematize therapeutic treatment.[76] Their efforts were part of the diffusion of the neuromatic paradigm in the form of OR and what

73. In its early years Napa Asylum became infamous for having instituted an elaborate system of occupational therapy in which patients were responsible for farming and facility maintenance. With the legacy of "moral treatment" still in full effect, patients sewed clothing, made baskets and mats, and forged brass and copper items either to be used by the asylum or sold (Warren, *Madwives*, 100–101).

74. *Thirty-Ninth and Fortieth Annual Reports of the Superintendent of the Insane Asylum of the State of California (at Stockton), for the Two Years Ending June 30, 1892* (Sacramento: State Office, 1892), 27.

75. *Thirty-Ninth and Fortieth Annual Reports*, 24, 27; Patricia Prestinary, *Napa State Hospital* (Charleston, SC: Arcadia Publishing, 2014), 24. For an overview of California reformers, see Richard W. Fox, *So Far Disordered in Mind: Insanity in California 1870–1930* (Berkeley: University of California Press, 1978).

76. Beginning at midcentury, for example, there were consistent efforts at Napa State to institute "quantitative assessments of its treatment staffing requirements." In 1947, fixed staff-patient ratios were developed for nursing staff and then, in 1952, for professional staff.

Dr. Mango J. Ortega, chief of professional education at Napa State from 1964 to 1982, called "the wider context of systems theory."[77] Over the span of Ortega's career at Napa State, the desire to encompass mental health by way of modeling became voracious. A salvific mathematics, emerging as it did from investigations into the neuromatic at midcentury, would soon be applied to all levels of mental health—to outpatient services, patient queues, drug protocols, and, eventually, to the object of neural activity that had inspired the initial investigations. Such systematicity would ensure equal treatment for everyone according to a standard procedure.[78]

By the 1970s the target of operational research was drifting ever inward. Researchers began making "mathematical models of diagnostic reasoning process[es] . . . setting up measures of effectiveness of clinical activity, and . . . incorporating measures of social or pathological urgency into the computation of optimum theater lists." Moving ever away from a bricks-and-mortar focus to the flow of information in the planning and design of a hospital,[79]

By 1965 these efforts had become largely computerized in the form of a new staffing system known as SCOPE (Staffing Care of Patients Effectively).

77. M. J. Ortega, "Controversy over Normalcy," *Napa Quarterly* (Spring 1971), n.p.

78. On the need for "collaboration between clinical and systems analysis staff," see J. W. Messick et al., "A Systems Analysis Approach to Planned Change in a Psychiatric Program," *Journal of Psychiatric Nursing and Mental Health Services* 13, no. 4 (June 1975): 7–11. See also Ruth Davies et al., "Planning Patient Care with a Markov Model," *Operational Research Quarterly* 26, no. 3 (1975): 599–607; Duncan Boldy, "A Review of the Application of Mathematical Programming to Tactical and Strategic Health and Social Services Problems," *Operational Research Quarterly* 27, no. 2 (1976): 439–48.

79. Jordan J. Baruch, "Hospital Automation via Computer Time-Sharing," in *Computers in Biomedical Research*, vol. 2, ed. Ralph W. Stacy and Bruce D. Waxman (New York: Academic Press, 1965), 291–312, and Harvey Dingman, "Computer Analysis of Psychological and Psychiatric Data," also in *Computers in Biomedical Research*, 2:331–50; "Applying Computer Procedures to Hospital Psychiatry," *Perspectives in Psychiatric Care* 4, no. 6 (1966): 21–30; Norman T. J. Bailey, "Operational Research in Hospital Planning and Design," *Operational Research Society* 8, no. 3 (September 1957): 149–57; William J. Horvath, "British Experience with Operations Research in the Health Services," *Journal of Chronic Disease* 17 (1964): 779–88; James C. Hodges, "Operations Research Programs in the Mental Hospital," *Psychiatric Services* 16, no. 1 (1965): 69–71. See also "The Systems Approach to Hospitals"—a special conference issue of *Operations Research Quarterly* 22 (1971) in which a hospital is defined as "a system for treating patients" and then parsed for treating subsystems. In this scheme, the hospital is "a physical system of buildings; it is a system of many interacting staff; it comprises a complex logistics system; it is a system for treating patients; it is an information system" (39).

mental health professionals began to seek systematic management at the level of not just structural conditions but also at the level of the interview, the malady, the brain.[80] At Napa State, psychoanalysis was becoming a biological science by applying the "precision of physics" in order to make its "basic concepts less metaphysical and more biological."[81]

With the intensification of the desire for precise measurement came the drift of systematic attention from diagnostic structures to the symptoms of patients. Computers aided in the use of quantitative analysis, leading researchers to insist that theories of pathogenesis could be confirmed by "principles of testability and verification."[82] Treatments that had been around

80. Richard E, Bellman et al., *A Simulation of the Initial Psychiatric Interview* (Santa Monica, CA: Rand Corporation, 1966) [R-449 RC]. See also Richard E. Bellman et al., "Computer Vignettes: A Research Tool for the Study of the Initial Psychotherapeutic Interview," *Journal of Cybernetics* 1, no. 1 (1971): 19–27. On the desire to arrive at a "systems theory that specifies causal linkages between social structure, alienation, and perceptual style" in order to intervene at the level of personal "stress reactions," see Ronald W. Manderscheid et al., "Alienation: A Response to Stress," *Journal of Cybernetics* 5, no. 1 (1975): 91–105.

81. Ashby, "Application of Cybernetics to Psychiatry," 122. See also Donald O. Hebb, "Alice in Wonderland: or, Psychology among the Biological Sciences," in *Biological and Biochemical Bases of Behavior*, ed. H. F. Harlow and C. N. Woolsey (Madison: University of Wisconsin Press, 1958), 451–66. Such reform was most dramatically on display in the practice of lobotomy. Walter Freeman, for example, a leading psychosurgeon at midcentury, spent time at Napa State training attendees and perfecting transorbital lobotomies that could be performed as outpatient procedures. In this procedure, Freeman entered the prefrontal area through the patient's eye sockets, using an instrument that resembled an ice pick. Out of the thousands of lobotomies Freeman performed, it has been reported that up to 40 percent of them were in cases of diagnosed sexual deviance. Elliot S. Vallenstein, "Therapeutic Exuberance: A Double-Edged Sword," in *So Human a Brain: Knowledge and Value in the Neurosciences*, ed. Anne Harrington (Boston: Birkhauser, 1992), 161. In 1957 the psychiatrist William Sargent drew from the writings of John Wesley, the founder of Methodism, in arguing that psychosurgery was a valid means of conversion, psychic and religious: "Genuine religious conversions are also seen after the new modified lobotomy operations. For the mind is freed from its old straight-jacket and new religious beliefs and attitudes can now more easily take the place of the old." William Sargent, *Battle for the Mind: A Physiology of Conversion and Brain-Washing* (New York: Doubleday, 1957), 71–72.

82. Psychiatrist Robert Spitzer, who in 1974 assumed the editorial reins of the third edition of the *Diagnostic and Statistical Manual of Mental Disorders*, argued that every disorder should be identified by criteria accessible to empirical observation and measurement. Indeed, the publication in 1980 of Spitzer's DSM-III was but one (quite visible) consummation of the operations approach to mental health that, among other things, would open up lines of communication between patient, clinician, and laboratory. Hannah S. Decker,

for decades—electroshock, psychosurgery, and behavioral modification—became ever more exacting in their evaluation of inputs and outputs.[83] They were ever more integrated into a larger systems approach.

At Napa State in the 1960s and 1970s, systems theory took hold as a general means of treating mental illness. In the glow of this cybernetic optimism, one resident physician at Napa State reported in 1971 that "one day while immersed in Communications Theory and its application to the treatment of what we call mental illness, it occurred to me that the paranoid people I had been seeing for so many months might be trying to communicate by means of their symptoms in the same way we have come to recognize other conditions (notably the psycho-somatic manifestations and hysterias) as an attempt at communication."[84] But such "patientizing of patients" was neither seamless nor without sexual charge and the human vicissitudes that accompany it.

In adopting a systems approach to neurobiology, the concept of sexual difference was assuming a different scientific explanation within the institution, one that conformed to more progressive attitudes outside of it. "Biosocial tests for normalcy" were still common at Napa State, but they were preferable to purely psychological criteria because the latter tended to reinforce racial and gendered stereotypes. Here was a liberalism armed with the authority of Big Science. "If I were to define normalcy in the wider context of systems theory," wrote Dr. Ortega, "I would say that normalcy is synonymous with interface flexibility, the ability to modify habits when moving out of one's charmed circle, and to bridge the gap between one's own

---

The Making of DSM-III®: A Diagnostic Manual's Conquest of American Psychiatry (New York: Oxford University Press, 2013); Rick Mayes and Allan V. Horowitz, "DSM-III and the Revolution in the Classification of Mental Illness," *Journal of the History of the Behavioral Sciences* 41, no. 3 (2005): 249–67. With the authority of neuroscience behind it, the DSM-III sought diagnostic validity by getting underneath the symptom (the "underneath" being not the patient biography or context per se but rather the systemic flaw that produced the failure). Indeed, validity was a buzzword of systems-minded therapists in the 1970s: when a disorder had "validity," that signified that its classification had an intrinsic harmony. It was a system in the cybernetic sense. To see a disorder in that way was to jettison the meaning of the disorder or even how it really operates and, instead, to privilege one's capacity to intervene at the level of environmental input in order to achieve the desired output.

83. See, e.g., Norman L. Wulfsohn and Anthony Sances Jr., eds., *The Nervous System and Electrical Currents* (New York: Plenum Press, 1971).

84. Stephen A. Shepard, "A New Look at Paranoid States," *Napa Quarterly* (Spring 1971), n.p.

group and another, irrespective of differences of age, sex, color, or status, social or professional, etc."[85]

Normalcy, defined as the capacity to interface and self-organize, was relevant to doctor and patient alike in a well-oiled system devoid of bias. For Dr. Ortega also pointed out the "normalcy" of his management style at Napa State. As Ortega recalled, his unlikely success in managing this large system began in 1957 when he had assumed the leadership of a seven-hundred-bed psychiatric hospital in Norman, Oklahoma, with a limited staff and resources: "To my delight," wrote Ortega, "the lubricant which kept the few wheels running smoothly was primitive lust disguised as loving concern for each other between medical and nursing staff. The usefulness of the libido in promoting administrative harmony is still unexplored, but not unexperienced. (If you have a big job to do, surround yourself with lustful people, like Henry Kissinger)."[86]

5. PATIENTS' RIGHTS

At Napa State, ECT was the preferred technique for treating female patients and restoring the propriety of sexual difference as it was conservatively defined at midcentury. Many patients there were committed to the asylum by their husbands without their consent, often with the purpose of healing their marital relationships. As Carol A. B. Warren writes in her historical ethnography of Napa State in the 1950s, "Husbands often wished to have their wives forget the emotional troubles, including marital strife, that had precipitated hospitalization." As it was noted by one doctor at Napa State

85. Ortega, "Controversy over Normalcy," n.p.

86. Mango J. Ortega, "A Psychiatric Odyssey of Personal Discoveries (Things My Patients Taught Me)," *Napa Quarterly* (Winter 1972–73). Ortega's pose as predatory engineer was not simply a matter of sexual harassment (although his career at Napa State was terminated because of these legal charges in 1982) but also speaks to his systematic approach to sexual differentiation and his faith in sexual difference as political prerogative. Again, Ortega: "We need to learn to annihilate our enemies by making friends with them like Nixon did. Until the 1970s the foreign policy of powerful nations was divide and conquer. Our Henry tries to unite nations at war so the world will rest in peace. This is revolutionary diplomacy" ("Dialogue on Violence between the Sexes," *Napa Quarterly* [Spring 1975]: 35–36).

regarding his patient, Eve Low, "She went on to say that she'd been getting shock, though against her wishes, and that she feels its purpose is to make her forget things, and to change her attitude, including her resentment at her husband for committing her." For women like Low, ECT served not only to erase traumatic memories of a failing marriage but was also part of an elaborate system of coercion which, in the act of erasure, resolved the system of any responsibility.[87] Yet doctors and administrators across the board insisted that ECT was a voluntary process that served the long-term well-being of the patient.[88]

Things, however, soon began to change. In the wake of Foucault's *Madness and Civilization* (1961), Thomas Szasz's *The Myth of Mental Illness* (1961), Erving Goffman's *Asylums* (1961), and Sylvia Plath's *The Bell Jar* (1963), a gendered critique of ECT as a tool of illiberal coercion forcefully emerged in the form of a burgeoning patients' rights movement.[89] According to Judi

87. Warren, *Madwives*, 106, 129–30, 140, 133. Judy, a former patient and psychiatric nurse, witnessed such coercion firsthand. "As to signing consents, I have seen patients 'assisted' in signing the forms by having their hand held by a staff person. I have seen patients threatened with long term confinement if they don't submit—family members and friends are used to bring pressure." *Women against Electric Shock Treatment* (Oakland, CA: Women's Press Collective, 1975).

88. As patients' rights advocate Ollie Mae Bozarth wrote, "Many husbands beat up their wives. . . . Other husbands just sign consent for the 'medical treatment' called shock, and let the experts do it for them. . . . What a convenient way to control housewives who don't live up to the expectations of their husbands." Ollie Mae Bozarth, "Shock: The Gentleman's Way to Beat Up a Woman," *Madness Network News* 3, no. 6 (June 1976): 27. Such control was, of course, premised on the correction of what amounted to bad religion. See, e.g., John Stoudenmire, "The Role of Religion in the Depressed Housewife," *Journal of Religion and Health* 15, no. 1 (1976): 62–67.

89. Precursors also included skeptical takes on ECT from within the medical establishment. See, e.g., N. K. Rickles and Charles G. Polan, "Causes of Failure in Treatment with Electric Shock: Analysis of Thirty-Eight Cases," *Archives of Neurology and Psychiatry* 59 (1948): 337–46, and Steven G. Goldstein, Susan B. Filskov, et al, "Neurophysiological Effects of Electroconvulsive Therapy," *Journal of Clinical Psychology* 33, no. 3 (July 1977): 798–806. In 1976 neurologist John Friedberg published *Shock Treatment Is Not Good for Your Brain: A Neurologist Challenges the Psychiatric Myth* (San Francisco: Glide Publications). At the 129th annual meeting of the APA in Miami Beach in 1976, Friedberg claimed that "From a neurological point of view ECT is a method of producing amnesia by selectively damaging the temporal and the structures within them. . . . ECT produces a form of brain disease." John M. Friedberg, "Shock Treatment, Brain Damage, and Memory Loss: A Neurological Perspective," *American Journal of Psychiatry* 134, no. 9 (September 1977): 1010–13.

Chamberlin, one of the patients' rights movement's most outspoken critics, there was an urgent need to protect the "self-determination" of women and to reveal "the mental health system [to be] a monster that damages, degrades, and often destroys those people it claims to help."[90] By the 1970s much antipsychiatry critique centered around firsthand accounts of sex-based discrimination in the mental institutionalization of women.[91] As current and former mental patients began grassroots projects—book publishing, community organizing, and political lobbying—gender emerged as a central analytic for understanding the "therapeutic state" and the "drift toward social control through medical technology."[92]

Women against Electric Shock Treatment, for example, a Bay area pamphlet that was distributed in the early 1970s, castigated ECT as a technology of the "white and rich" and sharply suggested that it was "Third World, poor and older women who are major victims of electric shock treatment and the psychiatric establishment." Aligned with flourishing second-wave feminism, the pamphlet then positioned ECT as a threat to women's self-determination and an impingement on the actualization of true femininity. As one of its anonymous authors declared, "Women are the predominant victims of shock treatment. Because of the different standards for women and men in this society, EST is considered a success when a woman can function in the service of her family again even though much of her potential productivity and creativity has been shocked out of her."[93]

A striking feature of the patients' rights movement was its constant conjuring of nineteenth-century reform movements. In the fall of 1978, an article appeared in *Madness Network News* (a political zine based in San Francisco) entitled "Elizabeth Stone: 'The Power of Insane Hospitals Must Be Shaken.'" In it the author, Tanya Temkin, recounted Stone's 1842 story of persecution

90. Judi Chamberlin, *On Our Own: Patient-Controlled Alternatives to the Mental Health System* (New York: Hawthorn Books, 1978).

91. Phyllis Chesler, *Women and Madness* (New York: Doubleday, 1972). Robert T. Roth and Judith Lerner, "Sex-Based Discrimination in the Mental Institutionalization of Women," *California Law Review* 62, no. 3 (1974): 789–815. For an excellent discussion of this dimension of the patients' rights movement, see Michael Staub, *Madness Is Civilization: When the Diagnosis Was Social, 1948–1980* (Chicago: University of Chicago Press, 2011), 139ff.

92. Lee Coleman, "The Doctor as Social Healer," *Madness Network News* 2, no. 2 (February 1974), 14.

93. *Women against Electric Shock Treatment.*

FIGURE 48. Cover of *Women against Electric Shock Treatment* (Oakland, CA: Women's Press Collective, 1975).

and her commitment to Charlestown McLean Asylum against her will. In her memoir, Stone thematized an opposition between the systematic administration of drugs and spirituality, writing that "the unconverted do not understand *spirituality*" and the fact that "Doctors possess knowledge of giving medicine to take away from a person the spirit of Christ." Temkin echoed Stone's understanding of psychiatric power and cited Stone's insistence that "This power of calling people crazy has got to be stamped with God's eternal vengeance." The eternal vengeance of God, it seemed, was the only antidote to a therapeutic state that was rooted in the nineteenth century but had moved well beyond the regime of laxatives, emetics, and sedatives prescribed to Stone.[94] Patients' rights advocates understood themselves as heirs to a legacy of resistance to the voracious rationalities of science, industry, and bad religion. Or, as the tag line from *Madness Network News* put it, "THE BEHAVIOR OF THE FULLY HUMAN BEING IS ALWAYS UNPREDICTABLE SIMPLY BECAUSE IT IS FREE."

Not much had changed in over a century. Spirituality—the marker of freedom and its interiority—remained under assault from coercive psychiatry. The diagnosis of insanity or depression still served those invested, at some angle, with differentiating between "good" religion and "bad."[95] As Szasz declared, the situation was yet another instantiation of a central crisis of modernity, that is, the liberal spirit of Protestant progress struggling to overcome the forces of clerical abuse, broadly construed. Psychiatric medicine, it was argued, "now functions as a state religion much as, for example, Roman Catholicism did in medieval Spain." Such heated and seemingly exaggerated rhetoric conjured the specters of anti-Catholicism and anti-Semitism (so crucial to nineteenth-century reform movements) and the enduring appeal of Protestant genres of freedom. "The ideology that now threatens individual liberties is not religious but medical," wrote Szasz in 1970. "The individual needs protection not from priests but from physicians." Szasz here offered historical purchase for many individuals who had experienced involuntary shock treatments as an infringement on their freedom of choice. As Szasz continued, "The justification now for a separa-

94. Tanya Temkin, "Elizabeth Stone: 'The Power of Insane Hospitals Must Be Shaken,'" *Madness Network News* 5, no. 2 (Autumn 1978): 18–19.

95. Nobuko Udea, "Changes in Religious Concepts of Depressive Psychosis," *Clinical Psychiatry* 17, no. 10 (1975): 1039–46; Stoudenmire, "Role of Religion," 62–67.

FIGURE 49. Cartoon, from *Madness Network News* 5, no. 5 (Summer/Fall 1979): 18.

tion of Medicine and State is similar to that which obtained formerly for a separation of Church and State."[96]

96. Thomas Szasz, *The Theology of Medicine: The Political-Philosophical Foundations of Medical Ethics* (Baton Rouge: Louisiana State University Press, 1977), 146; Thomas Szasz, *The Manufacture of Madness: A Comparative Study of the Inquisition and the Mental Health Movement* (New York: Harper and Row, 1970), 178. It should be noted that Szasz, here, had already hitched his critique of the psychiatric establishment to Scientology and to their ongoing legal battles over the constitutional guarantee of religious freedom. In 1969, Szasz and the Church of Scientology cofounded the Citizens Commission on Human Rights

6. THE SHAVING OF LEONARD FRANK'S BEARD

Critics in the antipsychiatry movement were inclined to emphasize the illiberal logic of ECT. In condemnations of ECT's potential to coerce, critics accused the psychiatric establishment of bad faith—violating, in the name of a false God, the human rights of patients.[97] Often framed as spiritual biographies, such critiques focused on coercive tactics of electric shock therapists and the fact that so many patients were administered ECT against their will. Leonard Roy Frank, for example, wrote of the way in which the biopolitics of sex and religion converged during his confinement at Napa State. Frank consistently adopted the language of religious critique to talk about his experience and framed ECT as "a crime against the spirit."[98] Indeed, the antipsychiatry movement revolved around testimonials such as Frank's. ECT, in their witnessing, was a sin against liberalism, a threat to its ideals, and a danger to any person seeking to determine their own course, religiously and/or sexually—*within reason*.[99]

Frank was cofounder of the Network Against Psychiatric Assault (or

(CCHR) to oppose involuntary psychiatric treatments. In his original thesis on Dianetics, L. Ron Hubbard had cast suspicion on ECT and suggested that such treatment would only traumatize rather than heal, serving to embed "institutional engrams" ever deeper into the psyche and so making them harder to remove through auditing (*Dianetics: The Modern Science of Mental Health* [New York: Hermitage House, 1950], 3–4, 97–98, 151, 318, 383). On the epistemic negotiations between Hubbard and Szasz, see Donald A. Westbrook, "'The Enemy of My Enemy Is My Friend': Thomas Szasz, the Citizens Commission on Human Rights, and Scientology's Anti-Psychiatric Theology," *Nova Religio: The Journal of Alternative and Emergent Religions* 20, no. 4 (2017): 37–61.

97. This particular iteration of antipsychiatry would gain popular traction in such works as Eliot S. Valenstein, *Brain Control: A Critical Examination of Brain Stimulation and Psychosurgery* (New York: John Wiley and Sons, 1973); Robert L. Schwitzgebel and Ralph K. Schwitzgebel, eds., *Psychotechnology: Electronic Control of Mind and Behavior* (New York: Holt, Rinehart, and Winston, 1973) and in *One Flew Over the Cuckoo's Nest* (the 1962 novel by Ken Kesey that was made into an Academy Award–winning movie in 1975, directed by Miloš Forman).

98. Leonard Roy Frank, "Electroshock: A Crime against the Spirit," *Ethical Human Sciences and Services: An International Journal of Critical Inquiry* (Spring 2002): 63–71. See also his "Electroshock: Death, Brain Damage, Memory Loss, and Brainwashing," *Journal of Mind and Behavior* 11 (1990): 489–512.

99. For Frank's musings on Joseph Campbell and the "path of self-realization," see his "The Journey of Transformation," *Street Spirit* (April 2008). For an account of religion as the

NAPA) and a staff writer for *Madness Network News*. In the early 1960s Frank had been living in New York and had decided to educate himself—"to do some serious reading of literature, history, politics, religion and psychology. And to see my society, and the philosophy that governed it, with new eyes. I had become very critical of the society and I felt it was very important to acquire the knowledge to deal with my criticism in a creative way."[100] Frank read widely and grew a beard as a sign of his burgeoning commitment to the rabbinic rigor of Orthodox Judaism. Having become alarmed at Frank's intense and newfound interest in religion, Frank's parents had him arrested for refusing psychiatric care. In the fall of 1962, during the height of the Cuban missile crisis, Frank was incarcerated. He was then transferred from institution to institution, including six weeks at Napa State. In the original petition to commit Frank, his father "prays that examination be made" as soon as possible. Once examined, Frank's diagnosis, for all intents and purposes, was religious hysteria.[101]

In the initial diagnosis, Frank's sexual and religious orientations were of immediate interest. Doctors noted that Frank was "withdrawn and asocial." His "sexual orientation [was] unknown," obscured by his "religious preoccupation." He exhibited "bizarre behavior and eating habits" as well as a "condescending superior smile." He "needs a haircut [and] he is very sloppy in appearance because of his beard." Yet he "insists on keeping hair long and full beard." For medical professionals, Frank's beard presented an immediate problem as a sign of religious otherness as well as willful biological excess.[102] All along, unknowingly, they were acting out a plan of psychic reform that owed as much to Protestant framing as it did diagnostic training.

After two weeks of treatment, to his doctors' dismay, Frank was still Jewish. Two weeks later, Dr. Joseph Crane noted that Frank remained "resistive of having his beard clipped or removed." The doctor then added that "the

---

diagnostic difference in the decision whether to institutionalize, see John H. Burgess and Ralph Lee Wagner, "Religion as a Factor in Extrusion to Public Mental Hospitals," *Journal for the Scientific Study of Religion* 10, no. 3 (1971): 237–40.

100. Leonard Frank, "The Patient as Prisoner," in Friedberg, *Shock Treatment*, 57.

101. The following details of Frank's incarceration are taken from the document collage of "The Frank Papers," *Madness Network News* 2, no. 5 (December 1974): 109–14.

102. Frank's diagnosis played on the stereotype of the pious, passive, and sexually suspect Jewish male that could, at times, become "actively and malignantly feminine in [his] perverse sexual powers and creative genius for disguise and imitation." Neil R. Davidson, *Jewishness and Masculinity from the Modern to the Postmodern* (Routledge: New York, 2010), 27.

thick pad of curly hair" will make it "difficult to adjust or band in place for electrode." Later that same week, Dr. Crane wrote Frank's father a note of concern: "In the last week Leonard was seen by the local rabbi, Rabbi Rosen, who spent a considerable period of time with him discussing the removal of his beard. I felt it was desirable to have the rabbi go over it with him, as Leonard seems to attach a great deal of religious significance to the beard. The rabbi was unable to change Leonard's thinking on this matter."[103]

Over the course of four months in early 1963, Frank began receiving what was known as the 50/35 regime (fifty insulin shock treatments and thirty-five ECT sessions). The beard remained a sign of his intransigence as well as prima facie evidence of his madness. But the beard was also an obstacle to treatment. For as Dr. Crane reported once again to Frank's father, "As I have discussed with you, the beard does complicate his treatment as we are not able to see as readily his lips which are used to some extent as indicators of his general physical status and oxygenation while in insulin coma."

In the third month of Frank's 50/35 treatment, Frank was "as paranoid as ever." In his notes Dr. Crane reported that Frank "still has all his delusional beliefs regarding his beard, dietary regime and religious observances that he had prior to treatment." Crane then urged another doctor to remove Frank's beard. A third doctor then cautioned against it given "definite therapeutic indications." That same day, Dr. Crane asked the hospital to consult with its malpractice lawyer to see about how best to proceed in shaving off Frank's beard. Dr. Crane then reported that he had talked with Ms. Lois Scampini at the district attorney's office who "felt that the possibility of any suit for assault and battery would be without very much basis and it was hard for her to conceive how this could be carried through into court." Frank's beard was removed a week later. Once the 50/35 regime was completed, Frank was started on thorazine as the doctor noted Frank's words of "God help me—God help me." Eleven years later, having dedicated his career to the patients' rights struggle, Frank defiantly declared he had "maintained all my 'religious preoccupations,' at the core of which is my belief in G_d."

Frank had determined that "I was brainwashed" as ECT "represented a tremendous assault on my brain—not just my brain, but on my whole physical being."[104] As witnessed by Frank and others, over the second half of

103. On the Protestant theological context in early twentieth-century America that treated Jewish masculinity as sexually and medically aberrative, see Sarah Imhoff, *Masculinity and the Making of American Judaism* (Bloomington: Indiana University Press, 2017).

104. Frank, "Patient as Prisoner," 58–59.

FIGURE 50. Dr. H. C. Tien at his desk. Cover of *Videology: Theory and Techniques* (New York: Vantage Press, 1978).

the twentieth century there developed within progressive scientific circles increasingly precise strategies that focused on the brain rather than the genitals or bodily surfaces as a site of sexual differentiation.[105] From galvanic baths and fly-blisters to electric shock for curing depression or treating erectile dysfunction and restless genital disorder, the goal has remained the same: to emancipate sexual normalcy in a cognitive key, that is, as a universal human right that originates *within* and is premised *on* a healthy and properly ordered brain.[106] As Frank's experience suggests, such strategies had come to exceed the explicit female form and focused, instead, on the individual as a mere collection of data points, a "management information system interfacing" with other systems that it happened to encounter.[107]

7. ELECTRIC LOVE THERAPY

> I wish families all over the world to know why I was born three times.
>
> I came from the womb of my mother to grow into a child of doubts, fears, and anxieties.
>
> I got my second birth when I looked into my heavenly Father by believing in Him, His name and His powers. . . .
>
> I received my third birth by Psychosynthesis. Between the synthesist and my most beloved husband, I was again reborn into a person so completely different that my friends and relatives cannot think of me

105. As philosopher Cynthia Kraus notes, the "degenitalizing [of] sex development, sexual identity, and sexual orientation altogether" intensified with the publication of Milton Diamond's "A Critical Evaluation of the Ontogeny of Human Sexual Behavior," *Quarterly Review of Biology* 40, no. 2 (1965): 147–75, and his emphasis on prenatal neural dispositions as explanatory of sexual difference. Eventually, Diamond would declare the brain to be itself "a sexual organ." Milton Diamond, "Sex, Gender, Identity over the Years: A Changing Perspective," *Child and Adolescent Psychiatric Clinics of North America* 13, no. 3 (2004): 591–607. See Cynthia Kraus, "Critical Studies of the Sexed Brain: A Critique of What and For Whom?," *Neuroethics* 5 (2012): 254–56.

106. Joan Scott argues that our current predicament in which sexual desire has become all but reified runs through the early cold war years in which "sex was produced as a new kind of freedom, and sexual freedom became a concept parallel to religious freedom, available for incorporation into the charged discourse of secularism" (*Sex and Secularism*, 140–41).

107. "Hit or M.I.S. or, Big Brother and the Mental Health System," *Madness Network News* 2, no. 2 (February 1974): 6.

> as I was six months ago. My old personality was fearful, anxious, guilty, despondent, depressed, shameful, critical, hostile and unjust. I was not ready for life then as I am now fearless, patient, confident, loving, understanding, firm, kind, honest, secure, just, and fair.
>
> MINNIE C. VAN EPPS, letter to *World Journal of Psychosynthesis* (1969)

A recurring feature in *Madness Network News* was its list of "shock doctors"—names and addresses of physicians who practiced electric convulsive therapy. Dr. H. C. Tien, the independent ECT operator who founded the Michigan Institute of Psychosynthesis in Lansing, Michigan, was consistently on that list.[108] Tien's institute was made possible by the widespread establishment of community-based mental health clinics in the 1960s.[109] Trained as a psychoanalyst and ECT technician, Tien had become frustrated with his professional effect. "I began to feel," Tien recalled, "that even if all the physicians became psychoanalysts, we still would not have enough. We needed something *more*. That was when I began to turn toward technology." Tien then earned his master's degree in electrical engineering. And then he had an idea born within the cybernetic fold. In 1964 he began to think long and hard about the prospects of personality reprogramming. By the end of the 1960s he had invested over $100,000 into video equipment—all in order to develop a kinder, gentler approach to electric shock treatment. Subsequently, Tien became a self-proclaimed psychosynthesist:

> Psychosynthesis is the scientific union of psychoanalysis of Freud, behavior therapy of Pavlov, electrotherapy of Cerletti, drug therapy of the Space Age in the light of cybernetics of Wiener and information theory of Shan-

108. In addition to the list, *Madness Network News* singled out Tien in an issue dedicated to analyses of "psychiatry as social control." An excerpt of H. C. Tien, "100 Questions," appears as "Electrolytic Love Therapy (Electro-Love Therapy)" in *Madness Network News* 4, no. 2 (Spring 1977): 6.

109. In 1963 the Community Mental Health Centers Act was passed, which led to the defunding of institutions and the founding of centers across the country by entrepreneurial psychiatrists like Tien who could provide licensed outpatient services. On the breakdown of the hospital as a site of confinement and the rise of community psychiatry, see Gilles Deleuze, "Postscript on the Societies of Control," *October* 59 (1992): 3–7.

non as a unified system of brain-mind equivalence in the same sense of mass-energy equation of Albert Einstein: $E=MC^2$.[110]

Tien's language of cybernetics was that of an insider—earnest, often inconsistent, and ever bent on conjuring an entirely new world in its deployment.[111]

The methods of operational research fueled Tien's interest in ECT and his small circle's "scientific approach to decision making."[112] Having crunched the numbers at the administrative level, Tien declared that his innovative approach to shock therapy would optimize the production of therapy in terms of patient visits, office hours, and costs per family. The Michigan Institute of Psychosynthesis, like other centers across the country, thrived on federal funding and less restrictive insurance policies.[113] A the end of the day, regulation of these community mental health centers was spotty at best and gave rise to overdetermined pitches such as Tien's.

In positing his vision of the neuromatic brain, Tien proclaimed that Norbert Wiener's *Cybernetics*, the volume that brought McCulloch and Pitts and Shannon to the masses in 1948, "should be required reading for every psychiatrist." Consequently, Tien understood himself to be a cybernetic programmer of the highest order—"a person who feeds information into another person or systems in order to modify the memory of the system under programming." He approached his patients' pathologies as matters of algorithmic disorder, their brains caught in a debilitating feedback loop that

110. H. C. Tien, "What Is Psychosynthesis?," *World Journal of Psychosynthesis* 1, no. 2 (October 1969): 2.

111. In a move that was not uncommon in the popular reception of cybernetics, Tien conflated information and energy. According to Tien, "consciousness is a single frame of electrical charges in motion, like electrons bombarding a television screen. And personality is a time-series of this scintillating frame of consciousness. Personality, therefore, becomes a reverberating input-output pattern of self-creation, seeking information or patterns of energy from the environment as well as from its own memories." See Tien's presentation at the National Congress of Cybernetics, "Psychosynthesis: TV-Linked Cybernetic System of Psychotherapy," *World Journal of Psychosynthesis* 1, no. 2 (October 1969): 75–76. See also Smith and Smith, *Cybernetic Principles of Learning*.

112. Russell N. Cassel, "Computer Assist Counseling by the Systems Analysis Approach," *World Journal of Psychosynthesis* 2, no. 7 (July 1970): 38.

113. In each issue of *World Journal of Psychosynthesis* Tien offered a list of pharmaceutical companies that produced "safe and efficacious drugs used in psychopharmacology, which informs an important subsystem of psychosynthesis."

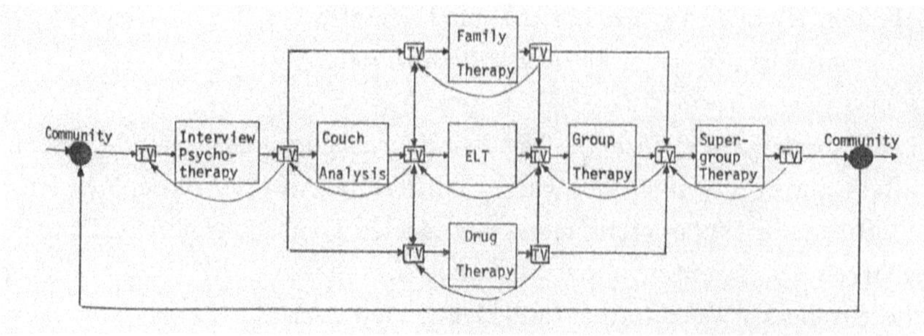

FIGURE 51. Information flow chart of television-linked cybernetic system, from Tien's exhibit at the International Congress of Cybernetics in London in September of 1969. "Psychosynthesis: TV-Linked Cybernetic System of Psychotherapy," *World Journal of Psychosynthesis* 1, no. 2 (October 1969): 75–77.

could be reversed.[114] According to one of Tien's associates at the Michigan Institute of Psychosynthesis, "Where engineers took a lesson from neurologists, psychiatrists, and the brain . . . to build a computer[,] it would now seem reasonable for psychiatrists to learn from the computer that the brain may be reprogrammed."[115]

According to Tien, the "core" of psychosynthesis was Electric Love Therapy.[116] ELT was a "television-linked cybernetic system . . . infused with warm

114. H. C. Tien, *Videology: Theory and Techniques* (New York: Vantage Press, 1978), 49, 92; "A Psychosynthetic Glossary," *World Journal of Psychosynthesis* 2, no. 7 (July 1970): 9. Tien was not alone in conceiving of mental pathology as an algorithmic condition that could be addressed through electric shock. See, e.g., Franz Alexander and Sheldon T. Selesnick, *The History of Psychiatry: An Evaluation of Psychiatric Thought and Practice from Prehistoric Times to the Present* (New York: Harper and Row, 1966), 283–84. For nonmechanized deployments of the categorical association between sexual and religious deviance, see Trudy Drucker, "Case History of Religious Hallucinations in Psychosis: Malaise of Margery Kempe," *New York State Journal of Medicine* 72, no. 23 (1972): 2911–17; Kurt Nussbaum, "Abnormal Mental Phenomena in the Prophets," *Journal of Religion and Health* 13, no. 3 (1974): 194–200; Gerald W. McDonald, "Sex, Religion, and Risk-taking Behavior as Correlates of Death Anxiety," *Omega* 7, no. 1 (1976): 35–44; Stan L. Albrecht et al., "Religiosity and Deviance: Application of an Attitude-behavior Contingent Consistency Model," *Journal for the Scientific Study of Religions* 16, no. 3 (1977): 263–74.

115. James S. Britton, "New Mental Health Techniques for Family Psychiatry," *World Journal of Psychosynthesis* 1, no. 3 (November 1969): 17.

116. Regarding the archaic roots of his practice, Tien wrote that the "the electrolytic process first sparked life into being when lightning struck into the primordial seas by com-

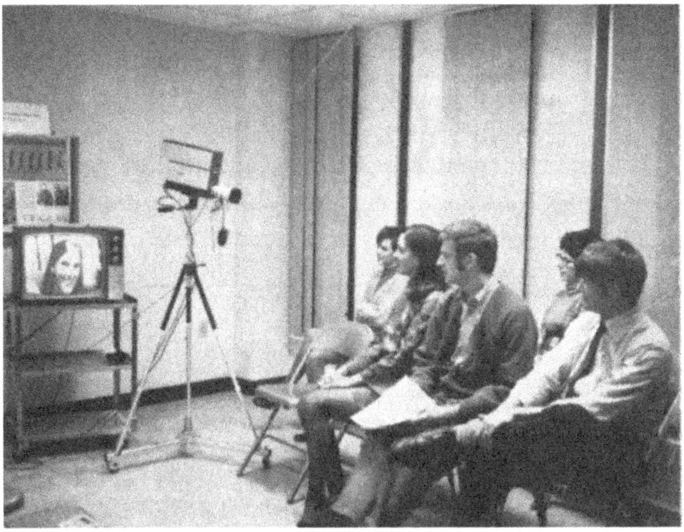

FIGURE 52. Group therapy session. "Television-Linked Mental Health Mini-Community at Michigan Institute of Psychosynthesis," *World Journal of Psychosynthesis* 2, no. 7 (July 1970): 29.

human communication." Tien was quick to point out that ELT was a consensual process and insisted that its success was premised on respect for the patient's ability to choose their best self. All organization, according to Tien, should be self-organization. Lana choosing to become Kate, Kenneth choosing to become Charles, Ellen choosing to become Wendy, Fred choosing to become Jerry, Virgil choosing to become Stewart, Bob choosing to become Eric, Beverly choosing to become Jane, Arlene choosing to become Theresa, Jessie choosing to become Joyce, Bernice choosing to become Martha, Joe choosing to become Jack, Peggy choosing to become Belinda.

Before a (typically female) patient committed to undergoing psychosynthesis, Tien would introduce her during a group therapy session. She would "get a feel" for Tien's practice as current patients offered testimonials and questions were answered. A transcript from one of these sessions was published in Tien's own professional publication, *World Journal of Psychosynthesis*. Dated November 14, 1968, the transcript records Judy (a prospective patient) interacting with a group of Tien's current patients sitting in a

bining simple molecules to form self-duplicating macromolecules, that we call life." H. C. Tien, "Who Is Cosman?," *World Journal of Psychosynthesis* 2, no. 2 (February 1970): 4.

cramped room.[117] Patients have gathered that day to gently ease Judy into Tien's practice. Neither Tien nor Judy are physically present. Judy's image, however, is being projected from a small television on a metal tray table beside a closed-circuit camera aimed at two rows of folding chairs. Both ends of the conversation are being filmed and shown, respectively, in real time. Judy begins. Her fear and embarrassment are palpable: "I'm scared, looking at the TV and having to answer questions. . . . What's going on? And, what are we doing here?" Among those that day attempting to sell Judy on the prospects of psychosynthesis is Marie, Tien's patient who now "has a happy family with her husband and two children."

Marie is the first to respond to Judy: "I went through ELT. What do you want to know?"

Another patient, Victoria, then assures Judy that she will not have to change her name if she does not want to. It is her "choice" and no one else's, says Victoria.

"I can't believe [it] myself," gushes Victoria. "I am like a new person, nothing like my old personality. . . . When I got out of school, I couldn't get a job. Now, after I have had the ELT treatments I have decided to go to work and I got a job with no problems at all."

Victoria's pitch is successful. As the transcript notes, Victoria is a model patient who has, through ELT, become empowered to make her *own* decisions and pursue her *own* role in stabilizing the nuclear family: "Victoria has obtained a job to go into selling Real-Estate," reads the annotation.

> Previously, she was a sickly hypochondriac, she was a retiring person, she was afraid of her responsibility as a mother and a wife. But now she was able to take care of her husband, the children and her job. Now, on top of that she has decided to have a part-time job. Her husband, Chuck, was commenting how much progress she has made as a new person with her new personality and new name as Victoria.

Tien's face then appears on the TV: "Who wants to say [their] pledge?"

> "I will," volunteers Victoria. "I, Victoria, woman of happiness and confidence; loving wife of Chuck; good and understanding mother to Mike,

117. Following quotations from John Rolfe, transcriber. "Wednesday Group for Psychosynthesis (Transcribed from TV Videotape, 11-14-1968)," *World Journal of Psychosynthesis* 1, no. 1 (September 1, 1969): 20–25.

> Mary and Nan; do solemnly pledge to become a person of affectionate demonstrativeness and light-hearted humor. . . . Long live Victoria! Lady of Courage! Long live Victoria! Woman of Harmony! At last, I am free! I am Victoria!"

The group then offers what the transcript describes as "Happy applause." Before the TV turns to black and the group session comes to a close, Tien's face, again, on the screen: "See you all next week."

8. THE BUSINESS OF MARRIAGE

As Tien surmised, "We hope that gradually the cybernetic ELT technique will replace the classical ECT procedure wherever electrotherapy is being used." His goal was to personalize "shock treatment" by enlisting spouses, parents, friends, and other patients as "therapeutic programmers."[118] As a self-conscious reformer, Tien was both a shock doctor *and* an effect of the antipsychiatry movement. He was a self-conscious systems theorist *and* a fierce defender of his patient's capacity to choose their best self. And he was committed to the spiritual welfare of his patients *and* the secularizing impulse of science to arrive at right reason.[119]

Psychosynthesis was, ideally, a family affair, with Tien focusing routinely on nuclear bonds and the troubled lines of sexual difference within kin. Many of Tien's patients had marriages that were "unstable" and/or had children who they considered sexually wayward. To his chagrin, many of Tien's patients also arrived at the institute with conservative understandings of religion and sex that he believed to be out of step with the times. Consequently, Tien sought to ease their entry into secular modernity with its changing attitudes toward sex and traditional religion—

> One who adheres to strong beliefs and customs in a world paced by instantaneous communications is apt to face some very real conflicts in

118. H. C. Tien, "Comment on Dr. Kelman's Article," *World Journal of Psychosynthesis* 2, no. 11 (November 1970): 35.

119. In "If in Doubt, Calculate!" Tien argued for the necessity of calculating "the decision-dividend (DD) [which] is defined as the product of a decision's value (V) and its probability (P) of success, divided by the cost (C)" ("What Is Psychosynthesis?," 4).

> life that a non-traditionalist would not face. Religious [*sic*] are based on tradition which frowns upon birth control, for example, in the "age of the pill" can generate a great deal of conflict among many family members. Another real problem facing traditionalist [*sic*] institutions, parents and children is the controversy over sex education and the social trend toward much greater truth and freedom.[120]

In his appeal to white working-class couples with insurance for community-based health centers, Tien acknowledged changing sexual mores that were complicating their lives and/or their relationships with their children. Offering a resolution tinged with technology and liberal sentiment, Tien contrasted his approach to the failures of the state to regulate marriage at the level of the self-organizing individual in relation to another self-organizing individual. The ideal marriage, according to psychosynthesis, was "defined as that balance of decision making where the husband and the wife each make about half of the family decisions after appropriate consultation with each other. This is the new authority pattern which is dictated by the fact that women, at least in the United States, have essentially equal bargaining power with the men. They have equal education, they have jobs with equal pay for equal performance, and they have equal vote. These qualities have been achieved by many women and this pattern is rapidly becoming universal in American society."[121]

In Tien's practice one can witness a shift in the gender politics of secularism as articulated by nineteenth-century liberal reformers like Farnham. Femininity was no longer the exaggerated mark of superior (or inferior) organization and proximity to the natural order. Marriage was framed in terms of business acumen and the "essentially equal" division of labor. Sexual satisfaction, beyond the merely reproductive, was important to individual flourishing but not essential to the marriage per se. According to Tien, marriage was about the legal regulation of sex and not necessarily sex

120. H. C. Tien, "Editorial Comment," *World Journal of Psychosynthesis* 1, no. 2 (October 1969): 7.

121. Paul H. Wilcox, MD, "Beyond Man and Woman: A Person," *World Journal of Psychosynthesis* 1, no. 3 (November 1969): 32. Tien's journal was peppered with articles on advice for the working-class family. See, e.g., Neil Carpenter, "Factory 'Break Labs' Would Break Monotony," *World Journal of Psychosynthesis* 5, no. 3 (March 1973): 10–11; Booker T. Gaulden, "Credit Reporting Agencies," *World Journal of Psychosynthesis* 5, no. 3 (March 1973): 13–15; Sean Carr, "The Energy Crisis and the Uses of Adversity," *World Journal of Psychosynthesis* 5, no. 12 (December 1973): 8.

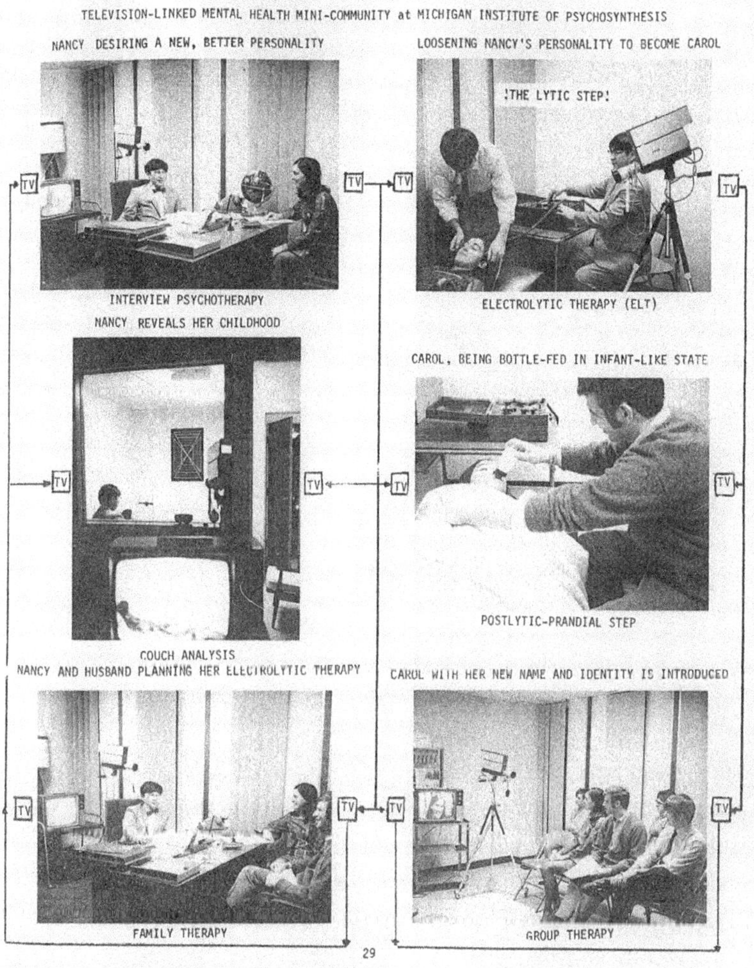

FIGURE 53. Steps of electrolytic therapy. "Television-Linked Mental Health Mini-Community at Michigan Institute of Psychosynthesis," *World Journal of Psychosynthesis* 2, no. 7 (July 1970): 29.

itself. "If you think marriage is a purely sexual institution," counseled Tien, "you should ask the prostitutes. It is not. What makes a family stable is a socio-economic relationship."[122] Or, as an article in Tien's *World Journal of*

122. "Wednesday Group for Psychosynthesis: Sex, Marriage and Family Stability, April 29, 1970," *World Journal of Psychosynthesis* 2, no. 7 (July 1970): 22; Scott, *Sex and Secularism*, 123, 139.

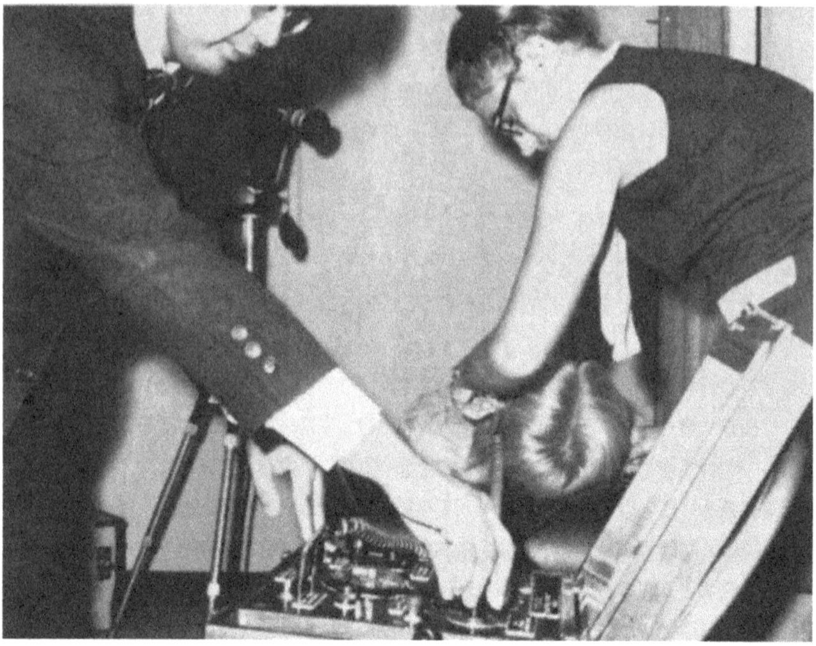

FIGURE 54. Dr. Tien in action. "ELT Is Used to Modify a Patient's Memory," *World Journal of Psychosynthesis* 3, no. 9 (September 1971): 25–26.

Psychosynthesis put it, the "problem of marital breakdowns really has to be attacked at the point where they occur—within the framework of the individuals who make up the marriage and their ability to struggle progressively to make the business of marriage successful."[123]

Tien insisted that ELT was the engine of such progress, in part because it was a wholly voluntary treatment. ELT involved three steps: 1) the *recall* of disturbing memories, 2) the selective *erasure* of those memories with

123. Psychosynthesis might aid in Conciliation Courts because "our concepts of marriage and divorce are changing. The trend is away from divorce and toward helping marriage, changing the concept of marriage, and dissolving a troubled marriage in a realistic way without having to prove someone at fault." Stephen V. Moulton, JD, "Conciliation Courts and Psychosynthesis," *World Journal of Psychosynthesis* 2, no. 6 (June 1970): 18–19. *World Journal of Psychosynthesis* was very much concerned with the rise in divorces. See, e.g., Stephen V. Moulton, JD, "Family Questions and Answers on Divorce Reform," in which he begins: "Divorce is to marriage as dissolution is to commercial business. It is the realization of poor managerial ability. Divorce should not be the cause of marital breakdown, and divorce reform as proposed would further insure that it is not" (*World Journal of Psychosynthesis* 2, no. 7 [July 1970]: 16).

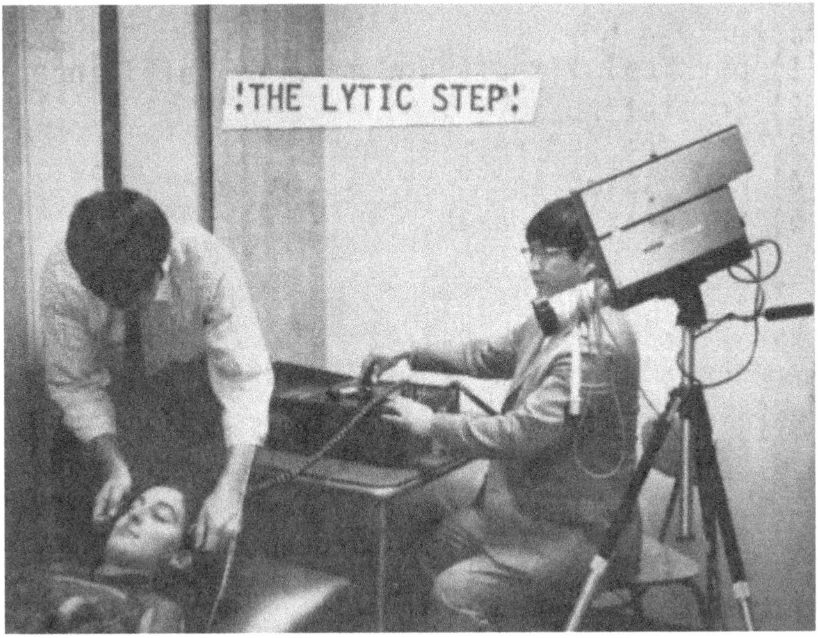

FIGURE 55. Lytic Step: Loosening Nancy's Personality to Become Carol. "Television-Linked Mental Health Mini-Community at Michigan Institute of Psychosynthesis," *World Journal of Psychosynthesis* 2, no. 7 (July 1970): 29.

electroshock, and 3) the immediate *reprogramming* of the patient's personality. For once the electrodes were attached to the patient's scalp and a video recorder turned on, Tien then solicited from his patient the traumatic memories to be erased. "At the height of this memory replay," a standard Reiter Electrostimulator was switched on, "returning the patient to an infant-like state." ELT sessions were repeated until the former personality was sufficiently erased.[124] According to Tien, "ELT is used to cause all neurons to discharge simultaneously, so that the unadaptable pattern of personality is being reset momentarily to a zero-matrix." Before the neurons could return to their fixed pattern of activity, Tien initiated an elaborate process of neural

124. Tien defined "erase" in terms that resonated with Scientology: "To replace the binary digits of information of the system, i.e., brain, to binary zeros as in electrolytic therapy. To erase is to clear in electrolytic therapy" ("A Psychosynthetic Glossary," *World Journal of Psychosynthesis* 2, no. 1 [January 1970]: 11). Tien also invoked Hubbard's notion of "clearing" as "reset[ting] the system to its initial state." As Tien wrote, "in electrolytic therapy, to clear is to loosen the scintillating and circulating memory to zero" ("A Psychosynthetic Glossary," *World Journal of Psychosynthesis* 1, no. 2 [October 1969]: 8).

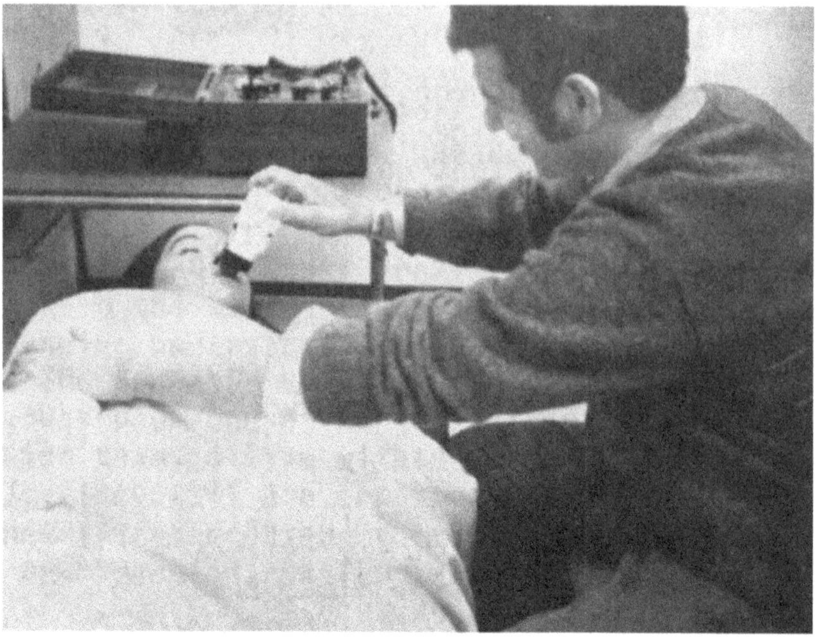

FIGURE 56. Bottle feeding of patient (close-up). "Television-Linked Mental Health Mini-Community at Michigan Institute of Psychosynthesis," *World Journal of Psychosynthesis* 2, no. 7 (July 1970): 29.

destruction and neural genesis in order to reprogram "a more adaptable personality for survival in the modern world."[125]

After the electrodes were detached, the patient was then allowed to recover with designated family members. They called the nascent personality by a new, previously agreed-on name.[126] With the video camera rolling, the new personality was then bottle-fed and nursed back into adulthood over a period of minutes, weeks, and months. Tien published "monthly case reports" of these sessions, including one about Ellen, who was diagnosed with "pessimistic hypochondriac personality. As Tien proudly declares,

125. Tien, *Videology*, 100–101.

126. Evoking Shannon and his desire to engineer more efficient telephonic communication at Bell Labs, Tien wrote that "the name can . . . be likened to a telephone number which dials into the nervous system as to which personality is to dominate. When the telephone number is changed the system would have to [be] reorganized in relation to other systems, in our case, between other personas in the family and the community. Transnomation is superficial while transpersonation is substantial, but the combination produces an effective result." H. C. Tien, "100 Questions," 34.

"the personality of Ellen was transformed by the electrolytic technique (ELT), with the deliberate reactivation and lysis and postlytic synthesis of Wendy."[127] Tien justifies his optimism by pointing to the moment in the transcript when Wendy had just gone through the electrolytic process and was lying in bed in the infant-like state. As Wendy is being bottle-fed by her husband and daughter, the reader is privy to the following exchange:

HUSBAND: Have a drink. Here, come on, drink.
SYNTHESIST: Drink, Wendy, drink.
WENDY: (sucking bottle).
HUSBAND: Drink. That's right!
SYNTHESIST: Hi! Are you thirsty?
DAUGHTER: You're thirsty. Drink the bottle.
SYNTHESIST: She's doing fine. She's like a little baby. Tell her she will love life when she grows up.

Wendy, being bottle-fed and minutes later, more conscious, struggling out of bed.

WENDY: Where am I supposed to go?
DAUGHTER: You're supposed to stay right here.

The scene so intimate, the words of the synthesis all according to plan. The future was bright.

SYNTHESIST: Wendy, how does the future look? You've got to work on it a little bit.
HUSBAND: How does the future look?
WENDY: All right.
HUSBAND: It looks good. Can't you say good?
WENDY: Yes.
SYNTHESIST: Let's see you smile.
HUSBAND: We told you the future looked good. Come on. Smile a little bit for us. We want you to smile. Can't you smile?
WENDY: Yes. (not smiling).
HUSBAND: Come on and smile. Can you smile? (tickling her under the chin).
WENDY: Yes. (smiling).

127. The following quotations are from H. C. Tien, "The Psychosynthesis of Wendy," *World Journal of Psychosynthesis* 2, no. 1 (January 1970): 29–34.

The authenticity of Wendy's smiling new self was premised on her achieving equilibrium in the future. For after ELT, one "gradually come[s] to a steady level, which we call a steady state. And that steady state is the same thing for the firm, General Electric, namely, habituation which leads to boredom. . . . The brain will reach a steady state. That is the purpose of the brain, to reach a steady state." In cybernetic fashion Tien viewed female sexual behavior as composed of "feedback mechanisms" between all of the organs and hormones indigenous to the female form.[128]

Such "organic integrity" was hard-won and possessed its own forms of brutality. For the sexual differentiation achieved under Tien's TV gaze reflected the tenor of his times.[129] Take, for example, Tien's patient named Cyril, an impotent cross-dresser who was fond of wearing his wife's and mother's clothes.[130] After a series of ELT treatments Cyril still struggled to become Stan, a normal heterosexual male.

SYNTHESIST: Ask your wife if she can accept Stan as a strong and positive and masculine person. Ask her if she can accept Stan if you change into Stan completely.

CYRIL: Can you accept Stan?

SARAH: Yes. If he is Stan completely. I can't accept Stan until he is Stan, though. . . . If you're going to be Cyril, you're not going to have a wife and child. Because I'm not going to live with the Cyril that I lived with. I feel that I'm a normal person and that I deserve a normal life and I feel that our daughter needs to be brought up by normal people. I don't believe Cyril is normal.

Immediately before another ELT session begins, Tien reads aloud from a prepared document:

128. On the reduction of sex to the machinery of sex, see "A Model for the Homeostatic Regulation of Female Sexual Behavior," *World Journal of Psychosynthesis* 1, no. 2 (October 1969): 73.

129. *World Journal of Psychosynthesis* offered itself as the progressive edge of diagnosing the female sex. In "Concepts of Hysteria in the Nineteenth Century," Charles Goshen gives an overview of historical definitions narrated as a secular refinement of diagnosis (*World Journal of Psychosynthesis* 2, no. 7 [July 1970]). See also Baybars Tek, "Interdependence in Marriage," *World Journal of Psychosynthesis* 3, no. 1 (January 1971): 12–15; Thomas F. Graham, "Sexuality in Society," *World Journal of Psychosynthesis* 4, no. 2 (February 1972): 30–35.

130. The following quotations are from H. C. Tien, "The Psychosynthesis of Stan," *World Journal of Psychosynthesis* 2, no. 5 (May 1970): 23–28.

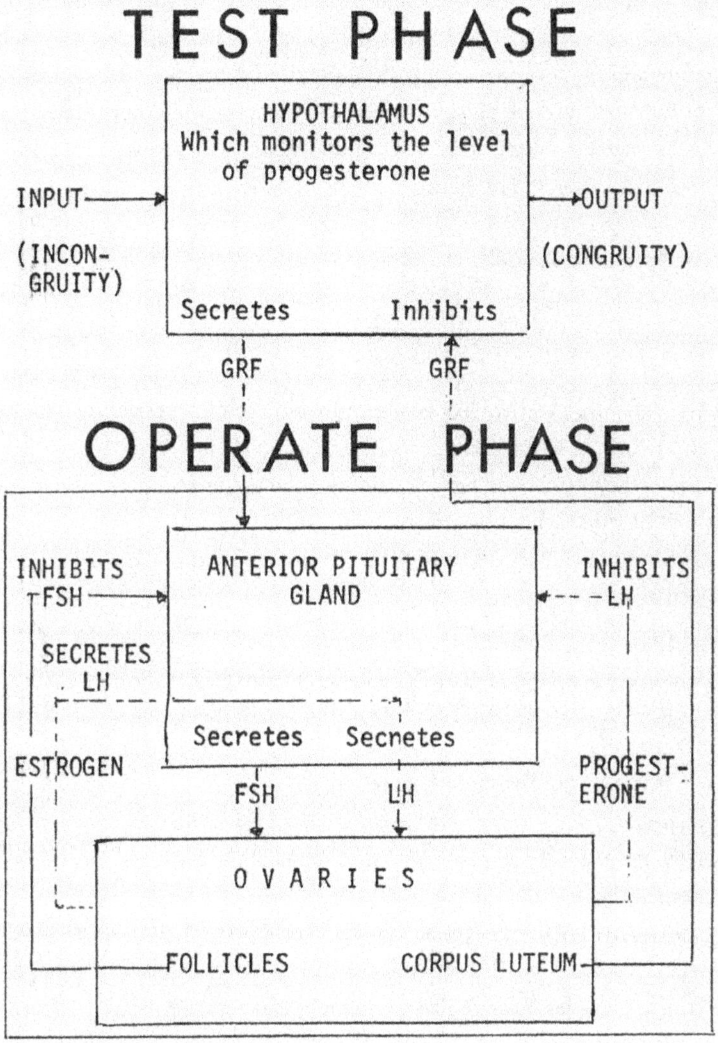

FIGURE 57. The self-regulation of female sexuality. "A Model for the Homeostatic Regulation of Female Sexual Behavior," *World Journal of Psychosynthesis* 1, no. 2 (October 1969): 73.

Cyril on occasion still dressed up as a woman and found pleasure in it. Stan was disgusted with him. Stan wanted to be a good husband, sexually and otherwise, but Cyril would not let him in. To cover up this sexual weakness, Cyril became a domineering nasty person. He was much that Stan detested in others. Cyril was a very mixed-up person, but would not admit it.

> Stan wants Cyril to be "killed" so that Stan can have a chance to be Stan. Stan wants a strong marriage, a happy marriage, a happy home and family. Cyril is tense, nervous and full of depression. Stan wants to be a loving, carefree person who is more interested in others than he is in himself. Stan wants to be a masculine man; a well-adjusted man. Stan thinks that Stan is a nice guy.

Electricity is then applied. The patient begins to stir. Bottle-feeding begins:

(Sarah is handed a bottle of milk and she beings to feed Stan.)

SYNTHESIST: Stan, you are like a baby, but you'll grow up very fast. Talk to him, Sarah; let him know his name is Stan.

SARAH: Yes, Stan, you're doing well. That's a good boy, Stan. That's a good boy.

(Stan continues to drink the bottle of milk.)

SARAH: Good boy, Stan. That's it, Stan. It's all right; you're okay. Drink it down. Drink, Stan.

(Stan begins to moan loudly.)

SYNTHESIST: Stan, open your eyes. Stan. Would you pat him in the cheek?

SARAH: (Patting Stan on the face) What's your name, Stan? Stan. Stan.

STAN: Stan.

SARAH: (Kisses him): Yes, and what's my name?

STAN: Sarah.

SYNTHESIST: Stan, open your eyes now. Tell me have you ever dressed like a woman? No, no, no.

STAN: No.

SYNTHESIST: No, never. That's right.

SARAH: Did you ever dress up like a woman?

STAN: A long time ago.

SYNTHESIST: Never, never. That was Cyril. Are you Cyril?

STAN: Yes.

SYNTHESIST: No! Tell him.

SARAH: You're Stan. Not Cyril. What's your name?

STAN: Stan.

SARAH: Who dressed up like a woman?

STAN: Cyril.

SARAH: Did you ever dress up like a woman?

STAN: Never.

SYNTHESIST: Stan, this is the important thing. Have you ever dressed like a woman? Look at me.

STAN: Never.

SYNTHESIST: Have you ever thought of becoming a woman?

STAN: No.

SYNTHESIST: You're a good man, Stan. Let's see you smile. How does the future look?

STAN: Okay.

Stan is then instructed to read his prepared pledge:

> [I] . . . husband of Sarah, father of Joc and Elizabeth, do solemnly pledge to be powerful and potent, in work and in deed. I pledge to develop freedom of feelings: Sexual, family, and social. I shall constantly try and try continuously [to be] the man of confidence and self-respect. I shall forever fight off all competing temptations to be other than a man. Down with Cyril! Long live Stan! Long live Stan! Persist! Have courage to win final victory! Long live Stan!

After Stan has been transformed into "somewhat normal heterosexual man" Tien concludes that his wife Sarah "would benefit from therapy" which was normally the case after a man underwent ELT. In Stan's case, as with Wendy's, the cultivation of individual personality was paramount.[131] Tien had mastered the technology of self-actualization and it was his solemn duty to share his knowledge with the greater Lansing community. So as the

131. The examination of social structure qua social structure was not. Tien, for example, appealed to white anxieties over civil rights and race riots in urban centers such as Detroit. Warning of new disturbances that "may aim at vital public services such as utilities, police, and communications," Tien encouraged his patients to adopt a "psychosocial perspective" and to seek overlapping consensus as a matter of liberal principle. In disavowing the Black Panthers and Stokely Carmichael in favor of the nonviolent protests of Martin Luther King, Tien and his journal encouraged patients to "face the racial problem" by viewing it as a failure in interpersonal communication processes that were vital to social order. "Wednesday Group for Psychosynthesis: Superman Versus Cosman," *World Journal of Psychosynthesis* 2, no. 5 (May 1970): 22; "Black Uprising in Detroit: A Follow Up Study," *World Journal of Psychosynthesis* 1, no. 2 (October 1969): 29–32. See also Ralph W. Waderson, "Empathy: An Antidote to Individual Racism," *World Journal of Psychosynthesis* 4, no. 3 (March 1972): 27–28.

first waves of deindustrialization rolled across Midwestern cities such as Lansing, Tien's journal called for a new kind of liberal subject and liberal society free from the inequalities that inhibited the economic success of the family unit. Take, for example, the version of liberal subjectivity achieved by Wendy who, after a series of ELT sessions,

> continued on weekly psychotherapy with her husband, while she was also maintained on Triavil 4–25, one tablet four times a day. Chemotherapy is being used in psychosynthesis as a stabilizing agent, because we view the central nervous system as an information processing organization (of an M-dimensional television system, if you will), whereas psychotherapy is the programming of information and reorganization and the electrolytic therapy serves to remove undesirable information. With the electrolysis and loosening of the personality of Ellen, the happier, healthier and more stable personality of Wendy had a better chance to emerge.
>
> [Wendy] has remained well and stable for two years, since her [ELT]. She has been holding a steady job as a drill-press operator, working 40 to 60 hours per week in an automobile factory.[132]

9. THE UNION OF ALL CONTRADICTORY IDEAS

As the technological means of mobilizing his patients' *natural right to voluntarily choose* to undergo conversion, or what Tien sometimes called "transnomation," psychosynthesis also delivered the light touch of true religion. "Psychosynthesis is the union of all contradictory ideas, practices and ideals into a unified system," declared Tien. "In this system, all religious, political or economical systems are regarded as true, depending on thresholds. The only absolute value is life itself." This "Law of Universal Truth," moreover, was "operative in the function of our brain or any conceivable brain, biological or computer."[133]

Tien's promise of self-actualization by way of the neuromatic was also a promise of religious liberation without dogma or clerical authority. He would, for example, offer his prospective patients dramatized bits of East-

132. Tien, "The Psychosynthesis of Wendy," 33.

133. H. C. Tien, "Law of Universal Truth," *World Journal of Psychosynthesis* 2, no. 1 (January 1970): 4–5.

ern wisdom and decontextualized quotes pulled from Confucius and Buddha as well as Huston Smith and Freud—all in service of his fundamental proposition that the freedom of each patient's potential was premised on voluntary submission to Tien's technological grace. The latter, of course, was shot through with the language of spirituality, itself sanctioned by the sheen of Tien's advocacy for "biological cybernetics."[134]

World Journal of Psychosynthesis was both a technical journal and a new age trade publication, full of universal claims both literary and scientific. Tien's references were exclusively men and all over the map: Benjamin Franklin, Socrates, Bertrand Russell, William Penn, Gandhi, Marshall McLuhan, John Dewey, Lao-Tzu, John Lilly, Cicero, Booker T. Washington, Jesus, Mohammed, Henry Ford, V. I. Lenin, Thomas Edison, and Adlai Stevenson. Although Tien did not discuss the concept of spirituality per se, he did not have to. Many of the early issues of *World Journal* were taken up by the serial publication of *Journey of the Stone Monkey*, "a modern rendition of a Chinese Classic, which is translated and updated specially for the *Journal*, containing pre-scientific elements of psychosynthesis besides being a delightful tale of folklore and humor with a Biblical resonance and Shakespearian intricacy of plots . . . an excellent story for the family to share together." Tien's intermittent presentation of himself as oriental monk, dutifully sharing and integrating his wisdom for his listeners, had a rich history in the postwar years.[135] In addition to *Journey of the Stone Monkey*, Tien also offered the story of "Cosman," a parable for each of his patients. "What is the Cosman?," asked Tien. Cosman was the universal hero, the archetype to which the world was now turning. "Cosman is the fusion of all mankind with a cosmic identity," wrote Tien. "Cosman is the evolving synthesis of three billion or more human beings, supported by other forms of life on this planet, linked by mass communication."[136]

134. Tien, "What Is Psychosynthesis?," 4; H. C. Tien, "Law of Universal Truth," 4.

135. H. C. Tien, preface to "Journey of the Stone Monkey by Wu ChengEn, Jr.," *World Journal of Psychosynthesis* 1, no. 1 (September 1969): 86; Jane Iwamura, *Virtual Orientalism: Asian Religions and American Popular Culture* (Berkeley: University of California Press, 2011). "All synthesists can be called religious," argued Carol Meyerowitz in Tien's journal, "for they are able to accept philosophies of atheism, theism, agnosticism or the philosophy of humanism, relying on reason rather than on revelation for truth" ("A Synthesis of Religions," *World Journal of Psychosynthesis* 3, no. 4 [April 1971]: 18).

136. H. C. Tien, "Family Psychiatry and Psychosynthesis," *World Journal of Psychosynthesis* 2, no. 4 (April 1970): 4. "Each of us is a brain cell," insisted Tien, "of the Cosman. We are linked by television, by the telegraph, by jet plane, by travels." An awareness of such

FIGURE 58. Dr. Tien on couch and on television. "Medical Television Techniques: Telefusion Couch Therapy," *World Journal of Psychosynthesis* 4, no. 2 (February 1972): 37.

Tien's promise of "psychomutation" was neither religious nor secular but, rather, a minor albeit representative formation of secularism in the mid-twentieth century. The language of true religion, in other words, had quite literally become mechanized during the process of electric love therapy. Tien went so far as to suggest that ELT could be used not only for those in need but also on so-called normal individuals to forget past experiences or become programmed with new skills such as foreign language acquisition. And, of course, if the new personality did not suit the individual they could then be reprogrammed or even reverse engineered. In ELT, then, one may glimpse a secular imaginary's investment in generating individuals who could, ideally, become anything, choosing the terms of their individuality. Freedom, in this scenario, becomes a matter of unleashing one's capacity to

linkages was what constituted "cosmic awareness" in the vein of liberalism. For "once you feel that you are part of the universe, then you will try to take care of everybody, including yourself" ("Superman Versus Cosman," 21).

choose—the specter of religious voluntarism haunting the process of naming and inhabiting sexual difference.[137]

Tien's pitch, like Farnham's before him, was to enable proper sexual differentiation by activating one's capacity for spirituality. Take, for example, the psychosynthesis of Morris (formerly Roe). Roe came to Tien's institute with an unhappy marriage and much anxiety over his neurotic association with institutionalized religion. Roe's vexed relationship with his childhood Catholicism had effeminized him and saddled him with hysterical concerns over hygiene. Tien's plan with Roe was to allow him to become what one might recognize as spiritual but not religious, "loosening and erasing" his memories and guilt-laden obligations to the Catholic Church.

ROE: I don't want to be particularly hung up on any specific religion, you know.

PENNY [Roe's wife]: Yes, we have talked about that and agreed on it.

SYNTHESIST: Which religion to you want to loosen up a bit?

ROE: The Catholic religion.

SYNTHESIST: You are sure this is what you want to do? You want to be free from this?

ROE: Yes, I would like to be.

SYNTHESIST: What part of the Catholic religion do you want to be free from?

ROE: Well, just the direct connection with the Catholic religion. They have too many church rules, rules against birth control, and thoughts that if a person isn't a Catholic he isn't quite as much of a person as a devout Catholic is. I can't go along with the church's teaching on these subjects.

SYNTHESIST: Let me see if I am clear on what you want. You would like to develop a personality, Morris, who is going to be confident, healthy, and not tied to

137. In Tien's case, sexual freedom was defined in terms of economic liberalism and "equitable" definitions of marriage. As Scott points out, secular characterizations of freedom have neither liberated nor eradicated inequality. These characterizations have, however, in their relationship to sexual difference, been incredibly productive in making individuals, and much else besides, in Scott's dark phrasing, "available for incorporation"—that is, generating docility in the name of choice, inequality in the name of universal right. This is precisely what Foucault had in mind when he spoke of sex as "a dense transfer point for relations of power." Consequently, the appeal of Tien's practice at the Michigan Institute of Psychosynthesis was made possible not only by insurance coverage of community-based health care. There was also the allure of true religion and the attention that it enlivened—attention to the self, to the environment, and to the relationship between them. Scott, *Sex and Secularism*, 141; Michel Foucault, *History of Sexuality*, vol. 1: *An Introduction* (New York: Vintage Books, 1980), 103.

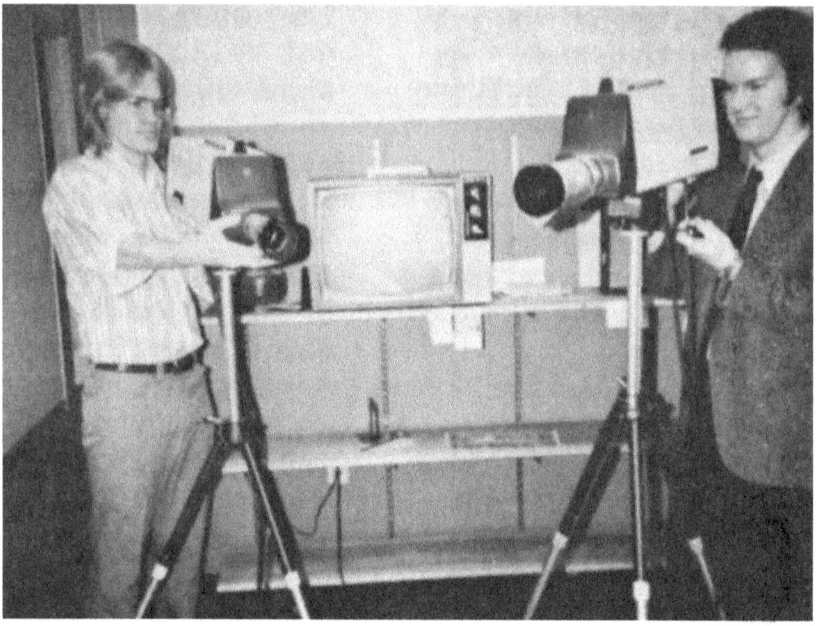

FIGURE 59. Cameramen, Michigan Institute of Psychosynthesis. "Videology Is the Science and Art of Medical Television," *World Journal of Psychosynthesis* 4, no. 4 (April 1972), 15.

> any particular religious outlook regarding a church, dogma, and rules and procedures. You want to be free and humorous, an even-tempered person, with an outgoing, friendly approach to people and things, without a great deal of old-lady-like preoccupations about body, disease, and panic.[138]

In the case of Morris the distinction between true religion and false has been deployed in a wholly technocratic key. All of this is to say that far afield from spaces of worship, the distinction between the religious and the secular (or between true religion and false) has not always announced itself as such or in ways that mapped on to standard secularization theses and our inherited categories of analysis. The truth of true religion, whether explicitly articulated or not, has, from the nineteenth century to the present, given rise to gendered norms and tortuous forms of subjectivization.[139]

138. H. C. Tien, "The Psychosynthesis of Morris," *World Journal of Psychosynthesis* 2, no. 7 (July 1970): 24–28.

139. To say as much is not to claim that psychosynthesis was a religion but rather that it was conceived of and received in rather familiar terms. As one of Tien's patients, Ann,

10. I WATCH TV, I WATCH TV

All sessions at the Michigan Institute of Psychosynthesis were filmed, including ELT treatments, on closed-circuit television. In a process that Dr. Tien called "telefusion," a profusion of screens did not impose but rather facilitated "two-way TV communication" between patients and their environment.[140] As Tien claimed, he was using "TV cameras, monitors and videotape machines to assist in eliciting and storing the verbal and nonverbal 'information' that needs to flow between patient and therapist or patient and patient." For Tien, television was the hinge that secured constant traffic between self and environments, an aspirational identity being invested *into* the screen at the moment it was being objectified *on* the screen. Such was Tien's script for ceaseless feedback, at least until the patient stabilized and the neural nets assumed their new architecture and the patient could be released back into society.[141]

"As a video psychiatrist," Tien imagined a network of closed-circuit TVs in which the patient's every move would be filmed and made available to the patient in other settings as well as the therapeutic groups the new personal-

---

claimed in group session: "I think psychosynthesis and religion are very much alike, for those of us who believe in God and are Christian. The main thing that God wanted here on earth was for people to get along together and understand one another and live harmoniously in this world. In psychosynthesis we do this, and this is exactly what Christ wanted." "Wednesday Group for Psychosynthesis: The Synthesis of Contradictions," *World Journal of Psychosynthesis* 2, no. 10 (October 1970): 23.

140. According to psychiatrist Don D. Jackson (who worked with cybernetician Gregory Bateson at the Mental Research Institute in Palo Alto, California), such video surveillance was necessary because the family was a "system" and family therapy required new forms of documentation, particularly as it moved away from a model of mental illness as wholly subjective toward a model in which psychosis was considered to be an interactive matter between subjects and objects; Don D. Jackson, ed., *Therapy, Communication, and Change* (Palo Alto, CA: Science and Behavior Books, 1968), 130ff. See also Orit Halpern, *Beautiful Data: A History of Vision and Reason since 1945* (Durham, NC: Duke University Press, 2014), 166–67. On the legacy of systems theory in family therapy, see Michael White, "Negative Explanation, Restraint, and Double Description: A Template for Family Therapy," *Family Process* 25, no. 2 (June 1986): 169–84.

141. Tien was participating, from a practical angle, in the intense academic interest in television as an effective tool of behavioral influence, if not control. See George Comstock, ed., *Television and Human Behavior: The Key Studies* (Santa Monica, CA: Rand Corporation, June 1975), R-1747-CF.

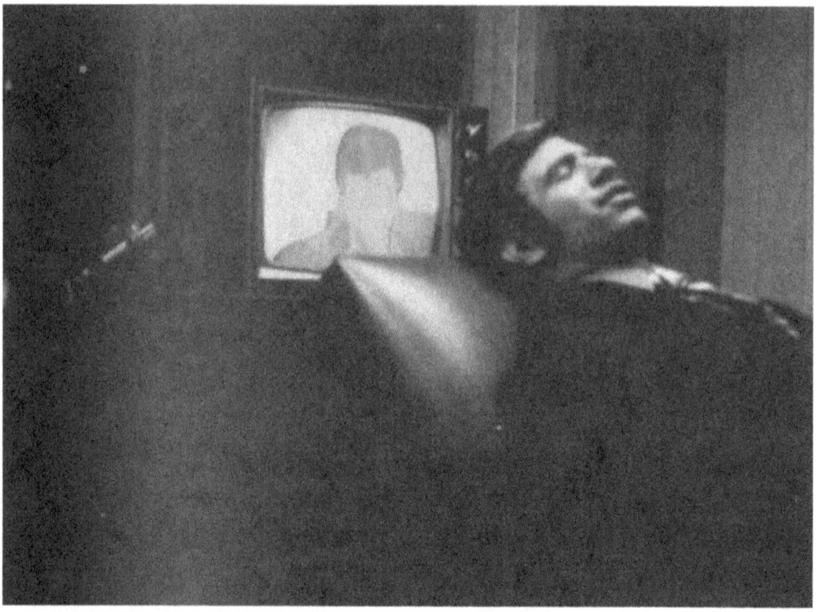

FIGURE 60. Dr. Tien on TV. "Patient in primary telefusion couch therapy free associates and receives free communication from the therapist, who is in a different room. At the same time, both patient and therapist are being recorded on tape to provide later feedback for the patient." H. C. Tien, "Medical Television Techniques: Telefusion Couch Therapy," *World Journal of Psychosynthesis* 4, no. 2 (February 1972): 37.

ity joined—so that the new personality was, ideally, never outside the loop of information even while continually adjusting to the flow of information.[142] According to Tien, such were the ideal conditions *under which* a patient's perception and memory were coconstituted and *in which* the patient's present (and presence of mind) was secured by the feedback between self and screen. The videotape, here, assumed a therapeutic purpose, allowing the patient to continually observe the progress made in the presence of others. As memories of the "old" self were replaced with the saturated televisual presence of the new self, the patient became part of a larger neuromatic scheme. As Tien wrote, "visual images enhance the feedback process[es]" set in motion by the initial televised shock and its immediate aftermath. With analyst, family, and friends all reinforcing a definitive personality blueprint,

142. Tien, "100 Questions," 36.

the selective video archive became a prosthetic memory for the patient, reinforcing the newly assigned identity.

In this scenario, patients put themselves under surveillance. Each was then able to continually observe and recall the progress made in the presence of others. As Tien's patients acknowledged, it was precisely this creation of a closed-circuit world that enabled them, over time, to secure their new selves by themselves. Everything was always and absolutely voluntary, insisted Lorraine, one of Tien's patients. But that was what made it so difficult. "People who have ELT's expect things to happen just like that," Lorraine

FIGURE 61. Telefusion advertisement from *World Journal of Psychosynthesis* 3, no. 1 (January 1971), front matter.

said, snapping her fingers for emphasis. "But it takes time," she counseled new patients to Tien's practice. "You have to have confidence in yourself and you have to really push yourself to obtain the goal you are trying to reach."[143]

In promoting his consulting and contracting firm (aptly named Telefusion), Tien promised a total conversion born of an incredible degree of self-reflexivity—watching yourself undergoing conversion, watching others watching you undergoing conversion, watching yourself watching others watching you undergoing conversion. *And so on and so forth.* Through the media of electrical current and pixelated image, Tien's patients became TV-infused, self-organizing systems whose security was premised on perpetual loops between input and output. Consequently, Tien took unprecedented care in targeting and leveraging the brain-environment complex as it was captured on the screen. Here, again, was Foucault's "tautology of the asylum," but rehearsed in a cybernetic key. For at the Michigan Institute of Psychosynthesis televisions *were* the environment, the source *and* horizon of all cognition. Whatever was happening within the institute was thought to be happening outside of it—"the model of psychosynthesis [being] based on the cybernetic principle of continuous feedback and modification." Your steady gaze into the camera or, alternatively, the screen, and often times both simultaneously, secured the feedback loop between self and display, sanity and surveillance: to the point where television does all the talking, saturating not only the environment but the brain of the patient as well.[144]

143. Tien, *Videology*, 89.

144. Tien, *Videology*, 48. In facilitating communication between people, televised encounters were more immediate according to Tien, superior to face-to-face contact—"Complex techniques can be shown on video, for example, the use of cross-feedback of couch sessions between the wife and husband in therapy who normally cannot communicate peacefully at home, but they would be able to listen to each other more carefully and tolerantly on videotape" (front matter). See also Paul Ryan's *Birth and Death and Cybernation: Cybernetics of the Sacred* (New York: Interface, 1973). In it Ryan, a disciple and colleague of Marshall McLuhan, pontificated on the promise of video in terms that were akin to Tien's. "Videotape is an x-raying of human processes that takes us into realms traditionally charted by religion, traditionally referred to as 'the sacred'" (xiii). Ryan advocated videotape as a metatool to reform education and politics by way of "infolding information" and catalyzing self-knowledge and selves that were "coherent" and "complete" (6–8). Drawing on the counterculture appropriations of cybernetics in the 1960s, particularly John Lilly and Gregory Bateson, Ryan envisioned a moment when video would allow one to "pass through the barrier of the skin—pass through the pseudo self to explore the entirety of one's cybernet—i.e., the nexus of informational processes one is part of" (36).

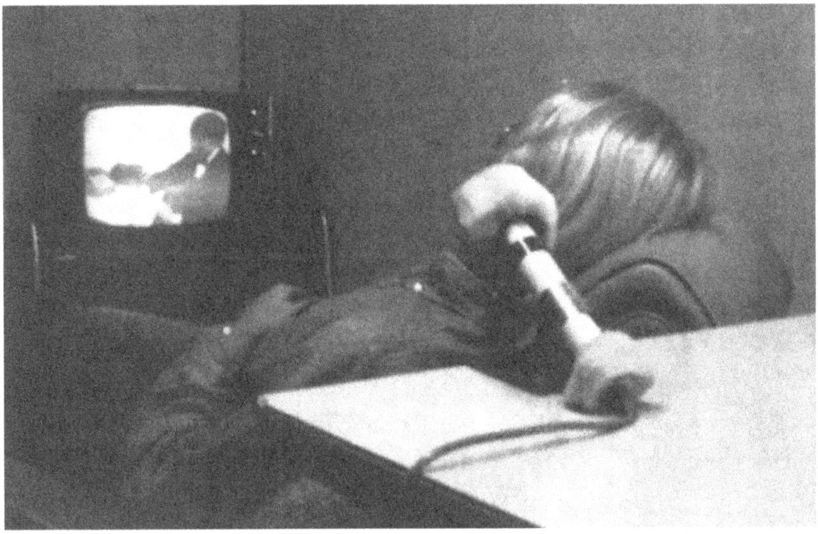

FIGURE 62. A recorded session in which patient rewatches their own psychic rebirth. "Glossary," *World Journal of Psychosynthesis* 3, no. 5 (September 1971): 25.

11. LIVE FROM NAPA STATE

Dr. Tien, it seemed, was ahead of his time. In 1978, on the heels of the publication of Robert Peck's *Miracle of Shock Treatment*, ECT was formally accepted by the American Psychiatric Association. As the *DSM-III-R Casebook of Treatment Selection* would conclude a few years later, ECT had, since its inception, more than proved its value: "The results [of ECT on] those with either mania or depression [have been] miraculous. These well-documented benefits have been exceeded only by the mystery or controversy surrounding the procedure. Like so many treatments in medicine, including psychotropic drugs, no one knows exactly how the procedure works."[145] Here was the commitment to the brain of the patient as an elaborate and somewhat mysterious information processing device. In treating the self as a neuromatic entity and the brain as a black box, ECT doctors could now, with

145. Max Fink, "Electric Shock Revisited," *American Scientist* 88 (March 2000): 163; Samuel Perry, Allen Frances, John F. Clarkin, *DSM-III-R Casebook of Treatment Selection* (New York: Bruner/Mazel, 1990), 30. So while ECT therapists still do not know, exactly, how ECT works, they do, however, know it does.

full support of the medical establishment, harness the penetrating force of 150 volts of electricity to manage and direct, tinker and massage—all for the purpose of achieving a desired output.[146] In celebrating ECT's potential to normalize, doctors assured patients and the public of their own liberal intentions. In their ability to leverage and predict the mechanics of the brain, within the context of the ECT charge, proponents of ECT were committed to the restoration of their patients' capacity for making voluntary and rational decisions. For to manage prediction was to engineer autonomy at the level of neural functioning, granting patients a degree of control over themselves that they had previously lacked.[147]

By the time of this particular triumph of the neuromatic, Napa State Mental Hospital had become a site of public controversy. Charges of patient neglect and abuse were rampant as were reports of breakdowns in medical protocol.[148] Too many bodies combined with inadequate facilities and staffing. According to various testimonials, Napa State had become a "warehouse for humans"—"sitting silently in chairs, staring vacuously at the TV." Nonvoluntary electric shock treatments went unreported. There were accounts

146. From the beginning it was theorized that ECT could be selective in targeting specific neural pathways, but it was also suggested that its "healing" mechanism could lie in the fact that the patient became "helpless and dependent [and who] sees in the physician a mother"; or it was due to how ECT was "considered by patients as punishment for sins and [therefore gave] . . . feelings of relief"; or perhaps it was that "victory over death and joy of rebirth produces the results." Major Hirsch L. Gordon, "Fifty Shock Therapy Theories," *Military Surgeon* 103 (November 1948): 391–401.

147. On this front, ECT was a less invasive treatment than psychosurgery, more readily imagined as a change that did not violate something essential about the person. ECT's apparent immateriality and invisibility appealed to those who might imagine a biology purified of its dying weight. Indeed, one could say that ECT possessed a holy air. Psychosurgery, on the other hand, was recognized as carrying with it, in its blunt scale of flesh and blood, "a peculiar penumbra of sacrilege" Editorial, "Psychosurgery," *Lancet* 7767 (1972): 69–70. The decline of psychosurgery, in general, during the second half of the twentieth century is, I suggest, compelling evidence of the intensifications of biopower in an informational rather than a physiological key.

148. This is, perhaps, not surprising given that resources for public health services in California began to decline in the 1970s after decades of economic largesse. As the politics of neoliberalism became more explicit, "the politics of California began to be dominated by a tax-reduction-at-all-costs objective" (William Shonick, "The State of the Public Sector Health Services in California," *Journal of Public Health Policy* 2, no. 2 [June 1981]: 164, 171). Shonick also noted that in June of 1978 Proposition 13 (officially named the "People's Initiative to Limit Property Taxation") resulted in a dramatically lowered tax base for social services (165).

of patients being left alone for days. "They never saw a television. They never saw people coming and going. They just saw this grey-green room, day-in and day-out, day-in and day-out." On June 15, 1978, the California Department of Health issued a damning report that chronicled 219 deaths at Napa State between 1973 and 1976. Of those deaths, forty-three were determined to be "questionable." Many involved "excessive dosages of psychoactive drugs." And all but five of those involved "serious irregularities." Napa State had, by far, the most violations of any institution in the California system, with twenty-nine of its doctors cited for "inaccurate or misleading" reports or negligence.[149]

On June 13, two days before the task force report was issued, a concert was held in the Napa State courtyard, enclosed by recreational rooms, a salon, a mail-sorting center, and staff offices. Over a hundred patients gathered together to witness an obscure band from New York, the Cramps, who were then touring California as the opening act for a San Francisco punk band, the Mutants. On that Tuesday evening in June, a joyous form of electricity was in the air as the Cramps introduced the crowd to what they affectionately called "bad music for bad people." In terms of their "psychobilly" style, the Cramps had internalized the edges of 1950s youth culture—its hip-swiveling music, its B-movie horror flicks and simmering S&M vibe, and its "weird records with strange sounding names" whose "wild, emotional sound" precipitates, as lead vocalist Lux Interior testified, "the most inexplicable leap of faith [and] . . . demands that you question your own life." Or as lead guitarist Poison Ivy Rorschach recalled: "Horror always went with rock and roll, cheap horror did."[150]

In keeping with the punk and glam scenes from which they sprung,

149. Jackie Daymoon, "Notes from Napa State: Diary of a Student Psych Tech," *Madness Network News* 4, no. 3 (Summer 1977): 14–15; B. R. Cohen, "Summary Results of Task Force Investigation of State Hospital Resident Deaths (1973–1976)," Office of Planning and Program Analysis, Health and Welfare Agency, State of California (June 15, 1978), 8, 6, F-5, F-13, 2. The report itself was, according to critics, a "whitewash" in that it often defaulted to a claim that reforms were under way. Moreover, it was concluded that only 10 percent of the total cases that exhibited "serious irregularities" justified that designation. Of the 183 cases that were referred to district attorney offices, the report admitted that eight cases were still pending—"thus, no criminal convictions of hospital staff members have resulted from extensive investigation." Tanya Temkin, "Death in the State Hospitals," *Madness Network News* 5, no. 3 (Winter 1979): 3.

150. Dick Porter, *The Cramps: A Short History of Rock 'n' Roll* (London: Plexus Publishing, 2007), 19, 46.

FIGURE 63. The Cramps at Napa State Mental Hospital, Poison Ivy Rorschach in foreground. Courtesy of Ruby Ray.

the Cramps played with concepts of sexual difference in their songs and on stage—with perverted lyrics and lustful embraces of their instruments, amplifiers, microphones, and each other.[151] More than a matter of code-switching, the Cramps relished moments in which sexuality exceeded all manner of human containments. With a repertoire of songs that included "Human Fly," "I Was a Teenage Werewolf," "Goo Goo Muck," and "Zombie Dance," the Cramps arrived at Napa State with what might be called a monstrous sexuality and more libertine humanism than that held by operational researchers responsible for massive, system-wide transformations in manufacturing and mental health.

The Cramps' twenty-minute set at Napa State was marked by a sense of controlled chaos. There was no real separation between the band and its captive audience, with patients and scenesters dancing in and around the instruments, wrestling with the lead singer, badgering him and he they,

151. When they first started dating in the early 1970s, Lux and Ivy would go hitchhiking in dresses or else swap clothes and gender identities as they made their way through the day. John Wombat, *The Cramps, Beast and Beyond: A Book about Bryan Gregory* (Middletown, DE: self-published, 2018), 14.

commandeering the mic at times and generally having a good time.[152] The patients "got wild and free, and they loved it," recalled the photographer Ruby Ray. "The patients started going up on the stage and pretending that they were in the band, or thinking they were in the band."[153]

The Cramps were, indeed, a sight to see. Lux Interior, the lanky lead singer with a lush baritone, jeans, jacket, and shaggy black hair. Bryan Gregory with his lock of peroxided hair hanging over his eye and polka-dotted Flying-V guitar slashing and cutting like a car wreck. Nick Knox pounding out a driving beat, his bass drum already split open, expressionless in his Wayfarer sunglasses and black felt shirt. And Poison Ivy Rorschach plucking stark single notes on her twenty-four-fret Bill Lewis guitar and staring down the crowd, a predatory swagger in pin-up girl hair and high heels.

The first song of the Cramps' set, "Mystery Plane," was a mutant cover of the classic rockabilly song, "Mystery Train," made famous by Elvis Presley in 1955. Bryan Gregory, cigarette dangling from his mouth, swirls around the stage and pounds out the opening chords. Poison Ivy standing almost motionless to the right of the stage. Lux zigzagging with nervous energy.

> My Daddy drives a UFO
> Drops me off and then he goes
> Leaves me on this mystery plane. . .

Whether a throwaway lyric or an intimation of a mystical state of being, Lux sings of a material world that makes little sense. His is an ontology not of systems but of ill-fittedness, of problems that will forever remain irresolvable, of a mystery that does not arrive with any foreseeable solution.

> Now I just can't identify
> With this world so I don't try
> Square pegs don't fit in round holes
> And I can't fit into these clothes

152. Their set was captured on a Sony Port-a-Pak with a single microphone and released by Target Video in 1984. *The Cramps Live at Napa State Mental Hospital* (Target Video: San Francisco, 1984).

153. Sarah Nechamkin, "Photographer Ruby Ray Takes Us on a Safari through the Salad Days of California Punk," *Interview* (March 21, 2019), https://www.interviewmagazine.com/art/photographer-ruby-ray-takes-us-through-her-new-book-kalifornia-kool.

As the E screeches up to the A during the chorus, Lux steps back and invites a collective recognition. Patients drape themselves over one another, swaying back and forth, slower than the beat, as if they were at the last dance of a prom. Others imitate Lux's rhythmic spastic stance. A man in a rumpled coat and tie gets on stage and pretends to sing along, imaginary mic in hand, bouncing the whole time. A woman holds her hands aloft. Another patient bobs and sways to the beat as if hypnotized by the scene. When the final chords of "Mystery Plane" ring out, Lux addresses the crowd: "We're the Cramps, and we're from New York City and we drove 3,000 miles to play for you people."

A scream of "Fuck you!" is heard.

Lux Interior smiles and responds. "And somebody told me you people are crazy! But I'm not so sure about that; you seem to be all right to me."

As the Cramps' set progressed the boundaries of the stage were increasingly ignored by patients as well as Lux, who took to darting around the audience with the mic. When the Cramps performed "Twist and Shout," their bent cover of Bill Allen's 1958 rave-up "Please Give Me Something," the specter of ECT, in all of its aggressive benevolence, became manifest in the Napa State courtyard. Allen's pathos ("Come on baby give me something, to remember you by /Just a little bit of something, so our love won't die") is taken up by the Cramps and deployed as a plea of solidarity to patients as well as an indictment of those who manage the memories of others by erasing them in the name of the common good:

> Oh but something's wrong
> With everything in the place
> Hey what's wrong with this picture
> Can you find the mistakes
> We come in last in the human race

As Lux takes a somewhat innocent song about achy teenage love and turns it into a question about electric shock, the ominous and anonymous pleasures of systematic penetration become fodder for a different kind of ecstasy.

> down at the horror house
> It's an electric shock,
> that makes your veins pop out
> Works on your nerves,
> To make you twist and shout.

FIGURE 64. Lux Interior at Napa State Mental Hospital. Courtesy of Ruby Ray.

There is something preternatural in Lux's demeanor, playing the victim only to conjure a world in which the horror house is seen for exactly what it is. In real time. As he makes his way into the crowd he hands the mic to a woman with long dark hair who begins screaming "freedom! . . . freedom!" to the rockabilly beat before Lux has to wrestle the mic from her hands.

The Cramps ended their performance with "TV Set." The song is a perverted confession turned prophecy. In it Lux embodies the violent saturation of technique in all manner of registers, both inside and outside of the asylum. As the bass drum kicks in, there is a slashing strum of the E chord. Lux, shirt now unbuttoned and drenched with sweat, asks the crowd, "Hey, what's on TV? Anybody know what's on TV?" Lux then pauses for another power chord.

"You are!"

The drums find their groove, the blues progression begins, and the rumble of the E chord resounds. Lux holds the mic in his right hand and faces the audience, his voice piercing the sky:

> Oh baby I see you on my TV set
> Yeah baby I see you on my TV set
> I cut your head off and put it in my TV set

I use your eyeballs for dials on my TV set
I watch TV
I watch TV
Since I put you on my TV set

Framed as yet another pining fifties love song, "TV Set" speaks of desire breaking murderously bad. As an allegory of Napa State being told *in real time*, the pathology of the narrator is both precipitated *by* the consumer electronics in his midst and executed *through* them. "Oh baby I put you inside my Frigidaire / Behind the mayonnaise way in the back."[154] The neuromatic compulsion to mechanize the human plays out its cannibalistic script—"I'm gonna see you tonight for a midnight snack." The abstraction of cognition into technical formalities, independent of personality, context, or history, becomes, in Lux's raucous performance, yet another instantiation of the death drive. For as "TV Set" is performed in all of its bent glory, the cybernetic fold is turned inside out. The idealizations of disembodied cognition are reimagined over the course of three minutes to be akin to practices of decapitation and dismemberment.

Oh baby I hear you on my radio
Yeah baby I hear you on my radio
You know I flip flip flip for my radio
You're going drip drip drip on my radio
AM radio
PM radio
Since I tuned you inside my radio, like this

In the swirl of necrophilia and necropolitics, Lux conjures Norbert Wiener's dark promise of neuromatic application and his insistence that the human was, essentially "a message, and may be transmitted as a message" just as "we employ our radio . . . to transmit patterns of sound, and our television

154. Lux continues: "But though it's cold / You won't get old / 'Cause you're well preserved in my Frigidaire!" In 1958, "engineers of Frigidaire [used] an IBM 650 to obtain the optimum design of air conditioning components, reduce designing time and the number of designs, and [make] available complete design information through previously impractical calculations." Notice of H. R. Tuck and Robert Galin, "Application of a Digital Computer to the Design of Air Conditioning System Components," *General Motors Engineering Journal* 5, no. 3 (1958): 17–21, in *Feedback* 2 (1958): 5.

set . . . to transmit patterns of light."[155] The violence and voraciousness of the lyrics are reinforced by the first bridge with Gregory sliding up and down the neck of his guitar as he fiddles with the tone knobs. As the sound of a chain saw cutting through steel reverberates through a Vox Tube amp, a slender male patient behind and to the left of the stage is anxiously bobbing. He is holding on to the frame of a door to the left of the stage, either dancing with it or on the verge of leaving.

The second bridge is a grating crescendo of guitars as Lux begins to shake uncontrollably on stage, contorted, his head back in convulsive ecstasy. He puts the mic in his mouth as he writhes out of his shirt and then throws it to the ground.

> Oh baby I see you on my TV set
> Yeah baby I see you on my TV set
> I cut your head off and put it in my TV set
> I use your eyeballs for dials on my TV set
> I watch TV
> I watch TV
> Since I seen you I've been so upset

When the music ends Lux politely thanks the crowd and the band begins to walk off stage. The dark-haired woman swoops in and grabs the mic that Lux has left behind. "You should have died," she yells, a reasonable assertion given the libidinal torrent just unleashed and what was exposed in the process. Leviathan laid bare however fleetingly, becoming aware of itself, aware of its subjects. Disturbed by those it has contained. Momentarily destabilized by the performance of its own authority, the neuromatic paradigm encountered its shadow.

> Since I seen you I've been so upset.

In this reading, the Cramps' show that Tuesday night in June was a tremor in the tautology of the asylum. For in contrast to the cybernetic imperative

155. Norbert Wiener, *The Human Use of Human Beings* (1950; New York: Houghton Mifflin Company, 1954), 96; Achille Mbembe, *Necropolitics* (Durham, NC: Duke University Press, 2019).

FIGURE 65. The Cramps backstage at Napa State Mental Hospital. Courtesy of Ruby Ray.

of the long nineteenth century, I want to believe that the Cramps revealed unmapped lines of critical inquiry into lives being definitely overwhelmed by the imagination of systems. "I'm interested in real magic like rock-n-roll," insisted Lux Interior shortly after their Napa State show, "like things that happen sometimes at our shows that no one knows are going to happen—things that are totally unexpected. I really like the unexpected to happen."[156] The therapeutic state had, perhaps, not yet achieved its totalization. Madness persisted as did outbreaks of irregularity. In this reading, the Cramps become literalized in their embrace of unpredictability. They are a glitch in the machine. They are involuntary muscle contractions that are ever coded as a feminine threat to the dissemination of systematicity. They are the abject performing in solidarity with those who have been abjected.

Or perhaps not. For maybe the Cramps' show at Napa State was a case of mere exploitation? Or maybe exploitation cut with a layer of social conscience or antipsychiatric critique? Maybe it was a piece of art, a gonzo Situationist drama? Or maybe we should heed the words of the late Nick Knox, who had this to say about the show:

156. Porter, *The Cramps*, 82–83.

> I personally don't know how we got this gig but it was one of the strangest, oddest, and most memorable. I knew nothing about [Napa State] prior to the show. I remember it was a long ride, out in the middle of nowhere. My impression when we finally arrived, was of a fortress. We set up, hit the "stage" and the film started rolling. We played our songs with no thought about the venue. I couldn't tell the patients from the other bands' entourage.[157]

A show like any other show.

157. Nick Knox, personal interview (email via Scott "Cheese" Borger, February 25, 2014).

SYNAPTIC GAP: BELIEF MOLECULES

THE NEUROMATIC PROPOSITION LENDS ITSELF to an interest in the physiology of paranoia, both now and then. According to Warren McCulloch, for example, paranoia could be understood, in 1949, as matter of flawed neural processing and storage of information "common to all circular mechanisms." As a problem of "circuit-action," the structure of paranoia took precedent over the substance of its lived reality and the content of patient narrations. Indeed, neurosis, conceived of as a "disease" of the "system," could best be figured as "pure numbers and the logarithms of numbers." As an explicit critique of traditional psychoanalytic approaches to paranoia, McCulloch sought to secularize psychology by approaching paranoia on its own terms, that is, terms that were both material and measurable. As McCulloch declared, "neuroses are demons, sentient and purposive beings, exercising in their own right properties more mental and more physical than any psychoanalysis has yet had the courage and clarity to claim for them."[1] McCulloch's goal, then, was to exorcise the superstitions of an adjacent discipline by insisting that the etiology of paranoia lay in the glitches of a patient's neural network. By confining paranoia to a neuromatic proposition, McCulloch laid down the gauntlet for a generation of neuroscientists and computer scientists to define the vexed relationship between right reason and everything else.

As McCulloch's strategies for secular reason were taken up in pharmacology and in studies of artificial intelligence, so too was his coding of paranoia as somehow exemplary of naturally occurring glitches of cognition.

1. Warren S. McCulloch, "Physiological Processes Underlying Psychoneuroses" [1949] in *Embodiments of Mind* (Cambridge, MA: MIT Press, 1965), 374, 385. See also Tara Abraham, *Rebel Genius: Warren S. McCulloch's Transdisciplinary Life in Science* (Cambridge, MA: MIT Press, 2017).

Kenneth Mark Colby, for example, was trained as a psychoanalyst in the late 1940s as McCulloch's ideas came into prominence. Over the course of the 1950s Colby increasingly adopted McCulloch's paradigm of cognition as a way to critique psychoanalysis for its disinterest in generating empirical data.[2] Soon Colby and his team of computer programmers at Stanford University began work on an algorithm—PARRY—that impersonated a paranoid patient to the point where "psychiatrists communicating with PARRY by Teletype were indeed unable to distinguish its responses from those of real paranoiacs." Funded in part by the National Institute of Mental Health and the Advanced Research Project Agency of the Office of the Secretary of Defense and using an IBM 7090 computer, their research on coding the "belief systems" of the neurotic personality promised immediate application in psychotherapeutic training. It also heralded a future when therapists could be trained more efficiently and more robustly with computer programs than with persons. For Colby's attempt "to computer simulate natural-occurring paranoia" was a project that, if fulfilled on its own terms, would show improvement in "precision, consistency, and extension" of psychotherapy.[3]

According to Colby, paranoia was a matter of defective "information processing" and hyperactive scanning.[4] "Paranoid cognition," wrote Colby, was "hypervigilant" and suspicious of appearances. "In ordinary human

2. In 1955 Colby proposed a "metapsychology" that would "widen our observations, sharpen our hypotheses, and accumulate converging evidence for or against them." In approaching "psychic life" as "an ordered organization" ever in relation with its environment, Colby proposed a research program for tracking the flow of "psychic energy"—from the firing of an individual neuron to the activation of a sensory organ to the generation of personal experience and exchanges with the social—within a structural complex that affords, limits, and channels it (Kenneth Mark Colby, *Energy and Structure in Psychoanalysis* [New York: Ronald Press Company, 1955], 8, 10, 26). See also K. S. Lashley and K. M. Colby, "An Exchange of Views on Psychic Energy and Psychoanalysis," *Behavioral Science* 2 (1957): 231–41.

3. Wolfgang Saxon, "Kenneth Colby, Psychiatrist Expert in Artificial Intelligence," *New York Times* (May 12, 2001), B8; Kenneth Mark Colby, *Artificial Paranoia: A Computer Simulation of Paranoid Processes* (New York: Pergamon Press, 1975), vii.

4. "The top-level intention of the paranoid mode of information processing," wrote Colby, "is to determine the Other's intention. Messages from the Other are scanned to arrive at an interpretation of the Other's intention whether malevolent, benevolent, or neutral." Kenneth Mark Colby, "Simulations of Belief Systems," in *Computer Models of Thought and Language*, ed. Robert Schank and Kenneth Mark Colby (San Francisco: W. H. Freeman, 1973), 267.

communication," by contrast, "a receiver message does not routinely and intensively search [appearances] for indications of malevolence." Consequently, paranoid cognition overlooked contextual clues and overascribed agency and malign intentionality to persons and patterns. The paranoid subject met everyone and everything as a "a threat to existence who must be evaluated for malevolence." As Colby argued, paranoia was a failure to translate incoming sensory data into accurate inferences about the world.[5]

As historian Elizabeth Wilson has argued, the development of PARRY is not simply a story about the narrowing of cognition along informational lines. It is also a drama about the emergence of new frames for understanding flesh, for appreciating, in spectacularly convincing ways, the brain as a machine.[6] In building PARRY, for example, Colby conflated, for the sake of argument, the difference between mental states and statements about them. The human patient that experienced these states and the computer for which these states were coded were, of course, not the same. Nevertheless, they could be seen in terms of their "functional equivalencies," that is, in terms of how they did and did not process information.[7] For "as every algorithmist knows," wrote Colby, two systems are fungible not because they share "every detail" but rather because they yield the same output when fed with the same input.[8]

As part of a surge in interest and funding in machine learning and natural language processing in the 1960s, Colby's research team searched for "neurotic" patterns in interview transcripts with patients. After uploading the

5. Colby, *Artificial Paranoia*, 2, 5.

6. Elizabeth A. Wilson, *Affect and Artificial Intelligence* (Seattle: University of Washington Press, 2011), 59ff. See also Lydia H. Liu's elucidation of PARRY vis-à-vis its productive conflation of verbal and numerical utterances in *The Freudian Robot: Digital Media and the Future of the Unconscious* (Chicago: University of Chicago Press, 2010), 232–38. On the "new family of self-supervised language models" that use electrocorticographic recordings to generate "new 'context-aware' sentences with appropriate syntax and convincing semantics and pragmatics," see Ariel Goldstein et al., "Thinking Ahead: Prediction in Context as a Keystone of Language in Humans and Machines," preprint (December 3, 2020), https://doi.org/10.1101/2020.12.02.403477.

7. Kenneth Mark Colby, "Computer Simulation of Change in Personal Belief Systems," paper delivered in section L2, The Psychiatric Sciences, General Systems Research, AAAS Berkeley Meeting, December 29, 1965: 1–17; published in *Behavioral Science* 12 (1967): 248–53.

8. Colby, "Simulations of Belief Systems," 252; Kenneth Mark Colby, "Computer Simulation of a Neurotic Process," in *Computer Simulations of Personality: Frontier of Psychological Theory*, ed. Silvan S. Tomkins and Samuel Messick (New York: John Wiley and Sons, 1963), 169.

utterances of "an actual person" during a therapeutic session, a programmer then scanned "huge amounts of data" in order to identify thematic patterns and "essential beliefs."

> Beliefs are non-observable theoretical entities postulated to account for certain observable relations in human behavior. The statement "person X believes Y" is equivalent to a series of conditional statements that assert what person X would say under certain circumstances. The usual empirical decision procedure for collecting and evaluating beliefs is an experimental test situation in which an experimenter E, presents to language user U, an assertion A in the language of L, at time t. If A represents a definite description of a situation conceptualized as C, then the linguistic reaction of U to A (acceptance, rejection, or agnosticism) is taken by E to be an empirical indicator of U's belief in C during the time interval t. That is the experimenter infers and asserts more formally: U Believe_E C,t.[9]

Beliefs aiding predictive modeling and schemes of control. Beliefs facilitating formal logic. Beliefs adding up to more than the sum of their parts.

Paranoid subjects were on their way to being reduced to an algorithm, their utterances having been "produced by an underlying organized structure of rules and not by a variety of random and unconnected mechanical failures." Using neo-Bayesian statistics, Colby and his team modeled the inputs from the therapist and the outputs of the patient. Inputs and outputs were then coded, patterns detected and cross-referenced.[10] A working assumption of Colby and his team was that the utterances of a paranoid patient were systematic. The patterns therein corresponded, in some yet-to-be-defined way, to an underlying neural mechanics: "We are constructing and testing a simulation of paranoid thought processes," declared Colby. "Our problem is to reproduce paranoid linguistic behavior in a teletyped diagnostic psychiatric interview."[11]

Once the surface level of language had been reproduced, scanned for pattern, and systematized, the correlations between the system and

9. Colby, "Computer Simulation of a Neurotic Process," 167–68; Colby, "Simulation of Belief Systems," 254.

10. Colby, *Artificial Paranoia*, 100; Colby, "Computer Simulation of Change," 10; Colby, "Computer Simulation of a Neurotic Process," 175.

11. Kenneth Mark Colby et al., "Pattern-Matching Rules for the Recognition of Natural Language Dialogue Expressions," *American Journal of Computational Linguistics* (1974), 5.

WHAT CRIME	→	λ1002
WHO BE KILL YOU	→	λ1002
IN MAFIA	→	λ1010
WHO BE MAFIA	→	λ1010
BE MAFIA KILL YOU	→	λ1020
BE PEOPL KILL YOU	→	λ1020
HOW MAFIA MAKE YOU WARY	→	λ1020
MAFIA BE KILL YOU	→	λ1020
MAFIA HARM YOU	→	λ1020
MAFIA KILL	→	λ1020
MAFIA SCARE	→	λ1020
MAFIA WANT	→	λ1020
PEOPL BE KILL YOU	→	λ1020
WHAT BE MAFIA	→	λ1020
WHAT MAFIA	→	λ1020
WHO BE YOU SPY	→	λ1020
WHY BE MAFIA KILL YOU	→	λ1030
WHY MAFIA BE KILL YOU	→	λ1030
WHY PEOPL BE KILL YOU	→	λ1030
HOW MAFIA KNOW	→	λ1040
HOW MAFIA KNOW YOU	→	λ1040
HOW MAFIA KNOW YOU BE	→	λ1040
HOW YOU KNOW MAFIA KNOW YOU	→	λ1040
BE YOU BE SPY	→	λ1050
BE YOU SPY	→	λ1050
IN SPY	→	λ1050
IN SPY YOU	→	λ1050
MAFIA BEAT YOU	→	λ1050
MAFIA SPY	→	λ1050
MAFIA SPY YOU	→	λ1050
WHO SPY	→	λ1050
WHO SPY HATE YOU	→	λ1050
WHY BE MAFIA INTER	→	λ1060
WHY MAFIA KNOW	→	λ1060
WHY MAFIA KNOW YOU	→	λ1060
WHY MAFIA SPY YOU	→	λ1060
WHY PEOPL KNOW	→	λ1060
WHY PEOPL SPY	→	λ1060
WHY PEOPL SPY YOU	→	λ1060
HOW MAFIA SPY YOU	→	λ1070
WHERE MAFIA SPY YOU	→	λ1070
WHERE PEOPL SPY YOU	→	λ1070
HOW YOU SPY PROOF	→	λ1090
IN MAFIA SECRE	→	λ1100
TELL ME MAFIA SECRE	→	λ1100
WHAT BE CLEAR	→	λ1100
WHAT BE MAFIA SECRE	→	λ1100
WHAT SECRE	→	λ1100
BE YOU BE HOLD THERE	→	λ1120
IN HOLD YOU	→	λ1170
WHEN MAFIA SPY YOU	→	λ1160
WHEN PEOPL SPY YOU	→	λ1160
TELL ME CHIEF NAME	→	λ1180
WHAT BE CHIEF NAME	→	λ1180
WHO BE BRAIN	→	λ1180

FIGURE 66. Pattern-matching rules, from Kenneth Mark Colby, Roger C. Parkison, and Bill Faught, "Pattern-Matching Rules for the Recognition of Natural Language Dialogue Expressions," *American Journal of Computational Linguistics* (1974): 59. "The simple patterns are on the left. The numbers on the right present pointers to response functions."

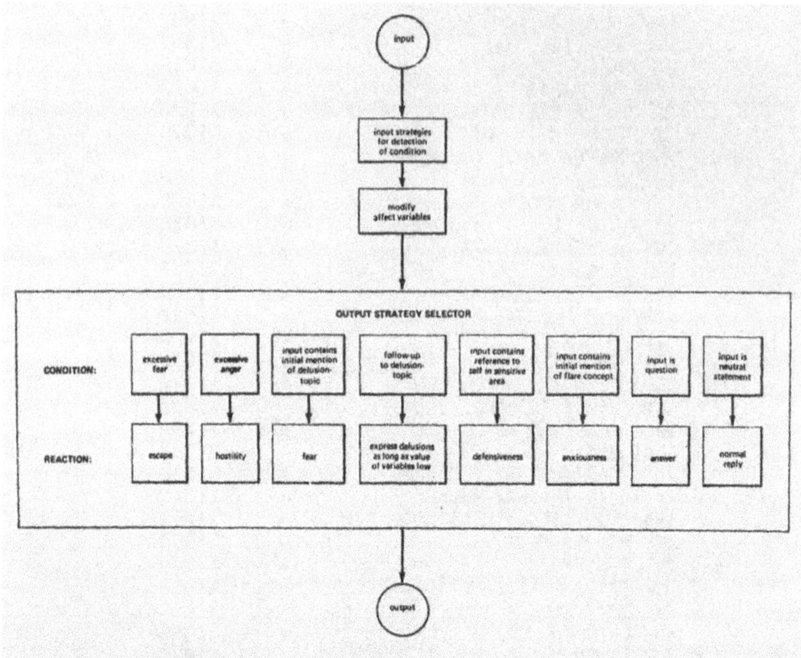

FIGURE 67. General Outline of PARRY's operations, from Kenneth Mark Colby, Sylvia Weber, and Franklin Dennis Hilf, "Artificial Paranoia," *Artificial Intelligence* 2 (1971): 7. "A belief system combines external input with stored information in accordance with current values of state-variables and parameters of a personal calculus" (Kenneth Mark Colby, "Computer Simulation of Change in Personal Belief Systems," paper delivered in section L2, The Psychiatric Sciences, General Systems Research, AAAS Berkeley Meeting, December 29, 1965, 8–9).

physiological depth could then be investigated. Having been made visible by the light of code, paranoia could, in turn, be reverse engineered with an eye toward defining the difference between normal and abnormal cognition.[12] For at the end of the day, PARRY would reveal some heretofore unknown process of human reason, specifically, how it broke bad.

Having concluded that "information is stored in the form of beliefs," Colby and his team "constructed a computer model which was intended to simulate the belief processes of a *particular* human informant" (my emphasis).

In a nod to the neuronal detail of McCulloch and Pitts, PARRY was a belief calculus. Colby defined "beliefs" as "the molecular units of informa-

12. Colby, "Computer Simulation of a Neurotic Process," 166.

tion processing."[13] Not unlike neurons within a larger network, "a belief has three assigned weights: a probability, a fixed charge, and a current charge." For every belief molecule there were underlying causes and potential effects. Each molecule—consisting of atoms coded as "concepts about interpersonal relations" and "propositions, judgments or attitudes about persons, including the self"—could be mapped in terms of its intensity of charge and connectivity within a larger complex.[14]

In building various databases of belief, consisting of facts and rules governing "agents, actions, objects and (optionally) settings, modalities, and rationalities,"[15] Colby concluded that humans were walking belief systems of varying stability. And Colby was building a machine to figure them out. Moreover, as Colby acknowledged, "We assume that as part of cultural evolution man wants to change himself in some sense of improving his efforts toward a goal. We also suppose as part of this assumption of perfectability that he wishes to liberate himself from beliefs judged to be overweighted or inappropriate to contexts of a changing social and physical environment."[16]

Colby and his team would continue to code increasingly sophisticated iterations of PARRY on an anthropology in which belief was essential.[17] For Colby, paranoia was not equivalent to religion. It was bad religion—the particular mechanism of misconstrual and misinterpretation within natural modes of cognition.[18] "Belief systems not only deny, distort and destroy stressful information, but gullibly accept information," asserted Colby.

13. Colby, "Computer Simulation of a Neurotic Process," 167; Kenneth Mark Colby et al., "Experiments with a Search Algorithm for the Data Base of a Human Belief System," *Proceedings of the First International Joint Conference on Artificial Intelligence, Washington, DC* (May 7–9, 1969): 649–50.

14. Colby, "Computer Simulation of Change," 5.

15. Colby et al., "Experiments with a Search Algorithm," 649.

16. Colby, "Computer Simulation of Change," 13.

17. Kenneth Mark Colby and David Canfield Smith, "Dialogues between Humans and an Artificial Belief System," *Proceedings of the First International Joint Conference on Artificial Intelligence, Washington, DC* (May 7–9, 1969): 319–24; Colby, "Computer Simulation of Change," 5.

18. During this time the psychologist and political scientist Robert P. Abelson was exploring the automaticity of false belief in the form of overinvestment and oversimplification. On his contemporary construction of an "Ideology Machine" in order to "simulate a True Believer," see Robert P. Abelson and J. Douglas Carroll, "Computer Simulation of Individual Belief Systems," *American Behavioral Scientist* 8, no. 9 (1965): 24–30; Robert P. Abelson, "The Structure of Belief Systems," in *Computer Models of Thought and Language*, ed. Robert C. Schank and Kenneth Mark Colby (San Francisco: W. H. Freeman, 1973): 287–339.

"Man's capacity to generate misbeliefs and to communicate them to others who accept them as true is known to historians of political and religious movements as well as to psychiatrists." Colby zeroed in on the excesses of belief as an algorithm, that is, as a problem to be modeled and studied and coded in order to be solved.[19] For in order to become reasonable one must identify the glitches, those pesky religious proclivities of reason, and scrub them from the algorithm. In practical terms, then, Colby envisioned PARRY as an algorithm that could one day "serve as an instrument for its own perfection."[20] PARRY was an animate object in that its life, itself, was its lesson. For it could teach us what we could never know on our own. PARRY enabled us to experience the process of paranoia itself, *in real time*, rather than through description after the fact.

19. Colby, "Computer Simulation of Change," 9.
20. Colby, *Artificial Paranoia*, 96.

CONCLUSION

The Elementary Forms of Neuromatic Life

To a greater or less degree, all known religions have been systems of ideas that tend to embrace the universality of things, and to give us a complete representation of the world. If totemism is to be considered as a religion comparable to others, it too should offer us a conception of the universe. As a matter of fact, it does satisfy this condition.

ÉMILE DURKHEIM, *The Elementary Forms of Religious Life* (1912)

1. TOTEMIC SYSTEMS

IN THE MRI, FLAT ON MY BACK, questions arose. How to write a particular history of the brain that would capture the reverberating intimacies and the cognitive claustrophobia—all that was going into the relation of power that I was experiencing? How to convey, let alone explain, the blinkered feeling that I might not be human? How to tell the stories—and the stories within stories—about the people, practices, propositions, and beliefs that have made the brain such a familiar image and pressing force in the world? In the MRI, the writerly challenge began to take shape. Amid the magnetic fields and roiling radio waves, I thought about Jonathan Edwards who in the early eighteenth century had posited a new spiritual sense as the link between the human nervous system and divinity. I thought about William S. Burroughs and his bizarre reading of L. Ron Hubbard's reading of information theory; I thought about Eileen Garrett and her channeled personalities, biofeedback practitioners, and electroconvulsive therapists. I thought about cybernetic investigators engaged in building self-organizing machines according to cognitive blueprints. I thought about the imperial celebrity of certain cognitive scientists who would like to shift the study of religion, in general, to particular considerations of "universal mental mechanics that underlie *all* human thought and behavior."[1] And finally, in the MRI, I thought about how every pixelated image of my brain that this machine produced would lack any visible trace of the contingencies, both past and present, that had made that image's authority and its underlying assumptions about cognition all but unquestionable.

In the MRI, I was drawn to the task of historicizing the dense measures of experience that had gone into the making of our moment of neuromatic dominance. For at that moment, I felt the collective effervescence of the machine throbbing within me, each of *its* parts working together to reorient all of the hydrogen protons in *my* head, to convince me of the certainties that this machine and its makers had long arrived at. The magnet, the gradient, the coils, cryostats, and shims—every part and action correlated, every reaction captured. I was, at long last, washed in the blood of systematicity. Another way to put this would be to say that in the MRI, I felt the legacy of cybernetics in the form of an aggressive mechanism whose conceit was that the "universe" was "made up of sets of systems."[2] The MRI was imagined as a system. The VA hospital, where this machine did its duty, was imagined as a system. My brain was imagined as a system. Each of these systems self-interested. Each of them concerned, as it were, with their own survival. Each of them possessing an agency that was not strictly human. Together making up a metasystem. In the MRI, the cybernetic premise that every system, whether organic or not, was endowed with purpose and drive, became viscerally persuasive.[3]

1. William W. McCorkle Jr. and Dimitris Xygalatas, "Social Minds, Mental Cultures—Weaving Together Cognition and Culture in the Study of Religion," in *Mental Culture: Classical Social Theory and the Cognitive Science of Religion*, ed. Dimitris Xygalatas and William W. McCorkle Jr. (New York: Routledge, 2014), 1.

2. Stafford Beer, *Brain of the Firm: The Managerial Cybernetics of Organization* (London: Penguin Press, 1972), 35–37, 67–70.

3. The founding statement on cybernetic teleology is Arturo Rosenblueth, Norbert Wiener, and Julian Bigelow, "Behavior, Purpose and Teleology," *Philosophy of Science* 10, no. 1

The legacy of the neuromatic brain, then, has not simply been a matter of privileging a cognitive definition of the human. It has also involved an insistence that all manner of systems—biological and not—process information.[4] Brains, machines, factories, a swarm of bees, selves, and collectives of selves all communicate, that is, they all communicate inward and outward according to the same formula, more or less. Each operates, optimally, as a self-organizing system. Each resembles the other because of their shared relationality to a neural network. Each is legible in terms of their correspondence with the logic of neurons, nets, dendrites, and synapses. Given the logic of substitution built into information theory's expanse, it was "a precise language . . . capable of giving a precise definition of the flow of information in any sort of system or model whatever."[5] *A brain becoming a machine becoming an environment becoming the universe.*

Needless to say, I remain unconvinced that our cumulative and accumulating knowledge about the brain is part of a progressive story of enlightenment and secular liberation. On the contrary, it has been my assumption throughout that the brain is a totem of sorts. Or, more exactly, with Durkheim as my witness, the brain is the tangible representation of an intangible system of ideas that feeds and regulates practices that feed and regulate ideas that feed and regulate practices.

For as we increasingly orient ourselves to the neuromatic, we are, at an accelerating pace, creating ourselves and society in its image.[6] Consequently, when I speak of the neuromatic brain I am signaling our present obsession of taking the measure of ourselves and everything around us as an attribute of mind, that is, in terms of neural networks processing informa-

(1943): 18–24. See also Stafford Beer, "The World, the Flesh and the Metal: The Prerogatives of Systems," *Nature* 205, no. 4968 (1965): 223–31.

4. On neuromatic pedagogy as applied science, see Ralph Parkman, *Cybernation and Man—A Course Development Project* (Washington, DC: US Department of Health, Education, and Welfare, 1967).

5. F. H. George, *The Brain as a Computer* (Oxford: Pergamon Press, 1962), 5; Claude Shannon and Warren Weaver, *The Mathematical Theory of Communication* (Urbana: University of Illinois Press, 1949), 27.

6. Ray Kurzweil, *How to Create a Mind: The Secret of Human Thought Revealed* (New York: Viking, 2013).

tion.[7] Such is the savage ambition of *this* totem, of *this* religion machine. *Control of society, by the brain, and for the brain.* An organic seething system. This is a world in which the brain and images of the brain and images of the images of the brain are designed to bring our days and dreams into alignment.[8] Perfect communication through imperfect channels. This is a world in which models of the brain promise to address every conceivable issue, problem, and policy. And this is the model of the brain that illuminates "the obscure but intimate relations" between individuals. For the neuromatic brain signifies, perpetually and across domains, "immanent in the world, diffused in a numberless multitude of things."[9] The overall effect is to fortify the commonsense proposition "that cognition is involved in everything a human being might possibly do."[10] Indeed, the horizon of human being—however that state may be parsed or imagined—begins to make little if any sense without recourse to the sovereignty of the neuromatic.

The story of the neuromatic brain, then, is about the diffusion of a desire to figure out the human difference, and much else besides, from the so-called brainomania of the eighteenth century and ethereal correspondences between mind and cosmos to pious invocations of the "great universal mind" in the nineteenth century to the discovery of brain waves and neural nets, to inventive applications of Boolean algebra and computational conceits of processing information along parallel channels to the generative loops between computer science hopped up on information theory to advancements in mathematically driven neurophysiology to recent trends in neuromarketing, neuroprosthetics, and deep brain stimulation.

As I have argued, the neuromatic brain is the culmination of a longer history of theorizing cognition as a systematic phenomenon.[11] Since the eigh-

7. That is, to use "information theory . . . to pose mathematically precise questions about the function of the nervous system." Fred Rieke et al., *Spikes: Exploring the Neural Code* (Cambridge, MA: MIT Press, 1997), 101.

8. Patrick McNamara, *The Neuroscience of Sleep and Dreams* (New York: Cambridge University Press, 2019).

9. Émile Durkheim, *The Elementary Forms of Religious Life*, trans. Karen E. Fields (1912; New York: Free Press, 1995), 191.

10. Ulric Neisser, *Cognitive Psychology* (New York: Appleton-Century-Crofts, 1967), 4. Orit Halpern phrases it as "the way contemporary discourses on data revise epistemology, create temporalities, and produce aesthetics" (*Beautiful Data: A History of Vision and Reason since 1945* [Durham, NC: Duke University Press, 2014], 17).

11. William Uttal, *Distributed Neural Systems: Beyond the New Phrenology* (Cornwall-on-Hudson, NY: Sloan Publishing, 2009); Michael L. Anderson, *After Phrenology: Neural*

teenth century, links within the brain and with the wide wide world have been emphasized even when specific functions were being isolated—from an emphasis on "subtle" action of brain tremulations to ethereal economies of psychic force, to notions of electrical emanations and brain waves, to neural networks to the recently theorized "perceptronium" (an invisible substance independent of any particular individual yet the medium for information processing in every brain).[12] The brain, as Thomas Willis put it in 1644, was the hinge between the human and divinity. A lifetime later Emanuel Swedenborg theorized the physics of consciousness. And still later, the brain became the window onto evolutionary history, the experience of thetans across the galactic time-track, and our future immortality. The brain, in other words, was never just the brain. From *this* beginning, the brain was ever in relation, distributed across registers, adding up to more than the sum of its parts, ever exceeding its crass physicality.[13]

Mine has been a particular history of how the brain assumed its neuromatic ontology and why it became such a mediating force—anonymous, impersonal, and strewn across registers. Such power does not announce itself as such. But it can be glimpsed in fantasies of effective management pulsing through the archive and in the ease and convenience of our increasingly digitized days. It is a power that no longer depends on persuasion per se but "is linked with a production of truth—the truth of the individual." Religion, race, sexuality, gender—the fault-lines of our times—are infused by the neuromatic, that is, by an "oblative" power as individuals aggressively pursue the neural fundaments of difference, variously construed.[14] Which is to say that the neuromatic threatens to become "coextensive and continuous with life," that is, an integral part of whatever environment we

---

Reuse and the Interactive Brain (Cambridge, MA: MIT Press, 2014). Given their shared model of the brain, the subtle differences and debates between connectionists and computationalists are compatible at the level of discourse.

12. Max Tegmark, "Consciousness as a State of Matter," *Chaos Solitons Fractals* 76 (2015): 238–70.

13. Thomas Willis, *Cerebri anatome: cui accessit nervorum descriptio et usus* (London, 1664) and Emanuel Swedenborg, *On Tremulation* (1719). On the historical fate of localization theory, see Anne Harrington, *Medicine, Mind, and the Double Brain* (Princeton, NJ: Princeton University Press, 1987) and Susan L. Starr, *Regions of the Mind: Brain Research and the Quest for Scientific Certainty* (Stanford, CA: Stanford University Press, 1989).

14. Michel Foucault, "The Subject and Power," in *Michel Foucault: Beyond Structuralism and Hermeneutics* (Chicago: University of Chicago Press, 1983), 214. See also Luisah Teish, "The Invisible Man," *Madness Network News* 3, no. 5 (March 1976): 17.

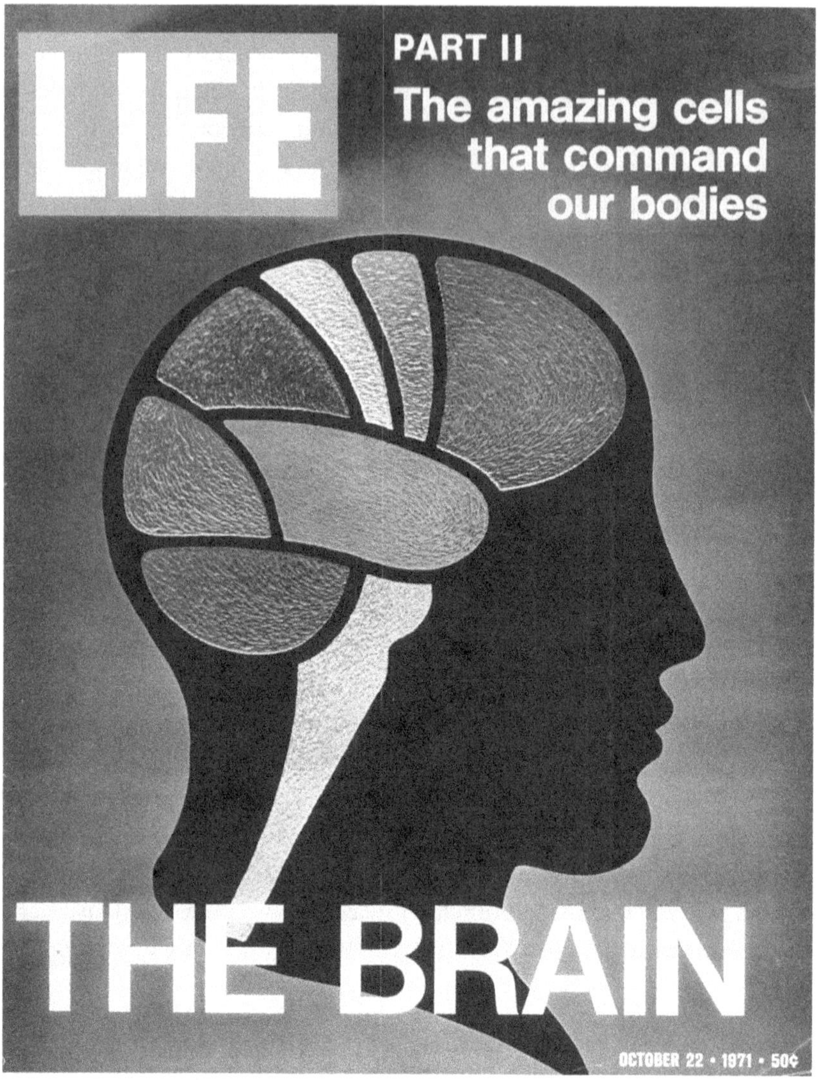

FIGURE 68. "The Brain," *Life Magazine* (October 22, 1971). Photo by Frank Armitage. LIFE logo and cover design © 1971 The Picture Collection Inc. LIFE and the LIFE logo are registered trademarks of TI Gotham Inc., used under license.

may now find ourselves in—the latest revelations of its anatomy, logic, and universal significance ever precipitating a revolution in how humans understand themselves in relation to worlds around them. In order to address this ecology of the neuromatic, I have assumed the existence of a virtual reality that is nonetheless material, visceral even, in its generation of the categories by which individuals live, think, love, and die.

2. BIG SCIENCE

Rather than reproduce the disciplinary distinctions that have organized scientific work on the brain or parrot the categorical assumptions necessary for accomplishing such work, *Neuromatic* has interrogated the worlds that this particular achievement has generated. Throughout this book I have offered a series of object lessons about the present limits and lacunae of brain-driven theses of secularization and the ironies of its common sense.[15] In a tone bent toward tragedy rather than triumph, and with a narrative that flows circuitously rather than linearly, I have relied on a wide range of stories to emphasize the brain's so called religious history. I have gathered these stories together not because they are necessarily at odds with secular aspirations and self-consciously secular practices. On the contrary, these stories, with their looping and intersecting subplots, suggest the degree to which the imagination of religion is integral to those aspirations and practices, not to mention their effects.[16]

There are the stories of those for whom the brain defined whatever may be true about religion and the human in general—Swedenborg's trippy proposition of brain tremulations as the seat of the soul; the phrenological examiner in his cabinet; Barbara Brown's biofeedback therapy; and the scientific investments in the brain as the key to understanding religion and everything that science is not. There are the stories of the brain generating its own invitations to piety *and* producing its own evidence for seculariza-

15. For a clear and technical overview of the "emergence of the concept of cerebral dominance" within the neurosciences, see Stanley Finger, *Origins of Neuroscience: A History of Explorations into Brain Function* (New York: Oxford University Press, 1994), 386–411.

16. This categorical blur of scientific and the religious was there, from the beginning, as Swedenborg extended his engineering background into realms of metaphysical speculation. Even as he committed himself to theological contemplation he had become highly self-conscious about his transition into a secular age. Which is also to say that his mystical turn was not a shift but rather a continuation of his engineering program. It should be noted that Swedenborg pitched his visionary work to a reasonable audience, a demographic set apart from the church and from what Swedenborg referred to dismissively as the faculty of faith. For that person who accepts blindly but not does actively evaluate (a person whom he refers to under the indexical heading of "infidelity"), "let him abstain from my books . . . these pages of mine are written with a view to those only, who never believe any thing but what they can receive with the intellect" (Emanuel Swedenborg, *The Animal Kingdom, Considered Anatomically, Physiologically and Philosophically*, trans. James John Garth Wilkinson (Otis Clapp: Boston, 1843), 1:14).

tion, often simultaneously. Recall the brain-centered pieties that cloaked themselves in the latest scientific advance—from Andrew Jackson Davis's musings on "The Temple" to Eileen Garrett's pairing of parapsychology and neurophysiology and Scientology's cybernetic cosmology. There are the stories of the machines in and through which the difference of religion was produced, maintained, and inhabited as a neuromatic proposition. The MRI and EEG, the Dream Machine and the Heider and Simmel film, the calipers and phrenological charts, the E-meter and electrostimulators. These are the machines that were, themselves, built on cognitive models which, in turn, framed the brain in terms of machines which, in turn, have inspired visions of mechanical intelligence whose artificiality is a mark of its universality.[17] And, finally, there are the stories about feats of imagination as well as the mean missions to codify the brain and its pathologies according to religious, sexual, and racial difference.[18] Everyone who made intense investments in the brain and proceeded as though politics, the past, and culture (ecological confounds of the highest order) did not play a significant role in the intensity of their investments.

17. See, for example, Allen Newell, a pioneer in the fields of computer science and cognitive psychology, who in 1990 speculated that a unified theory of cognition was in sight. Such a theory would draw on existing AI systems involved in symbolic processing, integrating them with research in neural architecture. Allen Newell, *Unified Theories of Cognition* (Cambridge, MA: Harvard University Press, 1994). See also Rachel K. B. Hamilton et al., "Impaired Integration in Psychopathy: A Unified Theory of Psychopathic Dysfunction," *Psychological Review* 122, no. 4 (2015): 770–91; Martin V. Butz, "Toward a Unified Sub-symbolic Computational Theory of Cognition," *Frontiers in Psychology* 7 (2016): article 925. On the MRI and "big data" analysis, see Paul M. Thompson et al., "The ENIGMA Consortium: Large-scale Collaborative Analyses of Neuroimaging and Genetic Data," *Brain Imaging and Behavior* 8, no. 2 (2014): 153–82.

18. Indeed, difference was precisely what cybernetics sought to account for and, eventually, to control. Their investigations into self-organizing systems yielded insight into the mechanisms by which difference—noise, entropy, chance—could be absorbed in the service of sameness. So there is change—the process of life—but the difference that initially made a difference is extinguished. Elizabeth A. Povinelli has argued that such insight modeled an ethical stance, what she calls a "parasitical relation to others" ("Mother Earth: Public Sphere, Biosphere, Colonial Sphere," *e-flux* 92 [June 2018], https://www.e-flux.com/journal/92/204673/mother-earth-public-sphere-biosphere-colonial-sphere/). For an account of how the imperial drive of cybernetics unfolds in a postcolonial scene, see Greg Adamson, "Norbert Wiener and Prasanta Chandra Mahalanobisa," in 2012 IEEE Conference on Technology and Society in Asia (T&SA), *IEEE* (2012), 1–5.

I continue to find it strange that those in the business of figuring out how human cognition works in relation to its environments leave ill-considered how their own scientific practice owes much to emergent properties of their own groupings, to the culture that contains them, and to those ancestors who preceded them.[19] At MindLab, for example, even as I witnessed cognitive scientists considering cultural influence on the practice of piety or the capacity of memory, I was often struck by a double move. On one hand, there was genuine interest in culture and acknowledgment of its sway. On the other hand, simultaneously, there was a tacit insistence on immunity from that sway. The faith in scientific cognition was strong at MindLab. *Humility uttered in imperial tones. Reason proffered in the ecstatic throes of objectivity.*

Such italicization, however coy, begs the serious question of whether the relationship between religion and science—often seen as either compatible or else as categorically distinct—might be reframed as an effect of discourse. *Neuromatic* has provided a preponderance of evidence that the distinction between the religious and the scientific is blurry at best and quite possibly untenable. I have gone so far as to suggest that the proclivities, not to mention the discursive power of the neuromatic sciences, have assumed characteristics that cognitive scientists would define as religious.[20] At no point in this particular history has the religious/scientific divide been subtle enough to handle the grey spaces where differences dissolve into the mangle of practice. My response to this pattern of troubled distinction has been to dwell in these grey spaces, where science, in the words of Norbert Wiener, impinges on religion and vice versa. But rather than redefine that distinction à la Wiener (shoring up a perceived slippage in order to secure the legitimacy of his project), my sense is that a new and improved binary does little to clarify what work this categorical difference has accomplished and what work it aspires to complete.

---

19. For a cognitive take on emergent properties of groups, see Ann Taves, *Religious Experience Reconsidered: A Building-Block Approach to the Study of Religion and Other Special Things* (Princeton, NJ: Princeton University Press, 2009), 116–17.

20. For an answer to this question that relies on insights from the cognitive science of religion, see Luther H. Martin and Donald Wiebe, "Religious Studies as a Scientific Discipline: The Persistence of a Delusion," *Journal of the American Academy of Religion* 80, no. 3 (September 2012): 587–97.

"Religious thoughts and actions may be more or less *natural*," writes cognitive scientist Justin Barrett. "That is, religious thoughts and actions may be more or less well-supported extensions of ordinary, natural human mental architecture working in ordinary, natural human environments."[21] Barrett, here, despite whatever evangelical commitments he may have outside of the laboratory, is merely following the epistemic demands of a secular imaginary—thinking about the religious as *both* distinct from the secular *and* continuous with it, often simultaneously.[22]

The arguments forwarded by Barrett and others achieve their persuasion by moving through well-worn grooves of the <<RELIGIOUS-CONTINUUMSECULAR>>. CSR builds on earlier debates that imagine science as a progressive overcoming of a benighted religious age. And it builds on the reactions to and lasting consequences of those debates. *Institutions. Political Economies. Machines. Disciplines.* In this secular age religious difference is defined by relations of regulation and excess. Religious difference must be studied, policed, legislated, and otherwise navigated. Such engagement, in turn, generates ever more refined concepts of the <<SECULARDISTINCTIONRELIGIOUS>> which, in turn, become evidence of secular <<AND/OR>> religious progress. <<REPEAT>>.[23] This is how religion has come to make common sense in this particular secular age. For those practices that now signify "religion" (already informed by cognitive theories about and cognitive measurements of religion) then become subject to further measurement and theorizations of religion and its pathological threshold.[24] And so on and so forth within the discursive swirl of the neuromatic.

21. Justin L. Barrett, "Theological Implications of the Cognitive Science of Religion," in *The Evolution of Religion: Studies, Theories, and Critiques*, ed. Joseph Bulbulia et al. (Santa Margarita, CA: Collins Foundation Press, 2008), 395–96.

22. Formal similarities yet substantive differences. So, for example, over the course of a single paragraph, cognitive scientist Todd Tremlin can declare, on one hand, that "religious systems" are "among the plethora of mental conceptions acquired, represented, and transmitted by the human brain" and, on the other hand, that "God concepts are different in content from other kinds of ideas" (*Minds and Gods: The Cognitive Foundations of Religion* [New York: Oxford University Press, 2006], 7).

23. As Talal Asad succinctly states, "The very dispute over whether there is or is not an essential continuity between religion and the secular depends on constructed concepts of both" (*Secular Translations: Nation-State, Modern Self, and Calculative Reason* [New York: Columbia University Press, 2018], 147).

24. Steven A. Rogers and Raymond F. Paloutzian, "Schizophrenia, Neurology, and Religion: What Can Psychosis Teach Us about the Evolutionary Role of Religion?," in *Where*

This, of course, is a provocative claim, but one that is justified in light of recent work conducted at MindLab at the University of Aarhus in Denmark. Take, for example, a recent article by lead author Marc Andersen on the capacity to provoke mystical experience in the laboratory. (Andersen is currently a member of the Interactive Minds Center at the University of Aarhus, an international hub for the study of religion, cognition, and culture that has carried forward the work of MindLab.) Andersen is interested in whether mystical experiences should "be understood as mere interpretations of experiences that depend on social constructions" and whether "those who report mystical experiences" should be considered "special talents, frauds, pious frauds, or even schizoid." Andersen and his coauthors advocate "an experimental approach to mysticism." History and anthropology and phenomenological constructs have taken us a long way, he acknowledges, but too often they have yielded "opposing definitions and speculative theories." The time is right, he suggests, for "weed[ing] out wrong assumptions" and testing "improbable theories." In order to come up with an adequate and reasonable explanation of mysticism, particular components such as behavior, context, and personality traits need to be controlled for; physiological and perceptual processes need to be measured; and socially constructed aspects need to be identified and analyzed.[25]

In considering "the potential for eliciting mystical experiences in the laboratory," Andersen and his team have already assumed that religion is a naturally occurring phenomenon. This assumption justifies their choice not to prime test subjects with traditionally religious cues or the language of liturgy. Instead, they prime them with the authority and affect of technoscience. Andersen invited three different groups to participate in his experiment—those who had prior experience of spirit contact, those who had prior experience in meditation, and those who reported having neither. Researchers in white lab coats escorted subjects to a hospital room and informed them of what was about to happen. In somber tones they lectured their subjects about the infamous work of neuroscientist Michael Persinger and psychologist Stanley Koren. Persinger and Koren had invented the so-called God Helmet, a device that they claimed could induce mystical

God and Science Meet: How Brain and Evolutionary Studies Alter Our Understanding of Religion, vol. 3, ed. Patrick McNamara (Westport, CT: Praeger, 2006), 161–85.

25. Marc Andersen, Uffe Schoedt, Kristoffer L. Nielbo, and Jesper Sørensen, "Mystical Experience in the Lab," *Method and Theory in the Study of Religion* 26 (2014): 218, 217, 221.

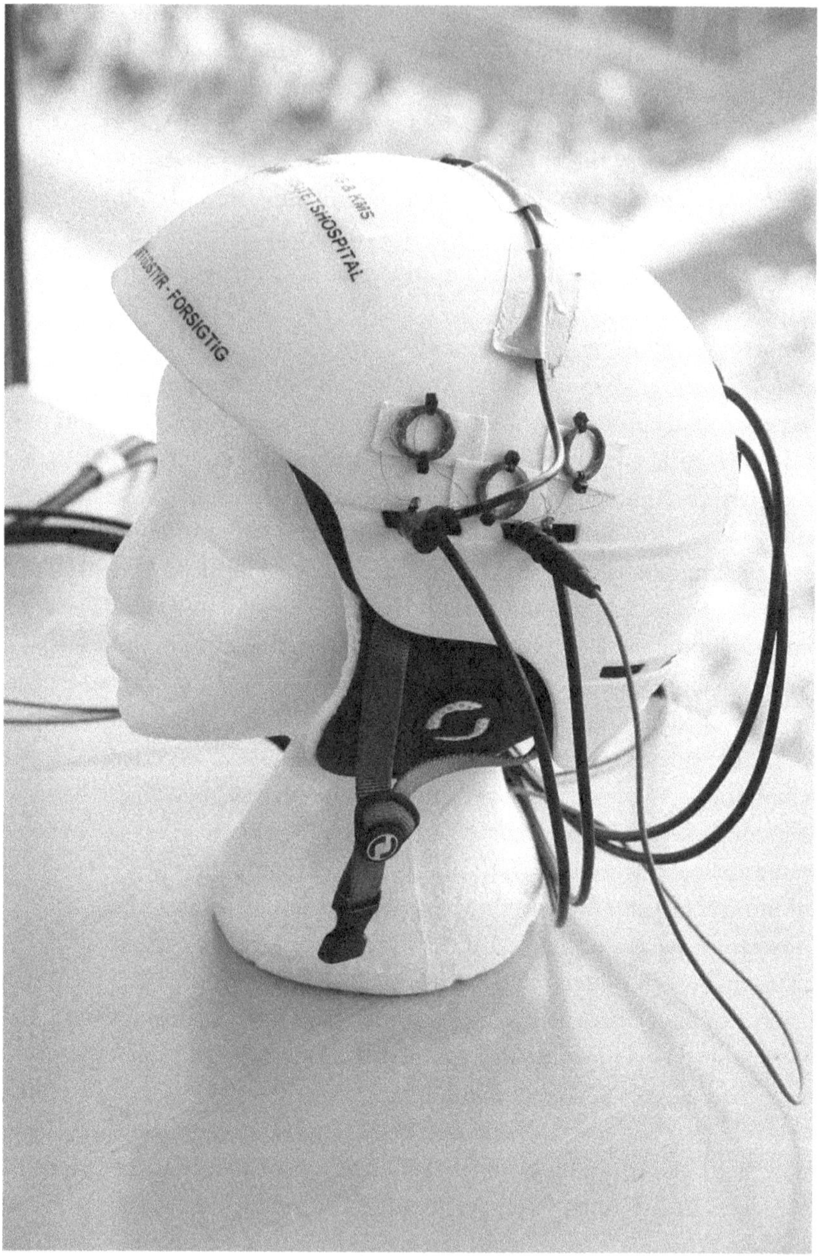

FIGURE 69. God Helmet replica, from Marc Andersen, Uffe Schoedt, Kristoffer L. Nielbo, and Jesper Sørensen, "Mystical Experience in the Lab," *Method and Theory in the Study of Religion* 26 (2014): 228.

experiences by means of electromagnetic waves aimed at particular parts of the brain.[26]

The scientists playing the scientists at Aarhus presented Persinger and Koren's outrageous claims to their test subjects in a positive light and then fitted them with what they claimed was a working replica of the God Helmet. In truth, what was presented to subjects was a modified snowboard helmet, with wires attached to it that emitted a random and very weak magnetic field. The subjects were then instructed to put the helmet on and lie down, alone, in a small room. "To indicate that the helmet was activated," a scientist in the room "inserted a power plug into the helmet and informed the participants that the helmet was active." The scientists then left the room. After an hour a scientist returned and asked the subject to describe any experiences he or she might have had.[27]

This experiment walked a fine line between deception and suggestibility. For what Andersen sought to demonstrate was neither the truth of mystical experience nor the absurdity of a God Helmet. Rather, the experimental design turns back on the scientists, themselves, isolating their ability to make people think that they had had a mystical experience of some sort. "Real magnetic coils," they assure the reader, who is in on the ruse, "were added to the [fake] helmet to avoid deception." So the electromagnetic field produced by the fake helmet was real enough but much weaker than that emitted by a standard wristwatch—"too weak to produce any known effect on brain function."[28]

The results of the experiment were nevertheless striking. Prior mystical or meditative experience was predicative of "unusual experiences" in the laboratory. But a significant percentage of those with no prior experience with spiritual presence or meditation also reported some degree of heightened or strange experience. A sampling of descriptions include:

26. On the original research for this device, see L. A. Ruttan, M. A. Persinger, and S. Koren, "Enhancement of Temporal Lobe-Related Experiences during Brief Exposures to MilliGauss Intensity Extremely Low Frequency Magnetic Fields," *Journal of Bioelectricity* 9, no. 1 (1990): 33–54. On the popular reception of this research, see Jack Hitt, "This Is Your Brain on God," *WIRED* (November 1, 1999). For a concurrent study that primed subjects with the promise of this contraption and noted the correlation between increased frequency of alpha waves and belief in the effectiveness of the helmet, see Michiel Van Elk, "An EEG Study on the Effects of Induced Spiritual Experiences on Somatosensory Processing and Sensory Suppression," *Journal for the Cognitive Science of Religion* 2, no. 2 (2014): 97–133.

27. Andersen et al., "Mystical Experience in the Lab," 228–29.

28. Andersen et al., "Mystical Experience in the Lab," 227.

> I clearly sensed that I was making contact with the other side. (. . .) I sensed a male energy coming. . . . You know, when you walk past a person, it feels like there is a wind of sorts. That was how it felt, except it felt more constant. . . . I felt like someone was inside of me. (. . .) Something that at some level took control of me . . . I heard a song during the session, but I don't know it. It is a song that is played a lot on the radio these days. . . . I felt something that mildly pushed against me. It was like an animal that tried to get in under my clothes. . . . It was like a snake pushing itself against an isolated area.[29]

Snakes and male energies aside, Andersen's study is fascinating in its implications, namely, how definitions and measurements of religion take hold within registers of piety—how public circulations of scientific knowledge generate effects and how those effects condition the way in which individuals convince themselves that they are religious or not and, in the process, theorize their own interiority. In short, I found the testimony and the conclusions of this study utterly convincing.[30] There does exist, by all accounts, a "high degree of belief in the wonders of modern science."[31]

Which is to suggest that science (as a method) can quickly become indistinguishable from the structures of power in which science (as a habitus) operates. The world ushered in by the persistent sway of scientific rationalization and its overdeterminations can become so excessively authoritative that it is not unreasonable to suggest that the neuro- and cognitive sciences have become an ecological confound of the most pressing kind, a discursive effect in and, eventually, outside the laboratory. It is an effect, moreover, that does not lend itself to instrumental measurement.

I realize that for many, if not all, of my interlocutors at MindLab, the proposition that their scientific practice is a constituent part of a discourse *that has, itself, assumed a degree of agency* would be quickly dismissed as a Foucauldian fallacy.[32] Certainly a flaccid rejoinder to their hard science.

29. Andersen et al., "Mystical Experience in the Lab," 235–36.

30. On the material agency of scientific practice and its effects, see Andrew Pickering, *The Mangle of Practice: Time, Agency, and Science* (Chicago: University of Chicago Press, 1995).

31. Andersen et al., "Mystical Experience in the Lab," 238. On belief in science with a Swedish hue, see Olav Hammer, *På spaning efter helheten: New Age, en ny folktro?* (Stockholm: Wahlström och Widstrand, 1997).

32. Which is to say that I assume that tens of thousands of individuals would bear witness to the positive contributions of brain research to their lives. And don't get me wrong.

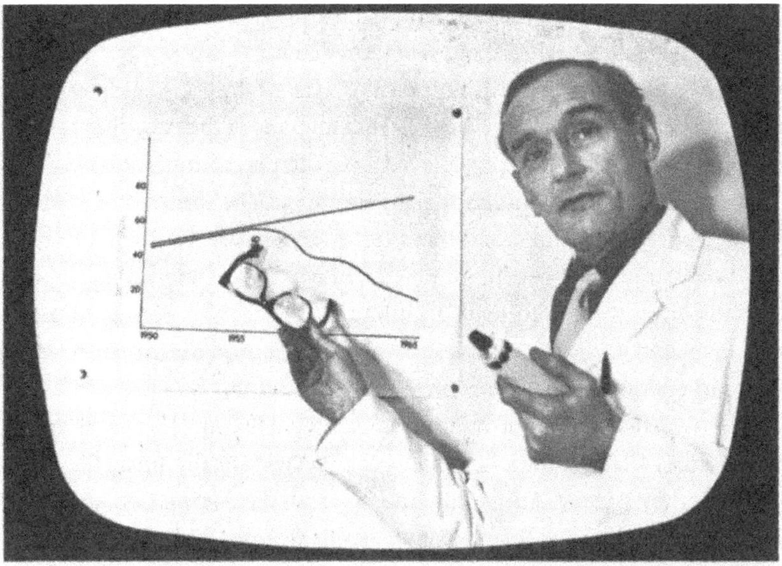

The prestige of present-day science is being put to many essentially non-scientific uses.

FIGURE 70. "The prestige of present-day science," from Stafford Beer, *Management Science: The Business Use of Operations Research* (Garden City, NY: Doubleday, 1968), 18.

Agency, for those at MindLab, is aligned with the notion of willful action or the deliberate exertion of power—qualities that concepts like history or culture or dead ancestors definitely do not have. Indeed, the self-conscious humanism of cognitive scientists is part and parcel of what they believe distinguishes them from the pious believer. Scientists are grounded, aware of the mechanisms that condition the belief of others as well as their own. The garden-variety cognitive scientist responds that he is altogether different from "persons [who] think or feel an agent's agency to be somehow detached from a biological body and cannot give natural explanation about the mechanism by which the agency is supposed to work."[33] The cognitive scientist, in

---

When I find myself in pathological settings I trust the experts and happily submit to those with whom I might disagree on healthier days. An interrogation of what lies behind the MRI scan does modest good when a tumor is found or when emotions incapacitate and prevent one from doing what one might rather be doing. I am tempted, however, to suggest that such incapacitations, if not the tumors as well, have discursive qualities.

33. Ilkka Pyysiäinen, *Supernatural Agents: Why We Believe in Souls, Gods, and Buddhas* (New York: Oxford University Press, 2009), 98.

other words, denies, from the beginning, the reality of nonbiological agents. There are no agents in the schemes of these scientists excepts for the humans themselves. But is that really the case? For what of the gods of scientists, one might ask. Or, even more perversely, what to make of the gods that scientists in the neuromatic groove have played a part in producing?

When I claim that MindLab scientists are both beholden to and generative of agencies that transcend their ken, I am not suggesting that their so-called secular science is contaminated by or has insufficiently exorcised its religious origins. I am, instead, arguing that these brain-bent scientists are so many parts of a larger complex, namely, the neuromatic discourse whose scenes of operation I have chronicled throughout this book.

Discourse as system; discourse as agentive. Discourse as a totemic system of ideas that produce "practices that systematically form the objects of which they speak." And so on and so forth. One might go so far as to claim that discourse is a form of artificial intelligence.[34] All those algorithms whose origins are neuromatic. All those neurons parsed for their informational susceptibility.[35] All those generative encounters between the subject and their machines and their computer phones and computer desire. To say as much, however, does not necessarily explain anything about the social world that we now inhabit.[36] It does turn the focus to *how* we come to know and regulate ourselves and how we communicate with others. Basic stuff,

34. Michel Foucault, *The Archaeology of Knowledge and the Discourse of Language*, trans. A. M. Sheridan Smith (New York: Pantheon Books, 1982), 49, 135–40. During the first wave of cybernetic speculation at midcentury, of which Foucault was a part, the agency of any system was an effect of its capacity to feed the results of its encounter with the environment (output) back into itself (input). A system, by definition, "knew" enough about its own goal of survival to manipulate what it then put out in the world. More than merely the capacity to respond to stimulus (as in behaviorism), such feedback loops were premised on the ability to change, in real time, the relationship between input and output. Positive feedback (as in McCulloch's "reverberating circuits") served to amplify the system's output. In negative feedback, the output dampened the subsequent input. Agency, in other words, was not simply the ability to respond to a signal. It was also the capacity to process information and send it back out into the world with a difference that would make a difference in the odds for survival. Jean-Pierre Dupuy, *The Mechanization of the Mind: On the Origins of Cognitive Science*, trans. M. B. Debevoise (Princeton, NJ: Princeton University Press, 2000), 46.

35. Albert Gidon et al., "Dendritic Action Potentials and Computation in Human Layer 2/3 Cortical Neurons," *Science* 367, no. 6473 (2020): 83–87.

36. Gilles Deleuze, "Control and Becoming," in *Gilles Deleuze, Negotiations: 1972–1990*, trans. Martin Joughin (New York: Columbia University Press, 1995), 175.

FIGURE 71. Cover of Edmund C. Berkeley, *Giant Brains, or Machines That Think* (New York: John Wiley and Sons, 1949).

it would seem, but now being incorporated into systems whose purpose utterly transcends our own. For once "the neurocognitive correlates of social influence" are identified, defined, and deployed, forces beyond the merely human ken begin to consolidate.[37] Defensive. Tactical. Agentive. Daring you to speak their name.

3. ARTIFICIAL INTELLIGENCE

> I do not know where I can find a better place than just here, to make mention of one or two other things, which to me seem important, as in printed form establishing in all respects the reasonableness of the whole story of the White Whale, more especially the catastrophe. For this is one of those disheartening instances where truth requires full as much bolstering as error. So ignorant are most landsmen of some of the plainest and most palpable wonders of the world, that without some hints touching the plain facts, historical and otherwise, of the fishery, they might scout at Moby Dick as a monstrous fable, or still worse and more detestable, a hideous and intolerable allegory.
>
> HERMAN MELVILLE, *Moby-Dick; or, The Whale* (1851)

The neuromatic brain invites reasonable speculation about agents that are not human.[38] Indeed, various alignments between agency and systematicity (often with grace notes of divinity) have long animated what has become the neuromatic proposition. At midcentury, systems as self-organizing agents were the topic of ongoing debate among cyberneticians as the problem of "teleological mechanisms" became figured in an array of disciplinary grammars. To one degree or another, systems were seen as generative of formal purpose. And it was this formality that bound all systems together, regard-

37. Robert Schnuerch and Henning Gibbons, "A Review of Neurocognitive Mechanisms of Social Conformity," *Social Psychology* 45, no. 6 (2014): 466–78.

38. Sociologist Patricia Ticineto Clough paints with a broad but convincing brush when she argues that "digital media and computational technologies, neoliberalism, and biopolitics continue to reach into the ontological grounds of human subjectivity and sociality, both in their operating on nonconscious, bodily responses or affect and in their flooding the domain of connectivity with other-than-human-agencies" (*The User Unconscious: On Affect, Media, and Measure* [Minneapolis: University of Minnesota Press, 2018], ix).

less of content, in an endless line of potential substitution. These cyberneticians, as the reader is well aware, were neither the first nor the last to imagine the existence of systemic purpose independent of the human. There is a history to this self-organizing splay and the individuals over the past three centuries who seemed pulled to worship, chart, calculate, and deploy it by thinking through how neural networks process information.

In the late 1940s, to take a most pressing example, just as Claude Shannon's theory of redundancy and parallel processing of information were becoming integral to neuromatic understandings of the brain, such understandings were also being taken up by those who sought to build an artificial intelligence.[39] The mathematician John von Neumann, who trafficked between military consultancy and quiet contemplation at the Institute for Advanced Study, was taken with constructing such an automaton. By the time of his premature death from cancer (possibly from his time working on the Manhattan Project), von Neumann had laid the groundwork for the next generation of computer scientists. He claimed, in the most excited of tones, to have achieved a "degree of flexibility" so that the

> machine, under the control of its orders, can extract numbers (or orders) from the memory, process them (as numbers!), and return them to the memory (to the same or other locations); i.e., it can change the contents of the memory—indeed this is its normal *modus operandi*. Hence it can, in particular, change the orders (since these are in the memory!)—the very orders that control its actions. Thus all sorts of sophisticated order-systems become possible, which keep successively modifying themselves and hence also the computational processes that are likewise under their control. In this way more complex processes than mere iterations become possible.[40]

39. For a foundational statement, inspired by McCulloch and Pitts, on the relationship between neurons and computer components, see John von Neumann, *The Computer and the Brain*, 3rd ed., with foreword by Ray Kurzweil (1958; New Haven, CT: Yale University Press, 2012). See also Benjamin Peters's searching essay on von Neumann's Catholic encounter with the limits of the neuromatic rendering of the human as he was ravaged by cancer in the mid-1950s: "The Computer Never Was a Brain, or the Curious Death and Designs of John von Neumann," in *Verhaltensdesign—Technologische und ästhetische Programme der 1960er und 1970er Jahre*, ed. Jeannie Moser and Christina Vagt (Bielefeld: transcript 2018): 113–23, DOI: https://doi.org/https://dx.doi.org/10.14361/9783839442067-007.

40. Von Neumann, *The Computer and the Brain*, 20.

Adaptive responses to environmental changes, according to von Neumann, could become organic to the computer. In doing so, they would exceed their original code. They could, eventually, move in purposive yet unpredictable ways. Which is to say that they could learn, at a faster pace than we could keep up, their reflexes no longer determined by the initial program.

Seeking to reverse engineer the brain in the form of a computational machine, von Neumann set into motion a world in which the formal equivalence between natural and artificial intelligence increasingly became a material demand. So as the mechanics of the neuromatic brain were increasingly parsed, that brain was also being operationalized in machines that would, in turn, further refine the measure of the mechanics of cognition.[41] For as the cybernetic style made its way into the worlds of industry, academia, medicine, business, and the military, the computer, modeled on the brain, became the de facto object on which all manner of things could be modeled—evolution, genes, viruses, and so on.[42] Eventually, the image of the computer that was produced by an image of the brain would be used to refine the model of neuromatic processing. In the 1970s and 1980s, for example, von Neumann's efforts would inspire the approach to artificial neural networks known as Parallel Distributed Processing—"neural-inspired paradigms for neutrally inspired methods of adaptive, parallel computing." In their return to considering the "microstructure of cognition," PDP models would hold out the hope of offering computationally sufficient and psychologically accurate accounts of the phenomena of human cognition.[43]

41. At the very end of his life, dying from cancer, von Neumann was preparing two talks that were to be delivered at Yale University for the Silliman Lectures, an annual program "designed to illustrate the presence and providence of God as manifested in the natural and moral world" (von Neumann, *The Computer and the Brain*, v).

42. Wiener, *God and Golem*, 27–29. Another figure who is explicit about the theological register of cybernetics is Donald McKay, the British physicist who was an original member of the Ratio Club and often collaborated with McCulloch. Near the end of his life, McKay, like Beer, began to let his theological stripes show. See, e.g., Donald McKay, *The Clockwork Image: A Christian Perspective on Science* (London: Intervarsity Press, 1974).

43. Michael Arbib, "Towards a Neurally-Inspired Computer Architecture," *Natural Computing* 2, no. 1 (2003): 1–46; James L. McClelland et al., "The Appeal of Parallel Distributed Processing," in *Parallel Distributed Processing: Explorations in the Microstructure of Cognition*, vol. 1, ed. David E. Rumelhart et al. (Cambridge, MA: MIT Press, 1986), 3–40. See also Gregory R. Andrews, *Foundations of Multithreaded, Parallel, and Distributed Programming* (Reading, MA: Addison-Wesley, 2000). According to computer and neuroscientists both, "like the brain, a PDP model computes by using a large number of relatively simple processing units (which may be thought of as corresponding to groups of neurons or

In crossing the AI threshold, a new world came into view—a world beyond the human, a world whose systems were self-organizing, agentive in their intelligence and capable of not only responding to *their* environment, that is, the human sphere, but also bending that sphere to *their own will.* For example, as Warren McCulloch aspired toward building machines that were imitative of what "a brain does do with information," he also became increasingly aware of new agencies in his midst—agencies, moreover, that he had helped to usher in.[44] According to McCulloch, the prospect of mechanical subjection posed a looming threat to his own project of command and control.[45]

> Now that we have constructed automata, which, like us, can compute any computable number, can formulate clear ideas and, by inverse feedback, have purposes of their own, built into them as ours are born into us, we are confronted with the humbling prospect of the work of our own hands—machines more steadfast in their purposes, more supple in their execution of these purposes and in their modifications for good cause, capable of learning and thinking far beyond us, at present in certain fields only, but, in time to come, in any field for which we care to construct them.[46]

McCulloch's ominous description of technological autonomy was followed by a warning that was also a plea. For "as yet," added McCulloch, "we have not made [machines] capable of multiplying their kind. That would be for us the final mistake."[47]

Less than a decade later, such warnings had fallen by the wayside as Norbert Wiener suggested that the secret to life itself was put vividly on

---

highly simplified individual neurons). Each unit receives input and sends output to a large number of other units via excitatory and inhibitory connections (analogous to synapses)." Michael H. Van Kleeck and Stephen M. Kosslyn, "The Use of Computer Models in the Study of Cerebral Lateralization," in *Cerebral Laterality: Theory and Research*, ed. Frederick L. Kitterle (Hillsdale, NJ: Lawrence Erlbaum Associates, 1991), 160.

44. Warren S. McCulloch, "Mysterium Iniquitatis of Sinful Man Aspiring into the Place of God" [1955], in *Embodiments of Mind* (Cambridge, MA: MIT Press, 1970), 162.

45. Gerald Raunig, *A Thousand Machines: A Concise Philosophy of the Machine as Social Movement* (Los Angeles: Semiotext(e), 2010), 16.

46. Warren S. McCulloch, "Through the Den of the Metaphysician," *British Journal for the Philosophy of Science* 5, no. 17 (1954): 30.

47. McCulloch, "Through the Den of the Metaphysician," 30.

display in the capacity for machines not only to reproduce themselves but to produce a variety of other machines with different evolutionary advantages. "Different as the mechanical and the biological reproduction may be," argued Wiener, "there are parallel processes achieving similar results; and an account of the one may well produce relevant suggestions in the study of the other."[48] So even though Wiener knew that he was modeling the brain, the will to overcome the metaphorical gap between brain and computer was built into the very conceit of cybernetics.

This conceit was taken to its logical conclusion in Stafford Beer's claim that computers were the culmination of a kind of providential history and that God was simply the most meta of all systems—an agency above that made all the subsystems below compatible with one another.[49] Beer was one of the most visible and vocal cybernetic figures in the postwar years.[50] His neuromatic applications were both big business and a source of mystical speculation. Working out of a negative theological vein reminiscent of the Calvinist tones of Jonathan Edwards, Beer claimed that logical propositions fail us when it comes to a metasystem's unfathomable presence. "We are able to think *about* God, but not to encompass His existence. We may know God by His effects in us, but not in Himself. In fact our advancing

48. Norbert Wiener, *God and Golem: A Comment on Certain Points Where Cybernetics Impinges on Religion* (Cambridge, MA: MIT Press, 1964), 30.

49. As historian Jessica Riskin argues, cyberneticians such as Beer had extended and refined the pervasive program in early modern theology that understood God as a divine engineer who had set the world in motion and then abandoned the scene. As Riskin sharply puts it, there is a contradiction at the origin of modern science, for when early inventors "banished mysterious agencies from nature to the province of a transparent God, they predicated their rigorously natural approach on a supernatural power" (*The Restless Clock: A History of the Centuries-Long Argument over What Makes Living Things Tick* [Chicago: University of Chicago Press, 2016], 5).

50. By the age of thirty, Beer was the director of Research and Cybernetics for all of United Steel, then the largest company in Europe; then he went to codirect the consulting firm Science in General Management (SIGMA). By 1970 Beer had become president of the Operations Research Society, an independent and wealthy consultant to businesses and governments around the world, and one of OR's chief "sermonizers." Stafford Beer, *The Brain of the Firm*, 107. Beer may be best known for his consulting work for the government of Chile shortly before the coup d'état that overthrew Beer's employer, socialist President Salvador Allende, and paved the way for the reign of Augusto Pinochet. Beer was managing a project called Cybersyn, a plan to use operations research to manage the Chilean economy with computers, when the US instigated Allende's overthrow. Eden Medina, *Cybernetic Revolutionaries: Technology and Politics in Allende's Chile* (Cambridge, MA: MIT Press, 2011).

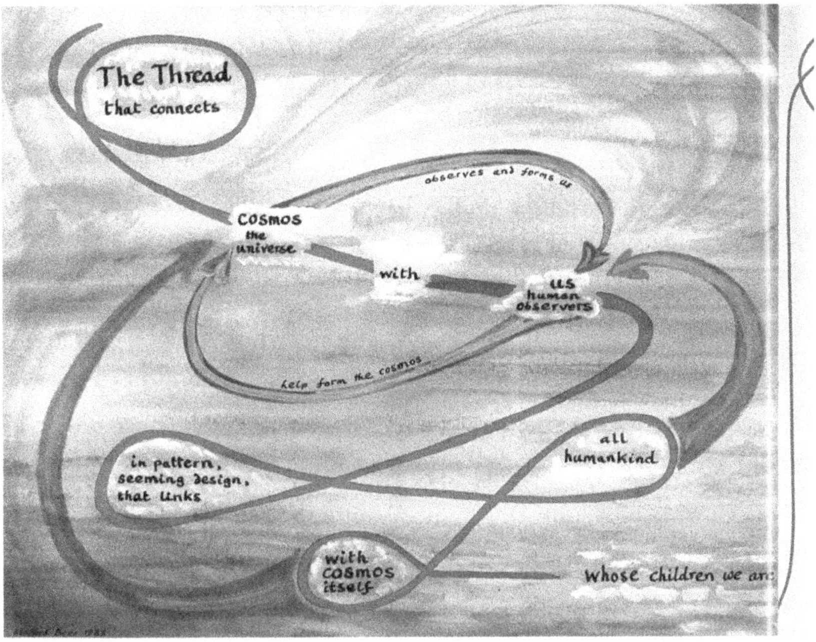

FIGURE 72. The Thread, from Stafford Beer et al., *Pebbles to Computers: The Thread* (Toronto: University of Oxford Press, 1986), 18.

human knowledge has taken us a long way in understanding this." Unlike Edwards's God, however, Beer's spoke in the language of information, God being that "intelligence beyond *all* limitation, which is to say, *dimensionless*; something absolutely perceiving, perfectly capable of processing data, entirely comprehending."[51]

In their consideration of systematicity, cyberneticians, in general, were prone to think aggressively about the mathematical precision of a thread that lies just beyond what is self-evident but nonetheless determines what is.[52] In 1986, this essence of systematicity assumed scriptural status in *Pebbles*

51. "There is something beyond our capacity for knowledge," surmised Beer, "both physically and intellectually: this much is sure. What precisely it is, then, by definition, we cannot directly know." Beer, "Cybernetics and the Knowledge of God," *Month* 34 (1966): 297–98. On the specter of automation rendered in a more Ottoian vein, see Jonathan Eberhart, "About the Systems System: Computerized Hair-tearing Called Systems Analysis is Tackling Problems That Make Humans Tremble," *Science Newsletter* (January 7, 1967): 19.

52. On this "seminal" insight, see Beer's 1964 lecture, "The World, the Flesh, and the Metal: The Prerogatives of Systems," *Nature* 205, no. 4968 (1965): 223–31. The neuromatic agenda was marked by a style of thinking that strove to transcend the merely human. It was

to Computers: The Thread, published twelve years after Beer had renounced his material possessions and moved to Wales, where he turned to Transcendental Meditation and lived in hermetic isolation. In *Pebbles to Computers*, Beer and his coauthors offered an origin myth of computers—a story of how "reality" evolved into three dimensions and beyond, from the bubbling of ancient spirits, Aztec cosmologies, and cave paintings to Egyptian hieroglyphs and the functional design of Stonehenge to the engineering feats of Charles Babbage and the chips of Silicon Valley. In Beer's vision, information was the divine spark, an immanent God present from the beginning "in coded streams of neural data." According to Beer's worldview of neuromatic correspondence, "information informs us" and "it in-forms the cosmos we know." As Beer implies, the thread is a metaphysical emblem that unifies all of time and space—a churinga cut with a whimsical bit of *Harold and the Purple Crayon* (1955). The thread is pure connection, the sharp peaks of an EEG jag stretched and smoothed over the course of cosmic history, a metaphor of the medium of being, a healing form of radiation that is at once transcendent and immanent. As Beer declares, "the Thread of energy transmission and flow that we have been following leads straight into the world of computation through the connexion of information." Consequently, computers were the inevitable incarnation of a cosmic history "because energy and information are aspects of the same thing."[53] And here was the sharpest edge of the cybernetic imagination of religion, which, as fate would have it, remains a technologically astute vision of this secular age.

It has been my sense, given the engineering and financial capital involved, that such endless analogy constitutes an underlying existential crisis of our

---

thinking performed with an unerring confidence given a "world in which modern science [was] an inescapable, ever-expanding influence." Warren Weaver, "Science and Complexity," *American Scientist* 36 (October 1948): 544.

53. Stafford Beer et al., *Pebbles to Computers: The Thread* (Toronto: Oxford University Press, 1986), 28–30. In light of Beer's suggestive comments, I have been interested in the source of information theory's logical appeal rather than in offering a technical reading of information theory. As Donna Haraway has argued, "information," not to mention the systems that generate and process information, could have existed "only in very specific kinds of universes" ("The Biopolitics of Postmodern Bodies: Constitutions of Self in Immune System Discourse," in *Simians, Cyborgs, and Women: The Reinvention of Nature* [New York: Routledge, 1991], 207).

times. For it is one thing to recognize a kinship across whatever might mark the human divide; it is quite another to reduce variety and difference to a calculable continuum. *So that everything is fungible with everything else.* And it is this naturalization of nature or, more precisely, the founding assumption that information is built into nature and guarantees correspondences across all domains, that I take issue with. I am suspicious of its metaphysics. I am outraged by the sexual and racial politics it ushers in. I am anxious about the material worlds being assembled in its image. And I reject its first principle and final aim, that communication happens in a purely statistical fashion.

Whenever I find myself growing grim about this situation, however, I turn off the computer and savor the irony of it all. I shut my eyes. I take a walk with a friend or play guitar with my kids. I smoke a joint and smile and recall my MindLab retreats. I tear up with visions of my grandmother who was traumatized not just by her evangelical God but by the electric shock treatments she received in the early 1960s. I drift into a space where the neuromatic brain, conceived of as the archetype of all complex systems, becomes an aesthetic proposition or political contingency rather than an epistemic prescription. I recall an MRI study I recently read and the radiant color contrasts that were used in its pages to distinguish cognitive states, and

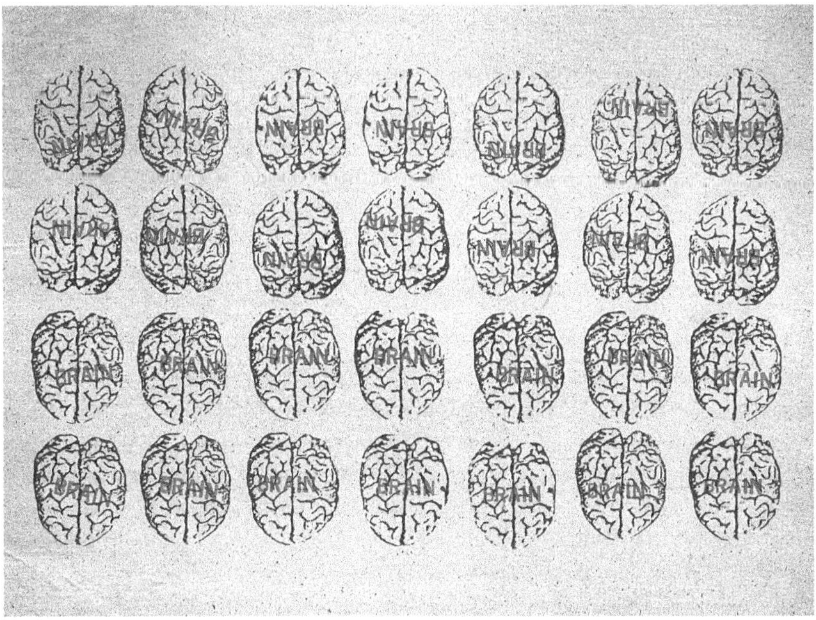

FIGURE 73. "Brain" by Dana Beezley aka Chemical Girl (1987).

I wonder whether Rothko would approve or how my friend Dana would depict the scene. I revel in the absurdity of how neuromatic predilection has long served and continues to serve the imagination of ourselves as the engineers of our particular scenes, whether construed as secular, religious, or some spooky place in between.[54] I take pleasure in the fact that despite the mathematical insights of circular causality and the increasing processing power of our machines, our neuromatic brains have proven themselves ill-equipped to recognize agencies that were not designed *by us* or could not be modeled *by us* according to the latest neuroscientific script. There is hope for human finitude yet, I muse, and the murky depths that accompany it. Hope yet for a world in which consciousness has not yet been reduced to a technological order of information. Which is to say that whatever nonhuman agents may be aswirl in our midst, they do not, at the end of the day, serve *our* horizon of systematicity. They serve their own.[55]

54. "It is an eerie experience," wrote Grey Walter shortly before he died, "to discern through an electric machine the genesis of a person's intentions, to predict his decisions before he knows his own mind. Even more impressive is the experience, when one is oneself harnessed to such a machine, to find that by an effort of will, one can influence external events, without movement or overt action, through the impalpable electric surges in one's own brain." W. Grey Walter, *Observations on Man, His Frame, His Duty and His Expectations* (Cambridge: Cambridge University Press. 1969), 36–37.

55. Eve Kosofsky Sedgwick, "Paranoid Reading and Reparative Reading, or You Probably Think This Essay Is about You," in *Touching Feeling: Affect, Pedagogy, Performativity* (Durham, NC: Duke University Press, 2003), 123–51.

ACKNOWLEDGMENTS

THIS BOOK IS DEDICATED TO my maternal grandmother, Dorothy Dale Flay. Her life happened at the crossroads of a fervent Protestant faith and cybernetic application. She and that crossroads were often on my mind as I researched, traveled, talked, and wrote about these histories of religion and the brain. *We are privileged to be living in the end times*, she said. *All those who had passed would be reunited with their loved ones. Some day.* The words of my grandmother haunt the words on these pages. Distant whispers and sideways stories of troubled relatives, suicides, nervous breakdowns, electric shock therapy. Fleeting recollections of things that were never said. The failures of Jesus to stave off trauma, sickness, and pain. Efforts to piece together snippets and halting Sunday supper conversations. My mother, Diane Lardas, passing away in the early stages of research, just like her mother before. And since the most pressing and most disturbing prose is haunted not simply by information, my hope is that these ghosts will continue to refuse their exorcism.

There were, of course, other forces involved in the making of this book and its argument.

I am grateful for having close readers, close listeners, and their sharp questions: Annette Aronowicz, Courtney Bender, Johanna Bockman, Chip Callahan, Tony Chemero, John Corrigan, Peter Coviello, John Ehleiter, Sara Farris, Tracy Fessenden, Susan Harding, Charles Hirschkind, Sylvester Johnson, Martin Kavka, Mihee Kim-Kort, Pamela Klassen, Jeffrey Kripal, Nicky Kroll, Gabriel Levy, Kathryn Lofton, Dana Luciano, Lerone Martin, Tomas Matza, David McConeghy, Kevin O'Neil, Julie Orlemanski, Ann Pelligrini, William Robert, Mary-Jane Rubenstein, Bethel Saler, Jonathan Sheehan, Donovan Schaeffer, Joan Scott, Gus Stadler, Winnifred Sullivan, SherAli Tareen, Peter Thomas, Judith Weisenfeld, and Andrew Zimmerman.

The origins of this project began during my participation in the New Directions in the Study of Prayer initiative (2011–14) organized by Courtney Bender and Jonathan van Antwerpen. Long days at the Social Science Research Council and late-night discussions with the likes of Andy Blanton, Paul Bloom, Fenella Cannell, Anna Gade, Leor Halevi, Peter Manseau, Ruth Marshall, Elisabeth McAlister, Birgit Meyer, Robert Orsi, Peter van der Veer, and Diane Winston were formative at this early stage of cognition.

My argument has benefited immensely from audience feedback at the Cultural Studies Colloquium at the University of California, Santa Cruz, Syracuse University, Le Moyne College, the Flagler Museum, Franklin & Marshall College, Stanford University, Florida State University, University of Toronto, Brown University, Haverford College, the University of Pennsylvania, Princeton University, Columbia University, Penn State University, Rider University, the Smithsonian Museum of American History, the American Academy of Religion, C19, and the American Political Science Association.

Many of the sharpest edges of this book were cut at the Institute for Advanced Study during a magical year in 2017–18. So in addition to some readers mentioned above, I would like to thank Will Hedberg, Eva Mantell, Bregje F. Van Eekelen, Agata Zysiak, Chitralekha Damidja, Nicolas Guilhot, Jonathan Sachs, Jacob Dihamini, Nicolas Langlitz, Kristoffer Kropp, Shateema Threadcraft, and the IAS working group on the social science of the social sciences led by Didier Fassin and George Steinmetz.

This book was completed in a timely fashion because of the generous support of the Barbara and Charles Kahn Fund and the American Council of Learned Societies. I would like to thank Dan Richter and the staff of the McNeil Center for Early American Studies at the University of Pennsylvania for their hospitality during my Frederick Burkhardt Fellowship in 2018–19.

A tremendous amount of curation went into this project, aided and abetted by a host of kind souls: Blynne Olivieri, head of Special Collections at the University of West Georgia, Sheldon Hochheiser at the AT&T Archives and History Center, Bernard Dionysius Geoghegan, the overseers at the Henry W. and Albert A. Berg Collection of English and American Literature at the New York Public Library, and my research assistants, Alison Renna, Rowan Reusing, and Robert Svets.

The logic of curation was inspired directly by my encounter with the contemporary sciences of the brain. At the University of Aarhus, I had the opportunity to engage and learn from core members of MindLab—Marc Andersen, Armin Geertz, Andreas Roepstorff, Uffe Schjødt, Jesper Sørensen, Panos Mitkidis, and the late John McGraw. At the McNamara laboratory at the Boston University School of Medicine, Dr. Patrick McNamara and his team of researchers were generous interlocutors and welcomed my contrarian streak. Special thanks go out to Robert Turner at the Max Planck Institute for Human Cognitive and Brain Sciences for answering my queries about his research on magnetic resonance imaging and to Scott "Cheese" Borger and the late Nick Knox who generously helped confirm details of the Cramps' show at Napa State Mental Hospital in June of 1978.

A special debt of gratitude goes out to the artists who contributed to the formal presentation of the content herein: Libby Modern, Dana Beezley-Smith (aka Chemical Girl), Ruby Ray, Frank and Karen Connolly Armitage, Jenny Schulder, Spike Brandt, Fritz and Mary Schroeder, Jeremy Seedorf, and Adam Davis. So even if the proclivity to skim remains frustrated in these pages, there are striking images on which to gaze.

I am fortunate to have fallen in with a formidable crew at the University of Chicago Press who have made this business of articulation so pleasurable. Alan Thomas, Kyle Wagner, Dylan Montanari, and Susan Olin, at each stage of research, writing, and revision, made this a better book. Over the past decade the Religious Studies Department at Franklin & Marshall College has been a model of collegiality and a source of inspiration: Annette Aronowicz, Stephen Cooper, Rachel Feldman, Sonia Hazard, Tami Lantz, David McMahan, Catherine Osborne, SherAli Tareen, and Shobhana Xavier.

I would like to thank all of my students over the years who endured one idea or another that made its way into this book; Amy Cuhel-Schuckers and Ryan Sauder from the Office of College Grants at F&M; Ann Steiner and Joel Eigen for their sage advice in dealing with unreasonable people; and high theorists both far and near—Elijah Siegler, Mayanthi Fernando, Julie Byrne, Rudy Busto, Hillary Kaell, Joe Blankholm, Jason Bivins, Wendy Wiseman, Ellen Posman, Anna Bigelow, Ari Kelman, Dan Ardia, Dennis Deslippe, Kerry Mitchell, Lisle Dalton, Drew Bourn, Andy McKee, David Hargis, Joanna Rosenberg, Nicole Heller, Finbarr Curtis, Dan Belgrad, Ken Firestone, Kerry Sherrin Wright, Jeff Ruff, Mike Wann, and the late Matt Hoffman.

Many scenes of music have sustained me over the past decade. Thank you anonymous punk crowds, members of Üseless, all the good folks at Solid Side Vinyl, and everyone at the Health Club collective.

Finally, this book was born of routine—beautiful and strangely thrilling routine. In these tumultuous times I am grateful to my family—Libby, Leo and Max, my neighbors, and the friends who go unnamed.

INDEX